Statistics and Computing

Springer

Statistics and Computing

Gentle: Numerical Linear Algebra for Applications in Statistics.
Gentle: Random Number Generation and Monte Carlo Methods.
Härdle/Klinke/Turlach: XploRe: An Interactive Statistical Computing Environment.
Krause/Olson: The Basics of S and S-PLUS.
Lange: Numerical Analysis for Statisticians.
Loader: Local Regression and Likelihood.
Ó Ruanaidh/Fitzgerald: Numerical Bayesian Methods Applied to Signal Processing.
Pannatier: VARIOWIN: Software for Spatial Data Analysis in 2D.
Venables/Ripley: Modern Applied Statistics with S-PLUS, 3rd edition.
Wilkinson: The Grammar of Graphics

W.N. Venables
B.D. Ripley

Modern Applied Statistics with S-PLUS

Third Edition

With 144 Figures

Springer

W.N. Venables
CSIRO Marine Laboratories
PO Box 120
Cleveland, Qld, 4163
Australia
Bill.Venables@cmis.csiro.au

B.D. Ripley
Professor of Applied Statistics
University of Oxford
1 South Parks Road
Oxford OX1 3TG
UK
ripley@stats.ox.ac.uk

Series Editors:

J. Chambers
Bell Labs, Lucent
  Technologies
600 Mountain Ave.
Murray Hill, NJ 07974
USA

W. Eddy
Department of Statistics
Carnegie Mellon University
Pittsburgh, PA 15213
USA

W. Härdle
Institut für Statistik und
  Ökonometrie
Humboldt-Universität zu
  Berlin
Spandauer Str. 1
D-10178 Berlin
Germany

S. Sheather
Australian Graduate School
  of Management
University of New South
  Wales
Sydney, NSW 2052
Australia

L. Tierney
School of Statistics
University of Minnesota
Vincent Hall
Minneapolis, MN 55455
USA

Library of Congress Cataloging-in-Publication Data
Venables, W.N. (William N.)
    Modern applied statistics with S-PLUS / W.N. Venables, B.D.
Ripley. – [3rd ed.]
        p.   cm. — (Statistics and computing)
    Includes bibliographical references and index.
    ISBN 0-387-98825-4 (hard adhesive : alk. paper)
    1. S-Plus.   2. Statistics—Data processing.   3. Mathematical
statistics—Data processing.   I. Ripley, Brian D., 1952–
II. Title.   III. Series.
QA276.4.V46   1999
005.369–dc21   99-18388

Printed on acid-free paper.

Production managed by Robert Bruni; manufacturing supervised by Joe Quatela.
Photocomposed copy prepared from the authors' PostScript files.
Printed and bound by Maple-Vail Book Manufacturing Group, York, PA.
Printed in the United States of America.

9 8 7 6 5 4 3 2 1

ISBN 0-387-98825-4 Springer-Verlag New York Berlin Heidelberg   SPIN 10717625

Preface

S-PLUS is a system for data analysis from the Data Analysis and Products Division of MathSoft, an enhanced version of the S environment for data analysis developed at Bell Laboratories (of AT&T and now Lucent Technologies). S-PLUS has become the statistician's calculator for the 1990s, allowing easy access to the computing power and graphical capabilities of modern workstations and personal computers.

The first edition of this book appeared in 1994 and a second in 1997. The S statistical system has continued to grow rapidly, and users have contributed an ever-growing wealth of software to implement the latest statistical methods in S. This book concentrates on using the current systems to do statistics; there is a companion volume which discusses programming in the S language in much greater depth.

Several different implementations of S have appeared. There are currently two 'engines' in use: S-PLUS 3.x and 4.x are based on version 3 of the S language whereas S-PLUS 5.x is based on S version 4. Furthermore, since 1997 the Windows version of S-PLUS has had a graphical user interface in the style of widespread Windows packages. Our aim is that this book should be usable with all these versions, but we have given lower priority to S-PLUS 3.x. Some of the more specialized functionality is covered in the *on-line complements* (see page 467 for sites) which will be updated frequently. The datasets and S functions that we use are available on-line, and help greatly in making use of the book.

This is not a text in statistical theory, but does cover modern statistical methodology. Each chapter summarizes the methods discussed, in order to set out the notation and the precise method implemented in S. (It will help if the reader has a basic knowledge of the topic of the chapter, but several chapters have been successfully used for specialized courses in statistical methods.) Our aim is rather to show how we analyse datasets using S-PLUS. In doing so we aim to show both how S can be used and how the availability of a powerful and graphical system has altered the way we approach data analysis and allows penetrating analyses to be performed routinely. Once calculation became easy, the statistician's energies could be devoted to understanding his or her dataset.

The core S language is not very large, but it is quite different from most other statistics systems. We describe the language in some detail in the early chapters, but these are probably best skimmed at first reading; Chapter 1 contains the most basic ideas, and each of Chapters 2 and 3 are divided into 'basic' and 'advanced' sections. Once the philosophy of the language is grasped, its consistency and logical design will be appreciated.

The chapters on applying S to statistical problems are largely self-contained, although Chapter 6 describes the language used for linear models that is used in several later chapters. We expect that most readers will want to pick and choose among the later chapters.

This book is intended both for would-be users of S-PLUS as an introductory guide and for class use. The level of course for which it is suitable differs from country to country, but would generally range from the upper years of an under-graduate course (especially the early chapters) to Masters' level. (For example, almost all the material is covered in the M.Sc. in Applied Statistics at Oxford.) Exercises are provided, but these should not detract from the best exercise of all, using S to study datasets with which the reader is familiar. Our library provides many datasets, some of which are not used in the text but are there to provide source material for exercises. (Further exercises and answers to selected exercises are available from our WWW pages.)

Both authors take responsibility for the whole book, but Bill Venables was the lead author for Chapters 1–4 and 6–8, and Brian Ripley for Chapters 5 and 9–14. The authors may be contacted by electronic mail at

```
Bill.Venables@cmis.csiro.au
ripley@stats.ox.ac.uk
```

and would appreciate being informed of errors and improvements to the contents of this book.

To avoid any confusion, S-PLUS is a commercial product, details of which may be obtained from `http://www.mathsoft.com/splus/`.

Acknowledgements:

This book would not be possible without the S environment which has been principally developed by Rick Becker, John Chambers and Allan Wilks, with substantial input from Doug Bates, Bill Cleveland, Trevor Hastie and Daryl Pregibon. The code for survival analysis is the work of Terry Therneau. The S-PLUS code is the work of a much larger team acknowledged in the manuals for that system.

We are grateful to the many people who have read and commented on draft material and who have helped us test the software, as well as to those whose problems have contributed to our understanding and indirectly to examples and exercises. We cannot name them all, but in particular we would like to thank Doug Bates, Adrian Bowman, Bill Dunlap, Sue Clancy, David Cox, Anthony Davison, Peter Diggle, Matthew Eagle, Nils Hjort, Stephen Kaluzny, Francis Marriott, Joseé Pinheiro, Brett Presnell, Charles Roosen, David Smith, Patty Solomon and Terry Therneau. We thank MathSoft DAPD and CSIRO DMS for early access to versions of S-PLUS and for access to platforms for testing.

Bill Venables
Brian Ripley
January 1999

Contents

Appendices

Typographical Conventions

Throughout this book S language constructs and commands to the operating system are set in a monospaced typewriter font `like this`. The character ~ may appear as ˜ on your keyboard, screen or printer.

We often use the prompts `$` for the operating system (it is the standard prompt for the UNIX Bourne shell) and `>` for S-PLUS. However, we do *not* use prompts for continuation lines, which are indicated by indentation. One reason for this is that the length of line available to use in a book column is less than that of a standard terminal window, so we have had to break lines that were not broken at the terminal.

Some of the S-PLUS output has been edited. Where complete lines are omitted, these are usually indicated by

```
    . . . .
```

in listings; however most *blank* lines have been silently removed. Much of the S-PLUS output was generated with the options settings

```
options(width=65, digits=5)
```

in effect, whereas the defaults are around 80 and 7. Not all functions consult these settings, so on occasion we have had to manually reduce the precision to more sensible values.

Chapter 1

Introduction

Statistics is fundamentally concerned with the understanding of structure in data. One of the effects of the information-technology era has been to make it much easier to collect extensive datasets with minimal human intervention. Fortunately the same technological advances allow the users of statistics access to much more powerful 'calculators' to manipulate and display data. This book is about the modern developments in applied statistics which have been made possible by the widespread availability of workstations with high-resolution graphics and computational power equal to a mainframe of a few years ago. Workstations need software, and the S[1] system developed at AT&T's Bell Laboratories and now at Lucent Technologies provides a very flexible and powerful environment in which to implement new statistical ideas. S is exclusively licensed to the Data Analysis Products Division of MathSoft Inc. who distribute an enhanced system called S-PLUS; we refer to the language as S and the environment as S-PLUS.

S-PLUS is an integrated suite of software facilities for data analysis and graphical display. Among other things it offers

- an extensive and coherent collection of tools for statistics and data analysis,

- a language for expressing statistical models and tools for using linear and non-linear statistical models,

- graphical facilities for data analysis and display either at a workstation or as hardcopy,

- an effective object-oriented programming language that can easily be extended by the user community.

The term *environment* is intended to characterize it as a planned and coherent system built around a language and a collection of low-level facilities, rather than the 'package' model of an incremental accretion of very specific, high-level and sometimes inflexible tools. Its great strength is that functions implementing new statistical methods can be built on top of the low-level facilities.

Furthermore, most of the environment is open enough that users can explore and, if they wish, change the design decisions made by the original implementors.

---

[1] The name S arose long ago as a compromise name (Becker, 1994), in the spirit of the programming language C (also from Bell Laboratories).

Suppose you do not like the output given by the regression facility (as we have frequently felt about statistics packages). In S you can write your own summary routine, and the system one can be used as a template from which to start. In many cases sufficiently persistent users can find out the exact algorithm used by listing the S functions invoked.

We have made extensive use of the ability to extend the environment to implement (or re-implement) statistical ideas within S. All the S functions that are used and our datasets are available in machine-readable form; see Appendix C for details of what is available and how to install it.

System dependencies

We have tried as far as is practicable to make our descriptions independent of the computing environment and the exact version of S-PLUS in use. This is not easy as there are two S 'engines' and two graphical interfaces in use. S-PLUS 3.x and 4.x are based on S3, version 3 of the S language, whereas S-PLUS 5.x is based on S4. All the systems have a command-line interface[2] but 4.x (which only exists on Windows, and includes versions 4.0, 4.5 and 2000) also has a configurable graphical user interface (GUI); see Appendix B.

Clearly some of the details must depend on the environment; we used S-PLUS 3.4 and 5.0 on Solaris to compute the examples, but have also tested them under versions 3.3, 4.5 and 2000 of S-PLUS for Windows, and S-PLUS 5.0 on Linux.

Where timings are given they refer to S-PLUS 4.5 running on a PC based on a 300MHz Pentium II CPU.

One system dependency is the mouse buttons: we refer to buttons 1 and 2, usually the left and right buttons on Windows but the left and middle buttons on UNIX (or perhaps both together of two).

Reference manuals

The basic references are Becker, Chambers & Wilks (1988) for the basic environment, Chambers & Hastie (1992) for the statistical modelling and first-generation object-oriented programming and Chambers (1998) for S4; these should be supplemented by checking the on-line help pages for changes and corrections as S-PLUS has evolved considerably since these books (even Chambers, 1998) were written. Our aim is not to be comprehensive nor to replace these manuals, but rather to explore much further the use of S-PLUS to perform statistical analyses.

---

[2] However the 'Standard Edition' of S-PLUS 4.5 or 2000 (as opposed to the 'Professional Edition') only has the GUI.

1.1 A quick overview of S

Most things done in S are permanent; in particular, data, results and functions are all stored in operating system files.[3] These are referred to as *objects*.

Variables can be used as scalars, matrices or arrays, and S provides extensive matrix manipulation facilities. Furthermore, objects can be made up of collections of such variables, allowing complex objects such as the result of a regression calculation. This means that the result of a statistical procedure can be saved for further analysis in a future session. Typically the calculation is separated from the output of results, so one can perform a regression and then print various summaries and compute residuals and leverage plots from the saved regression object.

Technically S is a function language. Elementary commands consist of either *expressions* or *assignments*. If an expression is given as a command, it is evaluated, printed and the value is discarded. An assignment evaluates an expression and passes the value to a variable but the result is not printed automatically. An expression can be as simple as 2 + 3 or a complex function call. Assignments are indicated by the *assignment operator* <-. For example,

```
> 2 + 3
[1] 5
> sqrt(3/4)/(1/3 - 2/pi^2)
[1] 6.6265
> library(MASS)
> mean(chem)
[1] 4.2804
> m <- mean(chem); v <- var(chem)/length(chem)
> m/sqrt(v)
[1] 3.9585
```

Here > is the S-PLUS prompt, and the [1] states that the answer is starting at the first element of a vector.

More complex objects will have printed a short summary instead of full details. This is achieved by the object-oriented programming mechanism; complex objects have *classes* assigned to them that determine how they are printed, summarized and plotted. This process is taken further in S-PLUS 5.x in which *all* objects have classes.

S-PLUS can be extended by writing new functions, which then can be used in the same way as built-in functions (and can even replace them). This is very easy; for example, to define functions to compute the standard deviation and the two-tailed *p*-value of a *t* statistic we can write

```
std.dev <- function(x) sqrt(var(x))
t.test.p <- function(x, mu=0) {
    n <- length(x)
    t <- sqrt(n) * (mean(x) - mu) / std.dev(x)
```

---

[3] These should not be manipulated directly, however.

```
                2 * (1 - pt(abs(t), n - 1))
        }
```

It would be useful to give both the t statistic and its p-value, and the most common way of doing this is by returning a list; for example, we could use

```
        t.stat <- function(x, mu=0) {
            n <- length(x)
            t <- sqrt(n) * (mean(x) - mu) / std.dev(x)
            list(t = t, p = 2 * (1 - pt(abs(t), n - 1)))
        }
        z <- rnorm(300, 1, 2)  # generate 300 N(1, 4) variables.
        t.stat(z)
        $t:
        [1] 8.2906
        $p:
        [1] 3.9968e-15

        unlist(t.stat(z, 1))  # test mu=1, compact result
                t       p
         -0.56308 0.5738
```

The first call to `t.stat` prints the result as a list; the second tests the non-default hypothesis $\mu = 1$ and using `unlist` prints the result as a numeric vector with named components.

Linear statistical models can be specified by a version of the commonly-used notation of Wilkinson & Rogers (1973), so that

```
        time ~ dist + climb
        time ~ transplant/year + age + prior.surgery
```

refer to a regression of `time` on both `dist` and `climb`, and of `time` on year within each transplant group and on age, with a different intercept for each type of prior surgery. This notation has been extended in many ways, for example to survival and tree models and to allow smooth non-linear terms.

Recent versions of S-PLUS on Windows provide an extensive graphical interface with menus and dialog boxes that can be used for simpler statistical analyses. We have chosen not to describe this system here (apart from a brief account in Appendix B) since it covers only a small part of the statistics we wish to cover and in any case a GUI should be intuitive and not need a detailed explanation.

1.2   Using S-PLUS

The different versions of S-PLUS require a number of ways to get started: please read the appropriate section of Appendix A for your system.

Bailing out

One of the first things we like to know with a new program is how to get out
of trouble. S-PLUS is generally very tolerant, and can be interrupted by Ctrl-
C[4] or Esc (versions 4.x and later on Windows). This will interrupt the current
operation, back out gracefully (so, with rare exceptions, it is as if it had not been
started) and return to the prompt.

You can terminate your S-PLUS session by typing

    q()

at the command line or from Exit on the File menu in a GUI version.

On-line help

There is a help facility that can be invoked from the command line. For example,
to get information on the function var the command is

    > help(var)

A faster alternative (to type) is

    > ?var

For a feature specified by special characters and in a few other cases (one is
"function"), the argument must be enclosed in double or single quotes, making
it an entity known in S as a character string. For example, two alternative ways
of getting help on the list component extraction function, [[, are

    > help("[[")
    > ?"[["

Many S commands have additional help for *name*.object describing their result:
for example, lm also has a help page for lm.object.

Further help facilities for some versions of S-PLUS are discussed in Appendix A.

1.3   An introductory session

The best way to learn S-PLUS is by using it. We invite readers to work through
the following familiarization session and see what happens. First-time users may
not yet understand every detail, but the best plan is to type what you see and
observe what happens as a result.

As explained previously you should first create a special directory and make
it your working directory for the session, then start S-PLUS.

The whole session takes most first-time users one to two hours at the appropriate leisurely pace. The left column gives commands; the right column gives
brief explanations and suggestions.

---

[4] This means hold down the key marked Control or Cntrl and hit the second key.

`library(MASS)`	A command to make our datasets available. Your local advisor can tell you the correct form for your system.
`?help`	Read the help page about how to use help.

`trellis.device()`
 	
On a monochrome **UNIX** system use
`trellis.device(color=F)`
	Turn on the graphics window. You may need to re-position and re-size to make it convenient to work with both windows. Try not to change the aspect ratio of the graphics window when you re-size it.

`x <- rnorm(1000)` `y <- rnorm(1000)`	Generate 1 000 pairs of normal variates
`truehist(c(x,y+3), nbins=25)`	Histogram of a mixture of normal distributions. Experiment with the number of bins (25) and the shift (3) of the second component.
`?truehist`	Read about the optional arguments.

`dd <- con2tr(kde2d(x,y))`
`contourplot(z ~ x + y,`
`    data=dd, aspect=1)`
	2D density plot. We convert its output into a form suitable for the Trellis visualization routines.

`wireframe(z ~ x + y,` `    data=dd, drape=T)`	Perspective plot with superimposed coloured levels.
`levelplot(z ~ x + y,` `    data=dd, aspect=1)`	Greyscale or pseudo-colour plot.
`x <- seq(1, 20, 0.5)` `x`	Make $x = (1, 1.5, 2, \ldots, 19.5, 20)$ and list it.
`w <- 1 + x/2` `y <- x + w*rnorm(x)`	w will be used as a 'weight' vector and to give the standard deviations of the errors.

`dum <- data.frame(x, y, w)`
`dum`
`rm(x, y, w)`
	Make a *data frame* of three columns named x, y and w, and look at it. Remove the original x, y and w.

`fm <- lm(y ~ x, data=dum)` `summary(fm)`	Fit a simple linear regression of y on x and look at the analysis.

```
fm1 <- lm(y ~ x, data=dum,
     weight=1/w^2)
summary(fm1)
```
Since we know the standard deviations, we can do a weighted regression.

```
lrf <- loess(y ~ x, dum)
```
Fit a smooth regression curve using a modern regression function.

```
attach(dum)
```
Make the columns in the data frame visible as variables.

```
plot(x, y)
```
Make a standard scatterplot. To this plot we will add the three regression lines (or curves) as well as the known true line.

```
lines(spline(x, fitted(lrf)),
     col=2)
```
First add in the local regression curve using a spline interpolation between the calculated points.

```
abline(0, 1, lty=3, col=3)
```
Add in the true regression line (intercept 0, slope 1) with a different line type and colour.

```
abline(fm, col=4)
```
Add in the unweighted regression line. abline() is able to extract the information it needs from the fitted regression object.

```
abline(fm1, lty=4, col=5)
```
Finally add in the weighted regression line, in line type 4. This one should be the most accurate estimate, but may not be, of course. One such outcome is shown in Figure 1.1.

You may be able to make a hardcopy of the graphics window by selecting the Print option from a menu.

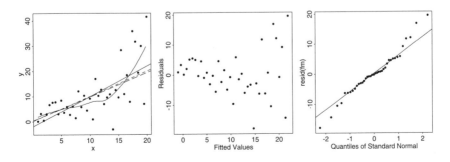

Figure 1.1: Four fits and two residual plots for the artificial heteroscedastic regression data.

```
plot(fitted(fm), resid(fm),
    xlab="Fitted Values",
    ylab="Residuals")
```
A standard regression diagnostic plot to check for heteroscedasticity, that is, for unequal variances. The data are generated from a heteroscedastic process, so can you see this from this plot?

```
qqnorm(resid(fm))
qqline(resid(fm))
```
A normal scores plot to check for skewness, kurtosis and outliers. (Note that the heteroscedasticity may show as apparent non-normality.)

```
detach()
rm(fm,fm1,lrf,dum)
```
Remove the data frame from the search path and clean up again.

We look next at a set of data on record times of Scottish hill races against distance and total height climbed.

```
hills
```
List the data.

```
splom(~ hills)
```
Show a matrix of pairwise scatterplots (Figure 1.2).

```
brush(as.matrix(hills))
```

Click on the Quit button in the graphics window to continue.
Try highlighting points and see how they are linked in the scatterplots (Figure 1.3). Also try rotating the points in 3D.

```
attach(hills)
```
Make columns available by name.

```
plot(dist, time)
identify(dist, time,
    row.names(hills))
```
Use mouse button 1 to identify outlying points, and button 2 to quit. Their row numbers are returned.

```
abline(lm(time ~ dist))
```
Show least-squares regression line.

```
abline(lqs(time ~ dist),
    lty=3, col=4)
```
Fit a very resistant line. See Figure 1.4.

```
detach()
```
Clean up again.

We can explore further the effect of outliers on a linear regression by designing our own examples interactively. Try this several times.

```
plot(c(0,1), c(0,1), type="n")
xy <- locator(type="p")
```
Make our own dataset by clicking with button 1, then with button 2 to finish.

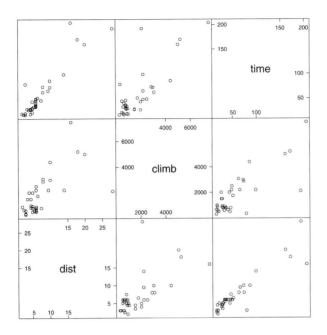

Figure 1.2: Pairs plot for data on Scottish hill races.

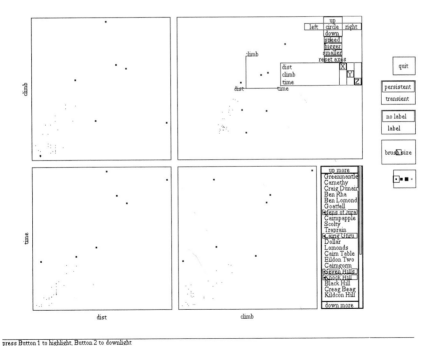

press Button 1 to highlight, Button 2 to downlight

Figure 1.3: Screendump of a brush plot of dataset hills (UNIX).

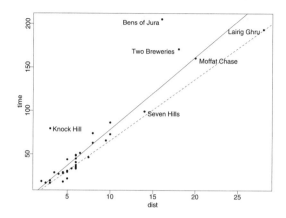

Figure 1.4: Annotated plot of time versus distance for `hills` with regression line and resistant line (dashed).

`abline(lm(y ~ x, xy), col=4)`
`abline(rlm(y ~ x, xy,`
`method="MM"),`
    `lty=3, col=3)`
`abline(lqs(y ~ x, xy),`
    `lty=2, col=2)`
	Fit least-squares, a robust regression and a resistant regression line. Repeat to try the effect of outliers, both vertically and horizontally.

`rm(xy)`	Clean up again.

We now look at data from the 1879 experiment of Michelson to measure the speed of light. There are five experiments (column Expt); each has 20 runs (column Run) and Speed is the recorded speed of light, in km/sec, less 299 000. (The currently accepted value on this scale is 734.5.)

`attach(michelson)`	Make the columns visible by name.
`search()`	The *search path* is a sequence of places, either directories or data frames, where S-PLUS looks for objects required for calculations.

`plot.factor(Expt, Speed,`
  `main="Speed of Light Data",`
  `xlab="Experiment No.")`
	Compare the five experiments with simple boxplots. (Prior to S-PLUS 5.x, just plot suffices.) The result is shown in Figure 1.5.

`fm <- aov(Speed ~ Run + Expt)` `summary(fm)`	Analyse as a randomized block design, with *runs* and *experiments* as factors.

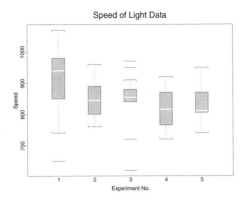

Figure 1.5: Boxplots for the speed of light data.

```
            Df Sum of Sq Mean Sq F Value    Pr(F)
    Run     19    113344    5965  1.1053 0.36321
    Expt     4     94514   23629  4.3781 0.00307
    Residuals 76   410166    5397
```

fm0 <- update(fm, . ~ . - Run)	Fit the sub-model omitting the non-sense factor, *runs*, and compare using
anova(fm0, fm)	a formal analysis of variance.

```
    Analysis of Variance Table
    Response: Speed

            Terms Resid. Df    RSS Test Df Sum of Sq F Value    Pr(F)
    1        Expt        95 523510
    2 Run + Expt        76 410166 +Run 19    113344  1.1053 0.36321
```

detach()	Clean up before moving on.
rm(fm, fm0)	

q()	Quit S-PLUS.

1.4   What next?

We hope that you now have a flavour of S-PLUS and are inspired to delve more deeply. We suggest that you read Chapter 2, perhaps cursorily at first, and the Sections 3.1–3, and 3.5, plus Appendix B if you are using S-PLUS 4.x. Thereafter, tackle the statistical topics that are of interest to you. Chapters 5 to 14 are fairly independent, and contain cross-references where they do interact. Chapters 7 and 8 build on Chapter 6, especially its first two sections.

Chapter 4 comes early, because it is about S not about statistics, but is most useful to advanced users who are trying to find out what the system is really doing.

On the other hand, those programming in the S language will need the material in our companion volume on S programming.

Chapter 2

The S Language

S is a language for the manipulation of objects. It aims to be both an interactive language (like, for example, a UNIX shell language) as well as a complete programming language with some convenient object-oriented features. In this chapter we are concerned with the interactive language, and hence certain language constructs used mainly in programming are postponed to Chapter 4.

This chapter divides into two parts. The first seven sections give an informal overview of the language which will suffice at first reading. Later sections (and Chapter 4) are more formal, and useful for answering "How do I do that?" and "Why did it do that?" questions.

2.1 A concise description of S objects

In this section we discuss the most important types of S objects and the simplest ways to manipulate them,

Naming conventions

Standard S names for objects are made up from the upper- and lower-case roman letters, the digits, 0–9, in any non-initial position and also the period, ' . ', which behaves as a letter except in names such as .37 where it acts as a decimal point. Some important conventions make use of a period in object names and users will become aware of these in due course. For example 'three dots', ..., is a reserved identifier, only used in the context of defining functions (see page 22). Other reserved identifiers include break, for, function, if, in, next, repeat, return and while.

Non-standard names are allowed using any collection of characters, but special steps have to be taken to use them. They are needed in some contexts but not in ordinary use and we suggest that they be avoided where possible. For example, so-called *replacement functions* must have names that end in the two characters " <- ", but they are not used explicitly.

In some (but not all) cases non-standard names may be handled by placing them within quotes. In particular this is true of function names (when used for function calls), object names to which a value is being assigned,[1] and argument or component names. Some examples are

```
> "style<-" <- function(p, value)    # a replacement function
        structure(p, style = value)
> Ttest(x1, x2)$"t-stat"  # one component of a result
[1] -3.6303
```

Note that S is *case sensitive*, so `Alfred` and `alfred` are distinct S names and that the underscore, ' _ ', is *not* available as a letter in the S standard name alphabet. (Periods are often used to separate words in names.)

Avoid using system names for your own objects; in particular avoid `c`, `q`, `s`, `t`, `C`, `D`, `F`, `I`, `T`, `diff`, `mean`, `pi`, `range`, `rank`, `tree` and `var`.

Language layout

Commands to S are either expressions or assignments. Commands are separated by either a semi-colon, `;`, or a newline. The `#` symbol marks the rest of the line as comments.

The S prompt is `>` unless the command is syntactically incomplete, when the prompt changes to `+`.[2] The only way to extend a command over more than one line is by ensuring that it *is* syntactically incomplete until the final line.

In this book we often omit both prompts and indicate continuation by simple indenting. When we do include prompts it is usually to separate input from output within a display.

An expression command is evaluated and (normally) printed. For example

```
> 1 - pi + exp(1.7)
[1] 3.332355
```

This rule allows any object to be printed by giving its name. Note that `pi` is the value of π. Giving the name of an object will normally print it or a short summary; this can be done explicitly using the function `print`, and `summary` will often give a full description.

An assignment command evaluates an expression and passes the value to a variable but the result is not printed. The recommended assignment symbol is the combination, " `<-` ", so an assignment in S looks like

```
a <- 6
```

which gives the object `a` the value 6. To improve readability of your code we strongly recommend that you put at least one space before and after binary operators, especially the assignment symbol. (We regard the use of " _ " for assignments as unreadable, but it is allowed. In S-PLUS 5.x " = " can almost always be used, but the exceptions are hard to get right.) Assignments using the right-pointing

---

[1] When S writes objects onto an external file in assignment form (using `dump`) the left-hand side of each top-level assignment is always placed in quotes, whether the name is standard or not.

[2] These prompts can be altered: see Section A.3.

combination " -> " are also allowed to make assignments in the opposite direction, but these are never needed and are little used in practice.

An assignment is a special case of an expression with value equal to the value assigned. When this value is itself passed by assignment to another object the result is a multiple assignment, as in

```
b <- a <- 6
```

Multiple assignments are evaluated from right to left, so in this example the value 6 is first passed to a and then to b.

It is useful to remember that the most recently evaluated non-assignment expression in the session is stored as the variable .Last.value[3] and so may be kept as a permanent object by a following assignment such as

```
keep <- .Last.value
```

This is also useful if the result of an expression is unexpectedly not printed; just print(.Last.value).

Vectors and matrices

The objects in S that most users deal with directly are vectors, functions or lists. Note that there are no scalars; vectors of length one are used instead. Vectors consist of numeric or logical values or character strings (and may not mix them). Normally it is unnecessary to be aware if the numeric values are integer, real or even complex, and whether they are stored to single or double precision; S handles all the possibilities in a unified way.

The simplest way to create a vector is to specify its elements by the function c (for concatenate):

```
> mydata <- c(2.9, 3.4, 3.4, 3.7, 3.7, 2.8, 2.8, 2.5, 2.4, 2.4)
> colours <- c("red", "green", "blue", "white", "black")
> x1 <- 25:30
> x1
[1] 25 26 27 28 29 30
> mydata[7]
[1] 2.8
> colours[3]
[1] "blue"
```

and the colon expression specifies a range of integers. The [1] indicates that the printout of the vector starts at element one. Individual elements of a vector are accessed by specifying them by number in square brackets, as in these examples. Character strings may be entered with either double or single quotes (in matching pairs), but will always be printed with double quotes.

Logical values are represented as T and F:

---

[3] If the expression consists of a simple name such as x, only, the .Last.value object is not changed.

```
> mydata > 3
 [1] F T T T T F F F F F
```

and may also be entered as TRUE and FALSE. Logical operators apply element-by-element to vectors, as do arithmetical expressions.

The elements of a vector can also be named and accessed by name:

```
> names(mydata) <- c('a','b','c','d','e','f','g','h','i','j')
> mydata
   a   b   c   d   e   f   g   h   i   j
 2.9 3.4 3.4 3.7 3.7 2.8 2.8 2.5 2.4 2.4
> names(mydata)
 [1] "a" "b" "c" "d" "e" "f" "g" "h" "i" "j"
> mydata["e"]
   e
 3.7
```

An assignment with the left-hand side something other than a simple identifier, as in the first of these, is called a *replacement*. They are disguised calls to a special kind of function called a *replacement function* (already mentioned in passing) that re-constructs the entire modified object. The expanded call in this case has the form

```
mydata <- "names<-"(mydata,
              c('a','b','c','d','e','f','g','h','i','j'))
```

Note that in the third command in the previous display the function names *extracts* the names of the components, and so is a completely different function from the replacement function of (apparently) the same name used in the first line.

More generally, several elements of a vector can be selected by giving a vector of element names or numbers or a logical vector:

```
> letters[1:5]
 [1] "a" "b" "c" "d" "e"
> mydata[letters[1:5]]
   a   b   c   d   e
 2.9 3.4 3.4 3.7 3.7
> mydata[mydata > 3]
   b   c   d   e
 3.4 3.4 3.7 3.7
```

(The object letters is an S vector containing the 26 lower-case English letters; the object LETTERS contains the upper-case equivalents.) We refer to taking *subsets* of vectors (although to mathematicians they are subsequences).

Elements of a vector can be <u>omitted</u> by giving negative indices, for example,

```
> mydata[-c(3:5)]
   a   b   f   g   h   i   j
 2.9 3.4 2.8 2.8 2.5 2.4 2.4
```

but in this case the names can not be used.

There are many special uses of vectors. For example, they can appear to be matrices, arrays or factors. This is handled by giving the vectors *attributes* or classes. All objects have two attributes, their mode and their length:

```
> mode(mydata)
[1] "numeric"
> mode(letters)
[1] "character"
> mode(sin)
[1] "function"
> length(mydata)
[1] 10
> length(letters)
[1] 26
> length(sin)
[1] 2
```

(The length of a function is one plus the number of arguments.)   Under S-PLUS 5.x it is important to note that every object has a *class* and this is much more important than the mode.

Giving a vector a dim attribute allows, and sometimes causes, it to be treated as a matrix or array. For example

```
> names(mydata) <- NULL      # remove the names
> dim(mydata) <- c(2, 5)
> mydata
        [,1] [,2] [,3] [,4] [,5]
[1,]   2.9  3.4  3.7  2.8  2.4
[2,]   3.4  3.7  2.8  2.5  2.4
> dim(mydata) <- NULL
```

Notice how the matrix is filled down columns rather than across rows. The final assignment removes the dim attribute and restores mydata to a vector.[4] A simpler way to create a matrix from a vector is to use matrix

```
> matrix(mydata, 2, 5)
        [,1] [,2] [,3] [,4] [,5]
[1,]   2.9  3.4  3.7  2.8  2.4
[2,]   3.4  3.7  2.8  2.5  2.4
```

The function matrix can also fill matrices by row,

```
> matrix(mydata, 2, 5, byrow = T)
        [,1] [,2] [,3] [,4] [,5]
[1,]   2.9  3.4  3.4  3.7  3.7
[2,]   2.8  2.8  2.5  2.4  2.4
```

---

[4] In S-PLUS 5.x this removes the names.

As these displays suggest, matrix elements can be accessed as `mat[m,n]`, and whole rows and columns by `mat[m,]` and `mat[,n]`, respectively. In S-PLUS 5.x such objects are of class `"matrix"`.

Arrays are multi-way extensions of matrices, formed by giving a `dim` attribute of length three or more. They can be accessed by `A[r, s, t]` and so on. Sections 2.8 and 4.3 discuss matrices and arrays in more detail.

Lists

A list is used to collect together items of different types. For example, an employee record might be created by

```
Empl <- list(employee="Anna", spouse="Fred", children=3,
             child.ages=c(4,7,9))
```

As this example shows, the elements of a list do not have to be of the same length. The components of a list are always numbered and may always be referred to as such. Thus `Empl` is a list of length 4, and the individual components may be referred to as `Empl[[1]]`, `Empl[[2]]`, `Empl[[3]]` and `Empl[[4]]`. Furthermore, since `Empl[[4]]` is a vector, `Empl[[4]][1]` is its first entry. However, it is more convenient to refer to the components explicitly by name, in the form

```
> Empl$employee
[1] "Anna"
> Empl$child.ages[2]
[1] 7
```

Names of components may be abbreviated to the minimum number of letters needed to identify them uniquely. Thus `Empl$employee` may be minimally specified as `Empl$e` since it is the only component whose name begins with the letter 'e', but `Empl$children` must be specified as at least `Empl$childr` because of the presence of another component called `Empl$child.ages`.

In many respects lists are like vectors. For example, the vector of component names is simply a `names` attribute of the list like any other object and may be treated as such; to change the component names of `Empl` to a, b, c and d we can use the replacement

```
> names(Empl) <- letters[1:4]
> Empl[3:4]
$c:
[1] 3
$d:
[1] 4 7 9
```

Notice that we select *components* as if this were a vector (and not `[[3:4]]` as might have been expected). (The distinction is that `[ ]` returns a list with the selected component(s) so `Empl[3]` is a list with one component, whereas `Empl[[3]]` extracts that component from the list.)

The concatenate function, `c`, can also be used to concatenate lists or to add components, so

```
Empl <- c(Empl, service = 8)
```

would add a component for years of service.

The function `unlist` converts a list to a vector:

```
> unlist(Empl)
 employee spouse children child.ages1 child.ages2 child.ages3
  "Anna"   "Fred"  "3"      "4"         "7"          "9"
> unlist(Empl, use.names = F)
[1] "Anna" "Fred" "3"    "4"     "7"     "9"
```

which can be useful for a compact printout (as here). (Mixed types will all be converted to character, giving a character vector.)

The function `c` has named argument `recursive`; if this is T the list arguments are unlisted before being joined together. Thus

```
c(list(x = 1:3, a = 3:6), list(y = 8:23, b = c(3, 8, 39)))
```

is a list with four (vector) components, but adding `recursive=T` gives a vector of length 26. (Try both to see.)

Factors

A factor is a special type of vector, normally used to hold a categorical variable, for example,

```
> citizen <- factor(c("uk","us","no","au","uk","us","us"))
> citizen
[1] uk us no au uk us us
```

Although this is entered as a character vector, it is printed without quotes. Appearances here are deceptive, and a special `print` method is used. Internally the factor is stored as a set of codes, and an attribute giving the *levels*:

```
> print.default(citizen)
[1] 3 4 1 2 3 4 4
attr(, "levels"):
[1] "au" "no" "uk" "us"
attr(, "class"):
[1] "factor"
> unclass(citizen)   # oldUnclass(citizen) on 5.x
[1] 3 4 1 2 3 4 4
```

Why might we want to use this rather strange form? Using a factor indicates to many of the statistical functions that this is a categorical variable (rather than just a list of labels), and so it is treated specially. As just one example, using a factor as the response variable signals to the function `tree` that a classification rather than regression tree is required (see Chapter 10).

By default the levels are sorted into alphabetical order, and the codes assigned accordingly. Some of the statistical functions give the first level a special status, so it may be necessary to specify the levels explicitly:

```
> citizen <- factor(c("uk","us","no","au","uk","us","us"),
      levels = c("us", "fr", "no", "au", "uk"))
> citizen
[1] uk us no au uk us us
Levels:
[1] "us" "fr" "no" "au" "uk"
```

Note that levels which do not occur can be specified, in which case the levels *are* printed. This often occurs when subsetting factors.[5]

A frequency table for the categories defined by a factor is given by the function `table`

```
> table(citizen)
 us fr no au uk
  3  0  1  1  2
```

If `table` is given several factor arguments the result is a multi-way frequency array, as discussed more fully on pages 105ff.

Sometimes the levels of a categorical variable are naturally ordered, as in

```
> income <- ordered(c("Mid","Hi","Lo","Mid","Lo","Hi","Lo"))
> income
[1] Mid Hi  Lo  Mid Lo  Hi  Lo

Hi < Lo < Mid
> as.numeric(income)
[1] 3 1 2 3 2 1 2
```

Again the effect of alphabetic ordering is not what is required, and we need to set the levels explicitly:

```
> inc <- ordered(c("Mid","Hi","Lo","Mid","Lo","Hi","Lo"),
      levels = c("Lo", "Mid", "Hi"))
> inc
[1] Mid Hi  Lo  Mid Lo  Hi  Lo

Lo < Mid < Hi
```

Ordered factors are a special case of factors that some functions (including `print`) treat in a special way.

The function `cut` can be used to create ordered factors by sectioning continuous variables into discrete class intervals. For example,

```
> erupt <- cut(geyser$duration, breaks = 0:6)
> erupt <- ordered(erupt, labels=levels(erupt))
> erupt
  [1] 4+ thru 5 2+ thru 3 3+ thru 4 3+ thru 4 3+ thru 4
  [6] 1+ thru 2 4+ thru 5 4+ thru 5 2+ thru 3 4+ thru 5
  . . . .
  0+ thru 1 < 1+ thru 2 < 2+ thru 3 < 3+ thru 4 < 4+ thru 5 <
      5+ thru 6
```

---

[5] An extra argument may be included when subsetting factors, as in `f[i, drop=T]`, to include only those levels that occur in the subset. Under the default, `drop=F`, the levels are not changed.

Note that the intervals are of the form $(n, n+1]$, so an eruption of 4 minutes is put in category 3+ thru 4.

Data frames

A data frame is the type of object normally used in S to store a data matrix. It should be thought of as a list of variables of the same length, but possibly of different types (numeric, character or logical). Consider our data frame `painters`:

```
> painters
            Composition Drawing Colour Expression School
 Da Udine            10       8     16          3      A
 Da Vinci            15      16      4         14      A
Del Piombo            8      13     16          7      A
Del Sarto            12      16      9          8      A
Fr. Penni             0      15      8          0      A
....
```

which has four numerical variables and one character variable. Since it is a data frame, it is printed in a special way. The components are printed as columns (rather than as rows as vector components of lists are) and there is a set of names, the `row.names`, common to all variables.

```
> row.names(painters)
[1] "Da Udine"        "Da Vinci"         "Del Piombo"
[4] "Del Sarto"       "Fr. Penni"        "Guilio Romano"
[7] "Michelangelo"    "Perino del Vaga"  "Perugino"
....
```

Furthermore, neither the row names nor the values of the character variable appear in quotes.

Data frames can be indexed in the same way as matrices:

```
> painters[1:5, c(2, 4)]
           Drawing Expression
 Da Udine        8          3
 Da Vinci       16         14
Del Piombo      13          7
Del Sarto       16          8
Fr. Penni       15          0
```

But they may also be indexed as lists, which they are. Note that a single index behaves as it would for a list, so `painters[c(2,4)]` gives a data frame of the second and fourth variables which is the same as `painters[, c(2,4)]`.

Variables that satisfy suitable restrictions (having the same length, and the same names, if any) can be collected into a data frame by the function `data.frame`, which resembles `list`:

```
mydat <- data.frame(MPG, Dist, Climb, Day = day)
```

although data frames are most commonly created by reading a file (see read.table on page 30).

There is a side effect of data.frame that needs to be considered; all character and logical columns are converted to factors unless their names are included in I() so, for example,

```
mydat <- data.frame(MPG, Dist, Climb, Day = I(day))
```

preserves day as a character vector, Day.

Compatible data frames can be joined by cbind, which adds columns of the same length, and rbind, which stacks data frames vertically. The result is a data frame with appropriate names and row names.

It is also possible to include matrices and lists within data frames. If a matrix is supplied to data.frame, it is as if its columns were supplied individually; suitable labels are concocted. If a list is supplied, it is treated as if its components had been supplied individually.

Coercion

There is a series of functions named as.xxx that convert to the specified type in the best way possible. For example, as.matrix will convert a numerical data frame to a numerical matrix, and a data frame with any character or factor columns to a character matrix. The function as.character is often useful to generate names and other labels.

Functions is.xxx test if their argument is of the required type. Note that these do not always behave as one might guess; for example, is.vector(mydata) will be *false* as this tests for a 'pure' vector without any attributes such as names. Similarly, as.vector has the (often useful) side effect of discarding all attributes.

2.2 Calling conventions for functions

Functions may have their arguments *specified* or *unspecified* when the function is defined. (We saw how to write simple functions on page 3.)

When the arguments are unspecified there may be an arbitrary number of them. They are shown as ... when the function is defined or printed. Examples of functions with unspecified arguments include the concatenation function c(...) and the parallel maximum and minimum functions pmax(...) and pmin(...).

Where the arguments are specified there are two conventions for supplying values for the arguments when the function is called:

1. arguments may be specified in the same order in which they occur in the function definition, in which case the values are supplied in order, and

2. arguments may be specified as `name=value`, when the order in which the
   arguments appear is irrelevant. The name may be abbreviated providing it
   partially matches just one named argument.

It is important to note that these two conventions may be mixed. A call to a
function may begin with specifying the arguments in positional form but specify
some later arguments in the named form. For example, the two calls

```
t.test(x1, y1, var.equal = F, conf.level = 0.99)
t.test(conf.level = 0.99, var.equal = F, x1, y1)
```

are equivalent.

Functions with named arguments also have the option of specifying *default
values* for those arguments, in which case if a value is not specified when the
function is called the default value is used. For example, the function `t.test`
has an argument list defined as

```
t.test <- function(x, y = NULL, alternative = "two.sided",
           mu = 0, paired = F, var.equal = T, conf.level = 0.95)
```

so that our previous calls can also be specified as

```
t.test(x1, y1, , , , F, 0.99)
```

and in all cases the default values for `alternative`, `mu` and `paired` are used.
Using the positional form and omitting values, as in this last example, is rather
prone to error, so the named form is preferred except for the first couple of arguments.

Some functions (for example `paste`) have both unspecified and specified arguments, in which case the specified arguments occurring after the ... argument
on the definition must be named exactly if they are to be matched at all.

The argument names and any default values for an S function can be found
from the on-line help, by printing the function itself or succinctly using the `args`
function. For example

```
> args(hist)
function(x, nclass, breaks, plot = TRUE, probability = FALSE,
         include.lowest = T, ..., xlab = deparse(substitute(x)))
NULL
```

shows the arguments, their order and those default values that are specified for the
`hist` function for plotting histograms. (The return value from `args` always ends
with NULL.) Note that even when no default value is specified the argument itself
may not need to be specified. If no value is given for `nclass` or `breaks` when
the `hist` function is called, default values are calculated within the function.
Unspecified arguments are passed on to a plotting function called from within
`hist`.

Functions are considered in much greater detail in the companion volume on
S programming.

2.3   Arithmetical expressions

We have seen that a basic unit in S is a vector. Arithmetical operations are performed on vectors, element by element. The standard operators `+ - * / ^` are available, where `^` is the power (or exponentiation) operator (giving x^y).

Vectors may be empty. The expression `numeric(0)` is both the expression to create an empty numeric vector and the way it is represented when printed. It has length zero. It may be described as "a vector such that if there were any elements in it, they would be numbers!"

Vectors can be complex, and almost all the rules for arithmetical expressions apply equally to complex quantities. A complex number is entered in the form `3.1 + 2.7i`, with no space before the `i`. Functions `Re` and `Im` return the real and imaginary parts. Note that complex arithmetic is not used unless explicitly requested, so `sqrt(x)` for x real and negative produces an error. If the complex square root is desired use `sqrt(as.complex(x))` or `sqrt(x + 0i)`.

The recycling rule

The expression `y + 2` is a syntactically natural way to add 2 to each element of the vector `y`, but 2 is a vector of length 1 and `y` may be a vector of any length. A convention is needed to handle vectors occurring in the same expression but not all of the same length. The value of the expression is a vector with the same length as that of the longest vector occurring in the expression. Shorter vectors are *recycled* as often as need be until they match the length of the longest vector. In particular a single number is repeated the appropriate number of times. Hence

```
x <- c(10.4, 5.6, 3.1, 6.4, 21.7)
y <- c(x, x)
v <- 2 * x + 1
```

generates a new vector v of length 10 constructed by

1. repeating the number 2 five times to match the length of the vector x and multiplying element by element, and

2. adding together, element by element, 2*x repeated twice, y as it stands and 1 repeated ten times.

Fractional recycling is allowed in S-PLUS 3.x and 4.x, with a warning, but in S-PLUS 5.x it is an error.

Some standard S functions

Some examples of standard functions follow.

1. There are several functions to convert to integers; `round` will normally be preferred, and rounds to the nearest integer. (It can also round to any number of digits in the form `round(x, 3)`. Using a negative number rounds to a power of 10, so that `round(x,-3)` rounds to thousands.) Each of `trunc`, `floor` and `ceiling` round in a fixed direction, towards zero, down and up, respectively.

2. Other arithmetical operators are `%/%` for integer divide and `%%` for modulo reduction.[6]

3. The common functions are available, including `abs`, `sign`, `log`, `log10`, `sqrt`, `exp`, `sin`, `cos`, `tan`, `acos`, `asin`, `atan`, `cosh`, `sinh` and `tanh` with their usual meanings. Note that the value of each of these is a vector of the same length as its argument. The function `log` has a second argument, the base of the logarithms, defaulting to e. However, in S-PLUS 5.x, `log` has only one argument, and `logb` must be used for 'log to base'.

   Less common functions are `gamma` and `lgamma` ($\log_e \Gamma(x)$).

4. There are functions `sum` and `prod` to form the sum and product of a whole vector, as well as cumulative versions `cumsum` and `cumprod`.

5. The functions `max(x)` and `min(x)` select the largest and smallest elements of a vector `x`. The functions `cummax` and `cummin` give cumulative maxima and minima.

6. The functions `pmax(x1, x2, ...)` and `pmin(x1, x2, ...)` take an arbitrary number of vector arguments and return the element-by-element maximum or minimum values, respectively. Thus the result is a vector of length that of the longest argument and the recycling rule is used for shorter arguments. For example,

   ```
   xtrunc <- pmax(0, pmin(1,x))
   ```

   is a vector like `x` but with negative elements replaced by `0` and elements larger than `1` replaced by `1`.

7. The function `range(x)` returns `c(min(x), max(x))`. If `range`, `max` or `min` is given several arguments these are first concatenated into a single vector.

8. Two useful statistical functions are `mean(x)` which calculates the sample mean, which is the same as `sum(x)/length(x)`, and `var(x)` which gives the sample variance, `sum((x-mean(x))^2)/(length(x)-1)`.[7]

9. `sort` returns a vector of the same size as `x` with the elements arranged in increasing order. See page 44 for further details.

10. The function `rev` arranges the components of a vector or list in reverse order. `duplicated` produces a logical vector with value `T` only where a value in its vector argument has occurred previously and `unique` removes such duplicated values.

---

[6] The result of `e1 %/% e2` is `floor(e1/e2)` if `e2!=0` and `0` if `e2==0`. The result of `e1 %% e2` is `e1-floor(e1/e2)*e2` if `e2!=0` and `e1` otherwise (see Knuth, 1968, §1.2.4). Thus `%/%` and `%%` always satisfy `e1==(e1%/%e2)*e2+e1%%e2`.

[7] If the argument to `var` is an $n \times p$ matrix the value is a $p \times p$ sample covariance matrix obtained by regarding the rows as sample vectors.

Table 2.1: Precedence of operators, from highest to lowest.

`$`	for list extraction
`[, [[`	vector and list element extraction
`^`	exponentiation
`-`	unary minus
`:`	sequence generation
`%%, %/%, %*%`	and other special operators `%...%`
`* /`	multiply and divide
`+ -`	addition and subtraction
`< > <= >= == !=`	comparison operators
`!`	logical negation
`&	&& ||`	logical operators
`~`	formula
`<-, _, ->`	assignment

11. Set operations may be done with the functions union, intersect and setdiff, which enact the set operations $A \cup B$, $A \cap B$ and $A \cap \overline{B}$, respectively. Their arguments (and hence values) may be vectors of any mode but, like true sets, they should contain no duplicated values.

Operator precedence

The formal precedence of operators is given in Table 2.1. However, as usual it is better to use parentheses to group expressions rather than rely on remembering these rules. They can be found on-line from help(Syntax).

Generating regular sequences

There are several ways in S to generate sequences of numbers. For example, 1:30 is the vector c(1, 2, ..., 29, 30). The colon operator has a high precedence within an expression, so 2*1:15 is the vector c(2, 4, 6, ..., 28, 30). Put n <- 10 and compare the sequences 1:n-1 and 1:(n-1).

A construction such as 10:1 may be used to generate a sequence in reverse order.

The function seq is a more general facility for generating sequences. It has five arguments, only some of which may be specified in any one call. The first two arguments, named from and to, if given specify the beginning and end of the sequence, and if these are the only two arguments the result is the same as the colon operator. That is, seq(2,10) and seq(from=2, to=10) give the same vector as 2:10.

The third and fourth arguments to seq are named by and length, and specify a step size and a length for the sequence. If by is not given, the default by=1 is used. For example,

```
s3 <- seq(-5, 5, by=0.2)
s4 <- seq(length=51, from=-5, by=0.2)
```

generate in both `s3` and `s4` the vector $(-5.0, -4.8, -4.6, \ldots, 4.6, 4.8, 5.0)$.

The fifth argument is named `along` and has a vector as its value. If it is the only argument given it creates a sequence 1, 2, ..., `length`(*vector*), or the empty sequence if the value is empty. (This makes `seq(along=x)` preferable to `1:length(x)` in most circumstances.) If specified rather than `to` or `length` its length determines the length of the result.

A companion function is `rep` which can be used to repeat an object in various ways. The simplest form is

```
s5 <- rep(x, times=5)
```

which will put five copies of `x` end-to-end in `s5`.

A `times=v` argument may specify a vector of the same length as the first argument, `x`. In this case the elements of `v` must be non-negative integers, and the result is a vector obtained by repeating each element in `x` a number of times as specified by the corresponding element of `v`. Some examples will make the process clear:

```
x <- 1:4        # puts c(1,2,3,4)              into x
i <- rep(2, 4)  # puts c(2,2,2,2)              into i
y <- rep(x, 2)  # puts c(1,2,3,4,1,2,3,4)      into y
z <- rep(x, i)  # puts c(1,1,2,2,3,3,4,4)      into z
w <- rep(x, x)  # puts c(1,2,2,3,3,3,4,4,4,4)  into w
```

As a more useful example, consider a two-way experimental layout with four row classes, three column classes and two observations in each of the twelve cells. The observations themselves are held in a vector `y` of length 24 with column classes stacked above each other, and row classes in sequence within each column class. Our problem is to generate two indicator vectors of length 24 that will give the row and column class, respectively, of each observation. Since the three column classes are the first, middle and last eight observations each, the column indicator is easy. The row indicator requires three calls to `rep`:

```
> colc <- rep(1:3,rep(8,3));  colc
> [1] 1 1 1 1 1 1 1 1 2 2 2 2 2 2 2 2 3 3 3 3 3 3 3 3
> rowc <- rep(rep(1:4, rep(2,4)), 3); rowc
> [1] 1 1 2 2 3 3 4 4 1 1 2 2 3 3 4 4 1 1 2 2 3 3 4 4
```

These can also be generated arithmetically using the `ceiling` function

```
> 1 + (ceiling(1:24/8) - 1) %% 3 -> colc; colc
> [1] 1 1 1 1 1 1 1 1 2 2 2 2 2 2 2 2 3 3 3 3 3 3 3 3
> 1 + (ceiling(1:24/2) - 1) %% 4 -> rowc; rowc
> [1] 1 1 2 2 3 3 4 4 1 1 2 2 3 3 4 4 1 1 2 2 3 3 4 4
```

In general the expression `1 + (ceiling(1:n/r) - 1) %% m` generates a sequence of length `n` consisting of the numbers 1, 2, ..., `m` each repeated `r` times. This is often a useful idiom.

Logical expressions

Logical vectors are most often generated by *conditions*. The logical operators
are <, <=, >, >= (which have self-evident meanings), == for exact equality
and != for exact inequality. If c1 and c2 are vector valued logical expressions,
c1 & c2 is their intersection ('and'), c1 | c2 is their union ('or') and !c1 is
the negation of c1. These operations are performed separately on each component
with the recycling rule applying for short arguments.

Logical vectors may be used in ordinary arithmetic. They are *coerced* into
numeric vectors, F becoming 0 and T becoming 1. For example, assuming the
value or values in sd are positive

```
N.extreme <- sum(y < ybar - 3*sd | y > ybar + 3*sd)
```

would count the number of elements in y that were farther than 3*sd from
ybar on either side. The right-hand side can be expressed more concisely as
sum(abs(y-ybar) > 3*sd).

The function xor computes (element-wise) the exclusive or of its two arguments.

The functions any and all are useful to collapse a logical vector. The function all.equal provides a way to test for equality up to a tolerance if appropriate.

The missing value marker, NA

Not all the elements of a vector may be known. When an element or value is 'not
available' or a 'missing value', a place within a vector may be reserved for it by
assigning the special value NA.

In general any operation on an NA becomes an NA. The motivation for this
rule is simply that if the specification of an operation is incomplete, the result
cannot be known and hence is not available.

The function is.na(x) gives a logical vector of the same length as x with
values that are true if and only if the corresponding element in x is NA.

```
ind <- is.na(z)
```

Notice that the logical expression x == NA is not equivalent to is.na(x). Since
NA is really not a value but a marker for a quantity that is not available, the first
expression is incomplete. Thus x == NA is a vector of the same length as x
all of whose values are NA irrespective of the elements of x. In the same way,
min(x) is NA if any element of x is the missing value.

The preferred way to set a value to missing in S-PLUS 5.x is to use is.na
on the left-hand side of an assignment, as in

```
> is.na(mydata)[6] <- T
> mydata
 [1] 2.9 3.4 3.4 3.7 3.7  NA 2.8 2.5 2.4 2.4
```

S functions differ markedly in their policy for handling missing values. Some have an option `na.rm=T` to remove missing values. Many will omit rows that contain a missing value after reducing the data matrix to only those columns needed for a calculation. For statistical functions the `na.action` argument may allow other possibilities. The default `na.action` is usually `na.fail` which causes the procedure to stop; the alternative `na.omit` implements the row-omission policy. In S-PLUS 2000, there is further option, `na.exclude` that likke `na.omit` omits rows but fills in residuals and fitted values with `NA`s.

Missing values are output as `NA`, and can be input as `NA` or by ensuring that a value is missing (for example, a field is blank). When character strings are coerced to mode numeric the result is `NA` unless the string parses as a number.

Elementary matrix operations

We have seen that a matrix is merely a data vector with a `dim` attribute specifying a double index. However, S contains many operators and functions for matrices; for example `t(X)` is the transpose function. The functions `nrow(A)` and `ncol(A)` give the number of rows and columns in the matrix A.

The operator `%*%` is used for matrix multiplication. Vectors that occur in matrix multiplications are if possible promoted either to row or to column vectors, whichever is multiplicatively coherent. (This may be ambiguous, as we show.) Note carefully that if A and B are square matrices of the same size, then A * B is the matrix of element-by-element products whereas A `%*%` B is the matrix product. If x is a vector, then

    x %*% A %*% x

is a quadratic form $x^T A x$, where x is the column vector and T denotes transpose.

Note that x `%*%` x seems to be ambiguous, as it could mean either $x^T x$ or $x x^T$. A more precise definition of `%*%` is that of an inner product rather than a matrix product, so in this case $x^T x$ is the result. (For $x x^T$ use x `%o%` x; see page 98.)

Matrices can be built up from other vectors and matrices by the functions `cbind` and `rbind`. Informally `cbind` forms matrices by binding together vectors or matrices column-wise, and `rbind` binds row-wise. The arguments to `cbind` must be either vectors of any length, or matrices with the same column size, that is, the same number of rows. The result is a matrix with the concatenated arguments forming the columns. If some of the arguments to `cbind` are vectors they may be shorter than the column size of any matrices present, in which case they are cyclically extended to match the matrix column size (or the length of the longest vector if no matrices are given). The function `rbind` performs the corresponding operation for rows. In this case any vector arguments, possibly cyclically extended, are taken as rows. Note that we have already seen `cbind` and `rbind` operating on data frames.

Further matrix operations are discussed in Sections 2.8 and 4.3.

	Price	Floor	Area	Rooms	Age	Cent.heat
01	52.00	111.0	830	5	6.2	no
02	54.75	128.0	710	5	7.5	no
03	57.50	101.0	1000	5	4.2	no
04	57.50	131.0	690	6	8.8	no
05	59.75	93.0	900	5	1.9	yes
	. . . .					

Figure 2.1: Input file form with names and row names.

2.4 Reading data

Large data objects will usually be read as values from external files rather than entered during an S session at the keyboard. The S input facilities are simple and their requirements are fairly strict and rather inflexible. There is a presumption by the designers of S that you will be able to modify your input files using other tools, such as file editors and the UNIX utilities sed and awk, to fit in with the requirements of S. There are some tools within S that can cater for non-standard situations, as we discuss in the following, but you may also be able to use the data import functions discussed in Section 2.9.

If variables are to be held in data frames, as we strongly suggest they should be, an entire data frame can be read directly with the read.table function. There is also a more general input function, scan, that is useful in special circumstances.

The read.table function

In order to be read into a data frame, an external file should have a standard form:

1. The first line of the file should have a *name* for each variable in the data frame. (Header lines can be ignored by setting the argument skip to skip an appropriate number of lines.)

2. Each additional line of the file has as its first item a *row name* and the values for each variable, separated by spaces, tabs or both. Character strings containing blanks must be contained within quotes, which are otherwise optional.

The first few lines of a file to be read as a data frame are shown in Figure 2.1. By default numeric items (except row names) are read as numeric variables and non-numeric variables, such as Cent.heat in the example, as factors. (This can be changed if necessary via the argument as.is.)

The function read.table is used to read in the data frame

```
HPrice <- read.table("houses.dat")
```

Price	Floor	Area	Rooms	Age	Cent.heat
52.00	111.0	830	5	6.2	no
54.75	128.0	710	5	7.5	no
57.50	101.0	1000	5	4.2	no
57.50	131.0	690	6	8.8	no
59.75	93.0	900	5	1.9	yes
....					

Figure 2.2: Input file form without row names.

It is often convenient to generate row names within S-PLUS. In this case the file should omit the row name column (as in Figure 2.2), and we must specify header=T in the call to read.table. The row names may be specified as an argument, row.names, to read.table, but to ensure the default row names, namely the row numbers, are generated we must specify row.names=NULL, as in

```
HPrice <- read.table("houses1.dat", header=T, row.names=NULL)
```

If the row.names argument is omitted, the first non-numeric column with all components different will be used as the row names, if one exists. This can be very puzzling if it is not anticipated.

The argument na.strings can be used to specify a character vector of input strings to map to NA. The skip argument may be used to specify a number of lines in the file to be omitted before reading data.

Note that data frames *must* have both row and column names so read.table supplies default values where necessary. The default names can be surprisingly unhelpful, so we recommend always supplying at least column names.

Data file names

UNIX users can use any legal file name for data files. So can Windows users, but they have to take care how they specify them, as backslashes (\) within names have to be doubled inside S, for example, as

```
"c:\\mywork\\splus\\sws\\file.dat"
```

For functions expecting the name of a file as a particular argument this can (usually) also be specified with slashes as

```
"c:/mywork/splus/sws/file.dat"
```

which can be easier to use, especially for users conversant with UNIX. The file-name clipboard may be used in Windows to refer to the Windows clipboard for input or output (but the file length is limited to 32 Kb).

The function scan

The simplest use of the scan function is to input a single vector from the keyboard:

```
> x <- scan()
1: 23.4 45.6 77.8 12.9
5: 20 10 11
8: 33
9:
> x
[1] 23.4 45.6 77.8 12.9 20.0 10.0 11.0 33.0
```

The default arguments specify that input is to come from the keyboard. Data items are entered as the prompt changes to n: where n is the index of the next item to be read. Reading is terminated by an empty input line (only from the keyboard) or by the end of file.

Data may be read as a vector from an external file using scan with the file name as argument. Suppose, for example, the file mat.dat contains

12	5	4	3
5	17	2	1
6	4	19	0
4	5	1	21

where each line is intended to become a row of an S matrix M. To read the matrix we use

```
M <- matrix(scan("mat.dat"), ncol=4, byrow=T)
```

It is not necessary to specify how many rows the matrix has since this is deduced from the number of columns and the total number of items read. The argument byrow=T indicates that the vector is to fill the matrix by rows rather than by columns.

Like read.table, scan also has an argument skip that allows a number of lines at the top of the input file to be omitted before reading data.

Another common way to use scan is similar to read.table but with more flexibility. Suppose the data vectors are of equal length and are to be read in parallel from a data file input.dat. Suppose that there are three vectors, the first of mode character and the remaining two of mode numeric. The first step is to use scan to read in the three vectors as a list, as follows

```
indat <- scan("input.dat", list(id="", x=0, y=0))
```

The second argument is a template list that establishes the mode of the three vectors to be read and the structure of the result, which is a list. If we wish to access the variables separately they may either be re-assigned to variables:

```
label <- indat$id; x <- indat$x; y <- indat$y
```

or the list itself may be attached to the search path, by

```
attach(indat)
```

As a final example, suppose we need to read in a large file with 50 numeric variables, and each case occupies 5 lines of the file `big.dat`. We are willing to label the variables X1, X2, ..., X50. The first step is to construct the template list:[8]

```
inlist <- as.list(numeric(50))          # a list of 50 zeros.
names(inlist) <- paste("X", 1:50, sep="")
```

To read the file we must also specify that each case will occupy more than one line of the data file:

```
datlist <- scan("big.dat", what=inlist, multi.line=T)
```

This is only possible using the `multi.line=T` argument with `scan`; multi-line files cannot yet be read with `read.table`.

There is a function `count.fields` that will count the number of fields on each line of a file, which can be useful in locating faulty lines in a large file.

2.5 Model formulae

Model formulae were introduced into S as a compact way to specify linear models, but have since been adopted for so many diverse purposes in S-PLUS that they are now best regarded as an integral part of the S language. The various uses of model formulae all have individual features which are treated in the appropriate chapter, based on the common features described here.

A formula is of the general form

```
response ~ expression
```

where the left-hand side, `response`, may sometimes be absent and the right-hand side, `expression`, is a collection of terms joined by operators usually resembling an arithmetical expression. The meaning of the right-hand side is context dependent. For example, in non-linear regression it is an arithmetical expression and all operators have their usual arithmetical meaning. In linear and generalized linear modelling it specifies the form of the model matrix and the operators have a different meaning. In Trellis graphics it is used to specify the abscissa variable for a plot, but a vertical bar, | , operator is allowed to indicate conditioning variables.

Some functions such as `lme` and `nlme` for mixed effects models may require several formulae as arguments.

It is conventional (but not quite universal) that a function which interprets a formula also has arguments `weights`, `data`, `subset` and `na.action`. Then the formula is interpreted in the *context* of the argument `data` which must be a list, usually a data frame; the objects named on either side of the formula are

---

[8] `paste` is discussed on page 35.

looked for first in data and then searched for in the usual way (described in detail in the Section 2.7). The weights and subset arguments are also interpreted in the context of the data frame.

We have seen a few formulae in Chapter 1, all for linear models, where the response is the dependent variable and the right-hand side specifies the explanatory variables. We had

```
fm <- lm(y ~ x,  data=dum)
abline(lm(time ~ dist))
fm <-  aov(Speed ~ Run + Expt)
fm0 <- update(fm, . ~ . - Run)
```

Notice that in these cases + indicates inclusion, not addition, and - exclusion.

In most cases we had already attached the data frame, so did not specify it via a data argument. The function update is a very useful way to change the call to functions using model formulae; it reissues the call having updated the formula (and any other arguments specified when it is called). The formula term " . " has a special meaning in a call to update; it means 'what is there already' and may be used on either side of the ~.

It is implicit in this description that the objects referred to in the formula are of the same length, or constants that can be replicated to that length; they should be thought of as all being measured on the same set of units. The other two arguments allow that set of units to be altered: subset is an expression evaluated in the context of data that should evaluate to a valid indexing vector (of types 1, 2 or 4 on page 39). The na.action argument specifies what is to be done when missing values are found by specifying a function to be applied to the data frame of all the data needed to process the formula. The default action is usually na.fail, which reports an error and stops, but some functions have more accommodating defaults.

Further details of model formulae are given in later chapters. Many of these involve special handling of factors and functions appearing on the right-hand side of a formula.

2.6 Character vector operations

The form of character vectors can be unexpected and should be carefully appreciated. Unlike say C, they are vectors of character strings, not of characters, and most operations are performed separately on each component.

Note that "" is a legal character string with no characters in it, known as the empty string. This should be contrasted with character(0) which is the empty character vector. As vectors, "" has length 1 and character(0) has length 0.

Character vectors may be created by assignment and may be concatenated by the c function. They may also be used in logical expressions, such as "ann" < "belinda", in which case lexicographic ordering applies using the ASCII collating sequence.

There are several functions for operating on character vectors. The function
nchar(text) gives (as a vector) the number of characters in each element of
its character vector argument. The function paste takes an arbitrary number
of arguments, coerces them to strings or character vectors if necessary and joins
them, element by element, as character vectors. For example

```
> paste(c("X","Y"), 1:4)
[1] "X 1" "Y 2" "X 3" "Y 4"
```

Any short arguments are re-cycled in the usual way. By default the joined elements are separated by a blank; this may be changed by using the argument
sep=string, often the empty string:

```
> paste(c("X","Y"), 1:4, sep="")
[1] "X1" "Y2" "X3" "Y4"
```

Another argument, collapse, allows the result to be concatenated into a single
long string. It prescribes another character string to be inserted between the components during concatenation. If it is NULL, the default, or character(0), no
such global concatenation takes place. For example,

```
> paste(c("X","Y"), 1:4, sep="", collapse=" + ")
[1] "X1 + Y2 + X3 + Y4"
```

Substrings of the strings of a character vector may be extracted (element-by-
element) using the substring function. It has three arguments

```
substring(text, first, last = 1000000)
```

where text is the character vector, first is a vector of first character positions
to be selected and last is a vector of character positions for the last character to
be selected. If first or last are shorter vectors than text they are re-cycled
in the usual way.

The first argument is coerced to character if necessary, so one way to generate
number labels padded with leading zeros to a constant width is to use a construction like the example on page 48.

For another example, the dataset state.name is a character vector of length
50 containing the names of the states of the United States of America in alphabetic
order. To extract the first four letters in the names of the last seven states:

```
> substring(state.name[44:50], 1, 4)
[1] "Utah" "Verm" "Virg" "Wash" "West" "Wisc" "Wyom"
```

Note the use of the index vector [44:50] to select the last seven states.

See Section 4.2 for further character manipulation functions.

2.7   Finding S objects

It is important to understand where S keeps its objects and where it looks for objects on which to operate. The objects that S creates at the interactive level during a session it (usually) stores permanently, as *files* in the .Data subdirectory of your working directory. (Subdirectory _Data under Windows.) On the other hand, objects created at a higher level, such as within a function, are kept in what is known as a *local frame* which is only temporary and such objects are deleted when the function is finished.

Each permanent object is held as a file using a *mapped* file name, possibly as part of a large file known as __BIG or __Objects. What is important to note is that when S is restarted at a later time, objects created in previous sessions are still available. This explains why we recommend that you use separate working directories for different jobs. For some users common names for objects are single letter names like x, y and so on, and if two projects share the same data subdirectory their objects easily become mixed up. However, data frames provide another convenient way of encapsulating and partitioning small collections of related objects within the same data directory.

When S looks for an object, it searches through a sequence of places known as the *search path*.[9] Usually the first entry in the search path is the data subdirectory of the current working directory. The names of the places currently making up the search path are given by invoking the function

```
search()
```

To get the names of all objects currently held in the first place on the search path, use the command

```
objects()
```

The names of the objects held in any place in the search path can be displayed by giving the objects function an argument. For example,

```
objects(2)
```

lists the contents of the entity at position 2 of the search path. It is also possible to list selectively, using the pattern argument of objects but as this differs between systems, please consult your on-line help. Finally, users of S-PLUS 4.x and 2000 can explore objects graphically; see page 459.

Conversely the function find(*object*) discovers where an object appears on the search path, perhaps more than once. For example (on one of our UNIX systems after library(treefix, first=T)) we have

---

[9] The official reference books often use the name *search list* but we find this use of 'list' potentially misleading.

```
> search()
[1] ".Data"
[2] "/usr/local/splus/library/treefix/.Data"
[3] "/usr/local/splus/splus/.Functions"
[4] "/usr/local/splus/stat/.Functions"
[5] "/usr/local/splus/s/.Functions"
[6] "/usr/local/splus/s/.Datasets"
[7] "/usr/local/splus/stat/.Datasets"
[8] "/usr/local/splus/splus/.Datasets"
> find("prune.tree")
[1] "/usr/local/splus/library/treefix/.Data"
[2] "/usr/local/splus/splus/.Functions"
```

(The format will differ between systems and S-PLUS versions. In particular, on
S-PLUS 5.x it will be like

```
> search()
[1] ".Data"         "treefix"        "splus"      "stat"
[5] "data"          "documentation"  "trellis"    "main"
> find("prune.tree")
[1] "treefix" "splus"
```

although the full path information is available via search("paths").)

In the preceding example the object named prune.tree occurs in two places
on the search path. Since the version from our treefix library occurs in position 2 before the S-PLUS version in position 3, ours is the one that will be found
and used. By this mechanism changes may be made to S-PLUS functions without
affecting the original version.

The 'places' on the search path can be of two main types.[10] As well as data
directories of (S created) files, they can also be S lists, usually data frames. In the
S literature any entity that can be placed on the search path is sometimes referred
to as a *dictionary*, or as a *database*. The directory at position 1 (normally .Data
or _Data) is called the *working database*. A library section is a specialized use
of a directory, discussed in Appendix C.2.

As we have noted, if several different objects with the same name occur on the
search path all but the first will be masked and normally unreachable. To bypass
this, the function get may be used to select an object from any given position on
the search path. For example,

```
s.prune.tree <- get("prune.tree", where = 3)
```

copies the object called prune,tree from position 3 on the search path to an
object called s.prune.tree in the working database.[11] Functions masked and
conflicts are available to help check if system objects are being masked by
user-defined objects, whether intentionally or not.

---

[10] There are further types of possible database that we have never had occasion to use.

[11] Note that the object name as the argument to get must be given in quotes.

Extra directories, lists or data frames can be added to this list with the `attach` function and removed with the `detach` function. Examples were included in the introductory session. Normally a new entity is attached at position 2, and `detach()` removes the entity at position 2, normally the result of the last `attach`. All the higher-numbered databases are moved up or down accordingly. Note that lists can be attached by `attach(alist)` but must[12] be detached by `detach("alist")`. If a list is attached, a copy is used, so any subsequent changes to the original list will not be reflected in the attached copy. When a list is detached the copy is normally discarded, but if any changes have been made to that database it will be saved unless the argument `save=F` is set. The name used for the saved list is of the form `.Save.alist.2` (and is reported by `detach`) unless `save` is a character string when the database will be saved under that name.

When a command would alter an object that is not on the working database, a copy must be made on the working database first. Earlier systems (S-PLUS 3.x and 4.x) did this silently, but S-PLUS 5.x does not and will report an error, so a manual copy must be made. Objects are usually altered through assignment with a replacement function, for example (page 165),

```
> hills <- hills
> hills$ispeed <- hills$time/hills$dist
```

To remove objects permanently from the working database the function `rm` is used with arguments giving the names of the objects to be discarded, as in

```
rm(x, y, z, ink, junk)
```

If the names of objects to be removed are held in a character vector it may be specified as a named argument. An equivalent form of the preceding command is

```
junk <- c("x", "y", "z", "ink", "junk")
rm(list = junk)
```

The function `remove` can be used to remove objects with non-standard names or from data directories other than position 1 of the search path. The objects to be removed must be specified as a character vector. A further equivalent to the preceding command is

```
remove(junk, where = 1)
```

To remove all the objects in the working directory use

```
remove(objects())
```

A considerable degree of *caching* of databases is performed, so their view may differ inside and outside the S session. To avoid this, use the `synchronize` function. With no argument, it writes out objects to the current working database. With a numerical vector argument, it re-reads the specified directories on the search path, which can be necessary if some other process has altered the database.

[12] In 3.x and 4.x. In 5.x `detach(alist)` *is* allowed.

It is not advisable to use operating-system commands to manipulate the data directories, especially on Windows where S-PLUS keeps summary files of the contents.

This is the end of the informal part of Chapter 2. You might like to take a rest, start skimming or move on to Chapter 3 at this point.

2.8   Indexing vectors, matrices and arrays

We have already seen how subsets of the elements of a vector (or an expression evaluating to a vector) may be selected by appending to the name of the vector an *index vector* in square brackets. We now consider indexing more formally. For vector objects, index vectors can be any of four distinct types.

1. **A logical vector.** The index vector must be of the same length as the vector from which elements are to be selected. Values corresponding to T in the index vector are selected and those corresponding to F omitted. For example,

    ```
    y <- x[!is.na(x)]
    ```

    creates an object y that will contain the non-missing values of x, in the same order as they originally occurred. Note that if x has any missing values, y will be shorter than x. Another example is

    ```
    z <- (x+y)[!is.na(x) & x > 0]
    ```

    which creates an object z and places in it the values of the vector x+y for which the corresponding value in x was positive (and non-missing). The system function unique uses the function duplicated to give x[!duplicated(x)].

2. **A vector of positive integers.** In this case the values in the index vector must lie in the set { 1, 2, ..., length(x) }. The corresponding elements of the vector are selected and concatenated, in that order, in the result. The index vector can be of any length and the result is of the same length as the index vector. For example x[6] is the sixth component of x and x[1:10] selects the first 10 elements of x (assuming length(x) $\geqslant$ 10, otherwise there will be an error). For another example we use the dataset letters, a character vector of length 26 containing the lower-case letters:

    ```
    > letters[1:3]
    [1] "a" "b" "c"
    > letters[1:3][c(1:3,3:1)]
    [1] "a" "b" "c" "c" "b" "a"
    ```

3. **A vector of negative integers.** This specifies the values to be *excluded* rather than included. Thus

```
> y <- x[-(1:5)]
```

drops the first five elements of x.

4. A vector of character strings. This possibility only applies where an object has a names attribute to identify its components. In that case a subvector of the names vector may be used in the same way as the positive integers in case **2.** For example,

```
> longitude <- state.center[["x"]]
> names(longitude) <- state.name
> longitude[c("Hawaii", "Alaska")]
  Hawaii  Alaska
 -126.25 -127.25
```

finds the longitude of the geographic centre of the two most western states of the USA. The names attribute is retained in the result.

A vector with an index expression attached can also appear on the left-hand side of an assignment, making the operation a *replacement*. In this case the assignment operation appears to be performed only on those elements of the vector implied by the index. For example,

```
x[is.na(x)] <- 0
```

replaces any missing values in x by zeros. Note that this is really a disguised call to the function "[<-", which copies the entire object. Replacing even one element of a large object can be a memory-expensive operation. For another example note that

```
y[y < 0] <- -y[y < 0]
```

has the same effect as y <- abs(y).

The case of a zero index falls outside these rules. A zero index in a vector of an expression being assigned passes nothing, and a zero index in a vector to which something is being assigned accepts nothing. For example,

```
> a <- 1:4
> a[0]
numeric(0)
> a[0] <- 10
> a
[1] 1 2 3 4
```

Zero indices may be included with otherwise negative indices or with otherwise positive indices but not with both positive and negative.

Another case to be considered is if the absolute value of an index falls outside the range 1, ..., length(x). In an expression this gives NA if positive and imposes no restriction if negative. On the left-hand side of a replacement, a positive index greater than length(x) extends the vector, assigning NAs to any gap, and a negative index less than -length(x) is ignored.

The functions `replace` and `append` are convenience functions; `replace(x, pos, values)` returns a copy of x with `x[pos] <- values` without affecting the original. The function `append(x, values, after=length(x))` appends `values` to x in the specified place, and returns a copy without changing x.

Array indices

An *array* can be considered as a multiply indexed collection of data entries. Any array with just two indices is called a *matrix*, and this special case is perhaps the most important.

A *dimension vector* is a vector of positive integers of length at least 2. If its length is k then the array is k-dimensional, or as we prefer to say, k-indexed. The values in the dimension vector give the upper limits for each of the k indices. Index ranges for S objects always start at 1 (unlike those of C which start at 0). Note that singly subscripted arrays do not exist other than as vectors.

A vector can be used by S as an array only if it has a dimension vector as its `dim` attribute. Suppose, for example, a is a vector of 150 elements. Either of the assignments

```
a <- array(a, dim=c(3,5,10)) # make a a 3x5x10 array
dim(a) <- c(3,5,10)          # alternative direct form
```

gives it the `dim` attribute that allows it to be treated as a $3 \times 5 \times 10$ array. The elements of a may now be referred to either with one index, as before, as in `a[!is.na(z)]` *or* with three comma-separated indices, as in `a[2,1,5]`.

To create a matrix the function `array` may be used, or the simpler function `matrix`. For example, to create a 10×10 matrix of zeros we could use

```
Zmat <- matrix(0, nrow=10, ncol=10)
```

Note that a vector used to define an array or matrix is recycled if necessary, so the 0 here is repeated 100 times.

It is important to know how the two indexing conventions correspond; which element of the vector is `a[2,1,5]` ? S arrays use 'column-major order' (as used by FORTRAN but not C). This means the first index moves fastest, and the last slowest. So the correspondence is

`a[1]`	$\longleftrightarrow$	`a[1,1,1]`,	`a[5]` $\longleftrightarrow$ `a[2,2,1]`,
`a[2]`	$\longleftrightarrow$	`a[2,1,1]`,	$\dots$ $\longleftrightarrow$ $\dots$,
`a[3]`	$\longleftrightarrow$	`a[3,1,1]`,	`a[149]` $\longleftrightarrow$ `a[3,5,9]`,
`a[4]`	$\longleftrightarrow$	`a[1,2,1]`,	`a[150]` $\longleftrightarrow$ `a[3,5,10]`

A formal definition of column-major order can be given by saying if the dimension vector is $(d_1, d_2, \ldots, d_m)$ then the element with multiple index $i_1, i_2, \ldots, i_m$ corresponds to the element with single index

$$i_1 + \sum_{j=2}^{m} \left\{ (i_j - 1) \prod_{k=1}^{j-1} d_k \right\}$$

The function `matrix` has an additional argument `byrow` which if set to true allows a matrix to be generated from a vector in row-major order.

Each of the dimensions can be given a set of names, just as for a vector. The names are stored in the `dimnames` attribute which is a list of (possibly NULL) vectors of character strings. For example, we can name the first two dimensions of `a` by

```
dimnames(a) <- list(letters[1:3],
                    c("i", "ii", "iii", "iv", "v"), NULL)
```

For a k-fold indexed array any of the four forms of indexing is allowed in each index position. There are two additional possibilities:

5. Any array index position may be empty.   In this case the index range implied is the entire range allowed for that index.

6. An array may be indexed by a matrix.   In this case if the array is k-indexed the index matrix must be an $m \times k$ matrix with integer entries and each row of the index matrix is used as an index vector specifying one element of the array. Thus the matrix specifies m elements of the array to be extracted or replaced.

So if `a` is a $3 \times 5 \times 10$ array, then `a[1:2,,]` is the $2 \times 5 \times 10$ array obtained by omitting the last level of the first index. The same sub-array could be specified in this case by `a[-3,,]`.

To give a simple example of a matrix index, consider extracting the diagonal elements of a square matrix `X`, that is, `X[1,1]`, `X[2,2]`, ..., `X[n,n]` in a vector, say, `Xii`, and then zeroing the diagonal.[13]

```
n <- dim(X)[1]              # same as nrow(X)
i.i <- matrix(1:n, n, 2)    # using the recycling rule
Xii <- X[i.i]               # extract the diagonal
X[i.i] <- 0                 # replace each by zero
```

Note that by default `a[2,,]` is not a $1 \times 5 \times 10$ array but a 5×10 array. Also `a[2,,1]` and `Xii` in the previous example are vectors and not arrays. In general if any index range reduces to a single value the corresponding element of the dimension vector is removed in the result. This default convention is sometimes helpful and sometimes not. To override it a named argument `drop=F` can be given in the array reference:

```
sua <- a[2,,]               # a   5x10 matrix
sub <- a[2,,, drop=F]       # a 1x5x10 array
```

Note that the drop convention also applies to columns (but not rows) of data frames, so if subscripting leaves just one column, a vector is returned.

However, this convention does not apply to matrix operations. Thus

---

[13] This example is artificial as there is a function `diag` that can be used for both purposes.

```
X <- matrix(1:30, 10, 3)
X[,1]
X %*% c(1,3,5)
```

result in a 10-element vector and a 10×1 matrix, respectively, which appears inconsistent. (One can argue for either convention, as reducing $1 \times n$ matrices to vectors is often undesirable.) The function `drop` forces dropping, so `drop(X %*% c(1,3,5))` returns a vector and `X[,1, drop=F]` returns a matrix. (Function writers have often overlooked these rules, which can result in puzzling or incorrect behaviour when just one observation or variable meets some selection criterion.)

The function `dim` can be used to find the dimensions of an array or matrix. For matrices two convenience functions `nrow` and `ncol` are provided[14] to access the number of rows and columns, respectively. There are further convenience functions `row` and `col` that can be applied to matrices to produce a matrix of the same size filled with the row or column number. Thus to produce the upper triangle of a square matrix `A` we can use

```
A[col(A) >= row(A)]
```

This is a logical vector index, and so returns the upper triangle in column-major order. For the lower triangle we can use `<=` or the function `lower.tri`. A few S functions want the lower triangle of a symmetric matrix in row-major order: note that this is the upper triangle in column-major order.

More generally there is a function `slice.index(A, k)` which generates an array like `A` with entries the value of the `k`th index. Thus if `M` is a matrix `row(M)`, for example, is the same as `slice.index(M, 1)`.

Array arithmetic

Arrays may be used in ordinary arithmetic expressions and the result is an array formed by element-by-element operations on the data vector. The `dim` attributes of operands generally need to be the same, and this becomes the dimension vector of the result. So if `A`, `B` and `C` are all arrays of the same dimensions

```
D <- 2*A*B + C + 1
```

makes `D` a similar array with its data vector the result of the evident element-by-element operations. However the precise rule concerning mixed array and vector calculations has to be considered a little more carefully. From experience we have found the following to be a reliable guide, although it is to our knowledge undocumented and hence liable to change.

- The expression is scanned from left to right.
- Any short vector operands are extended by recycling their values until they match the size of any other operands.

---

[14] Note that the names are singular: it is all too easy to write `nrows`!

- As long as short vectors and arrays *only* are encountered, the arrays must all have the same `dim` attribute or an error results.

- Any vector operand longer than some previous array immediately converts the calculation to one in which all operands are coerced to vectors. A diagnostic message is issued if the size of the long vector is not a multiple of the (common) size of all previous arrays.

- If array structures are present and no error or coercion to vector has been precipitated, the result is an array structure with the common `dim` attribute of its array operands.

S-PLUS 5.x has an additional rule: if any value found has length zero, the result has length zero. (Earlier versions would use the re-cycling rule and fill the result with missing values.)

Sorting

The S function `sort` at its simplest takes one vector argument and returns a vector of sorted values. The vector to be sorted may be numeric or character, and if there is a names attribute the correspondence of names is preserved. As a simple example consider sorting `mydata`:

```
> mydata
  a   b   c   d   e   f   g   h   i   j
2.9 3.4 3.4 3.7 3.7 2.8 2.8 2.5 2.4 2.4
> sort(mydata)
  i   j   h   f   g   a   b   c   d   e
2.4 2.4 2.5 2.8 2.8 2.9 3.4 3.4 3.7 3.7
```

Note that the ordering of tied values is preserved.

The second argument to `sort` allows partial sorting. The argument specifies an index or set of indices that represent the order statistics guaranteed to be correct in the result. The values between these reference indices will all be intervening values, but may not be in sorted order. For example, consider finding the median of a large sample from the $N(0, 1)$ distribution directly:

```
> x <- rnorm(100001)
> sort(x, partial=50001)[50001]
[1] 0.0028992
```

This is most often used to find quantiles, for which the functions `median` and `quantile` are provided.

A more flexible sorting tool is `sort.list`,[15] which produces an index vector that will arrange its argument in increasing order. Thus `x[sort.list(x)]` returns the same value as `sort(x)` and `x[sort.list(-x)]` is the vector x arranged in *decreasing* order. For example, to arrange the states of the USA from west to east according to their geographical centres we could use

---

[15] The use of "list" is unfortunate here since it has nothing to do with S list objects.

```
> latitude <- state.center[["y"]]
> names(latitude) <- state.name
> i <- sort.list(longitude)
> cbind(latitude = latitude[i], longitude = longitude[i])
                latitude longitude
        Alaska    49.250  -127.250
        Hawaii    31.750  -126.250
    ....
  Rhode Island    41.593   -71.124
         Maine    45.623   -68.980
```

Notice how under `cbind` the `names` attribute of the vector can become part of the `dimnames` attribute of the matrix.

A further sorting function is `order`. It takes an arbitrary number of arguments and returns the index vector that would arrange the first in increasing order, with ties broken by the second, and so on. For example, to arrange employees by age, by salary within age, and by employment number within salary, and print them together with their number of dependents, we might use:

```
m <- order(Age, Salary, No)
cbind(Age[m], Salary[m], No[m], Depndts[m])
```

Notice that `order` with a single argument is exactly equivalent to `sort.list`.

The function `rank` is related to sorting: it computes the ranks of the elements of a vector. This is not quite the same as the inverse of `sort.list`, as ties are averaged.

```
> shoes$B
 [1] 14.0  8.8 11.2 14.2 11.8  6.4  9.8 11.3  9.3 13.6
> rank(shoes$B)
 [1]  9  2  5 10  7  1  4  6  3  8
> rank(round(shoes$B))
 [1] 9.0 2.5 5.5 9.0 7.0 1.0 4.0 5.5 2.5 9.0
> sort.list(sort.list(round(shoes$B)))
 [1]  8  2  5  9  7  1  4  6  3 10
```

All four functions have an argument `na.last` that determines the handling of missing values. With `na.last=NA` (the default for `sort`) missing values are deleted; with `na.last=T` (the default for `sort.list`, `order` and `rank`) they are put last, and with `na.last=F` they are put first.

2.9 Input/Output facilities

In this section we cover a miscellany of topics connected with input and output.

Writing data to a file

A character representation of any S vector may be written on an output device
(including the session window) by the `write` function

```
write(x, file="outdata")
```

where the session window is specified by `file=""`. Little is allowed by way
of format control, but the number of columns can be specified by the argument
`ncolumns` (default 5 for numeric vectors, 1 for character vectors). This is sometimes useful to print out vectors or matrices in a fixed layout. For data frames,
`write.table` is provided:

```
> write.table(dataframe, file="", sep=",")
> write.table(painters, "", sep="\t", dimnames.write="col")
Composition      Drawing Colour  Expression      School
10       8       16      3       A
15       16      4       14      A
8        13      16      7       A
12       16      9       8       A
     ....
```

Omitting `dimnames.write="col"` writes out both row and column labels (if
any) and setting it to F omits both.

   We have found the following simple function (from library MASS) useful to
print out matrices or data frames.

```
write.matrix <- function(x, file="", sep=" ")
{
        x <- as.matrix(x)
        p <- ncol(x)
        cat(dimnames(x)[[2]],format(t(x)), file=file,
             sep=c(rep(sep, p-1), "\n"))
}
```

This produces a neatly formatted layout, and starts with the column labels (if any)
ready for reading in by `read.table` with `header=T`.

HTML output

The function `html.table` is similar to `write.table`, but formats the result as
an HTML table. There are variations on the HTML for tables accepted by Web
browsers, and this one seems to be in Netscape style.

Data import and export

S-PLUS 4.x and 5.x can import and export data frames from/to files in a variety of formats through the functions `importData` and `exportData`. The supported formats are those of a number of spreadsheets[16], databases, statistical (**SAS**, **SPSS**, **Stata**, **Systat**) and matrix-language (**GAUSS**, **MATLAB**) systems, as well as ASCII files (although there are many other ways to read and write these).

S-PLUS 4.x can also import and export data through a dialog box accessed from the File menu.

Reading non-standard data files

Occasionally it is helpful to use a different field separator from the default, which is any consecutive sequence of blanks, tabs or newlines (known as whitespace). For example, some spreadsheet programs' output files are written with fields separated by commas. The argument `sep=","` of `read.table` and `scan` will allow this. (In this case empty and blank fields will be read as missing values `NA`.) Using `sep="\t"` specifies that the fields are to be separated by a single tab character, thus allowing strings to be read that themselves contain blanks, and `sep="\n"` allows only the newline as a field separator, so every line will be read as a single field.

If input data are to be read from a file where data items occupy specific columns without guaranteed whitespace separators between items, the first option to consider is to modify the file so that there is separating whitespace. This makes the data easier to read and simpler to check directly from the data file if necessary, but is not an option if the items themselves may contain embedded blanks and tab characters. If it is necessary to use a file with fixed-width fields there is an argument, `widths`, to `scan` that allows fixed width input by specifying an integer vector of field widths. There is also a function `make.fields` that can be used to convert a file with fixed-width, non-separated, input fields into a file with separated fields. It should be noted that the `widths` argument of `scan` uses the `make.fields` function itself and a temporary file, so its use with very large files may pose a problem.

Another solution to handling complex or non-standard input is to read in each line of the file as a character string vector and extract the data items later. For example, suppose we wish to read 40 variables from each line of a file, `v40.dat`. Each variable occupies 2 columns, making 80 columns of data in all. Missing values are denoted by a blank entry. We may use

```
chdata <- scan("v40.dat", what="", sep="\n")
```

Specifying `sep="\n"` ensures that data items are separated only by the newline, and `what=""` establishes that the data to be read in are to be of mode `character` and so `chdata` is a character vector with each complete line of the data file as its elements.

To extract each column in numeric form we need to use a simple loop

---

[16] But not Excel 95 nor Excel 97 on UNIX.

```
dat <- list()                    # initially empty list
for(i in 1:40)
      dat[[i]] <- as.numeric(substring(chdata, 2*i-1, 2*i))
```

Coercion to numeric of a fully blank string results in a missing value, as required. (This also happens in the case of a string that does not parse correctly as a number.) To make the data more easily accessible it is a good idea to convert their representation to a data frame:

```
digits <- substring(1000+seq(along=dat), 3, 4)
names(dat) <- paste("V", digits, sep="")
dat <- as.data.frame(dat)
```

This device ensures the names are of constant length, so the first variable is V01 rather than V1. In turn this ensures that consecutive variables remain consecutive when the names are sorted. The function substring is discussed on page 35.

One fairly common problem in reading files is a failure to ensure that the end-of-file characters have been converted while transferring the data to the computer on which S-PLUS is used. All versions of S-PLUS will accept either LF (the UNIX norm) or CRLF (MS-DOS and VMS), but files transferred from Macintosh computers must be converted.

Executing commands from, or diverting output to, a file

If commands are stored on an external file, say, commands.q in the current directory, they may be executed at any time in an S session with the command

```
source("commands.q")
```

It is often useful to use options(echo=T) before a source file, to have the commands echoed. (If used in the file itself, it takes effect *after* the file is completed.) Similarly

```
sink("record.lis")
```

will divert all subsequent output from the session window to an external file record.lis. The command

```
sink()
```

restores output to the window once more.

Dumping S *objects*

Two functions are useful for externally manipulating or transmitting S objects. The function dump takes as its first argument a character vector giving the names of S objects to be written on an external file, by default the file dumpdata. These are written in *assignment form* so that executing the file dumpdata with the source command will re-create the objects.

```
dump(c("a", "x", "ink"), file="outdata") # dump objects
    ....
source("outdata") # re-create a, x and ink
```

The dumped objects are in a form that is fairly easy to read and to edit, but reading such a file with source can be slow, particularly for large numeric objects.

The function data.dump does a similar job to dump but the dumped objects may only be re-created using the companion function data.restore to read the file, and this operation is relatively fast. These two functions are intended for transmission of S objects between remote or incompatible computers, and unlike dump are guaranteed to preserve the storage mode (integer, single or double precision) of S objects.

General printing

The function cat is similar to paste with argument collapse="" in that it coerces its arguments to character strings and concatenates them. However instead of returning a character string result it prints out the result in the session window or optionally on an external file. For example, to print out today's date on our UNIX system:

```
> d <- date()
> cat("Today's date is:", substring(d,1,10),
            substring(d,25,28), "\n")
Today's date is: Sat May  8 1999
```

which is needed occasionally for dating output from within a function. Note that an explicit newline ("\n") is needed.

Other arguments to cat allow the output to be broken into lines of specified length, and optionally labelled:

```
> cat(1,2,3,4,5,6, fill=8, labels=letters)
a 1 2
c 3 4
e 5 6
```

and fill=T fills to the current output width.

The function format provides the most general way to prepare data for output. It coerces data to character strings in a common format, which can then be used by cat. For example, the print function print.summary.lm for summaries of linear regressions contains the lines

```
cat("\nCoefficients:\n")
print(format(round(x$coef, digits = digits)), quote = F)
cat("\nResidual standard error:",
    format(signif(x$sigma, digits)), "on", rdf,
    "degrees of freedom\n")
cat("Multiple R-Squared:", format(signif(x$r.squared, digits)),
    "\n")
cat("F-statistic:", format(signif(x$fstatistic[1], digits)),
    "on", x$fstatistic[2], "and", x$fstatistic[3],
    "degrees of freedom, the p-value is", format(signif(1 -
    pf(x$fstatistic[1], x$fstatistic[2], x$fstatistic[3]),
    digits)), "\n")
```

Note the use of `signif` and `round` to specify the accuracy required. (For `round` the number of digits is specified, whereas for `signif` it is the number of significant digits.) To see the effect of `format` versus `write`, consider:

```
> write(iris[,1,1], "", 15)
5.1 4.9 4.7 4.6 5 5.4 4.6 5 4.4 4.9 5.4 4.8 4.8 4.3 5.8
5.7 5.4 5.1 5.7 5.1 5.4 5.1 4.6 5.1 4.8 5 5 5.2 5.2 4.7
4.8 5.4 5.2 5.5 4.9 5 5.5 4.9 4.4 5.1 5 4.5 4.4 5 5.1
4.8 5.1 4.6 5.3 5
> cat(format(iris[,1,1]), fill=60)
5.1 4.9 4.7 4.6 5.0 5.4 4.6 5.0 4.4 4.9 5.4 4.8 4.8 4.3 5.8
5.7 5.4 5.1 5.7 5.1 5.4 5.1 4.6 5.1 4.8 5.0 5.0 5.2 5.2 4.7
4.8 5.4 5.2 5.5 4.9 5.0 5.5 4.9 4.4 5.1 5.0 4.5 4.4 5.0 5.1
4.8 5.1 4.6 5.3 5.0
```

There is a tendency to output values such as `0.6870000000000001`, even after rounding to (here) three digits. (Not from `print`, but from `write`, `cat`, `paste`, `as.character` and so on.) Use `format` to avoid this.

By default the accuracy of printed and converted values is controlled by the `options` parameter `digits`, which defaults to 7.

Changes in version 5.x

S-PLUS 5.x has a general notion of a *connection* to extend that of a file. Many of the types of connections will be familiar to UNIX devotees, for in UNIX they are just types of files, like pipes and fifos.[17] A (read-only) connection can also be an S character vector. All but the simplest uses of connections will be of interest only to system programmers, who are referred to Chambers (1998, Chapter 10) and especially the on-line help for the details.

Functions such as `scan` and `write` have an argument "file" that specifies (as a character string) the name of the file to be used; in many cases this can be replaced by a connection. Why would we want to do this? One reason is to keep a file open for further operations. Earlier versions of our function `ppinit` in library `spatial` had

---

[17] Also known as named pipes.

```
h <- scan(tfile, list(xl = 0, xu = 0, yl = 0, yu = 0, fac = 0),
          n = 5, skip = 2)
pp <- scan(tfile, list(x = 0, y = 0), skip = 3)
```

which opened the file, skipped two lines, read the third, closed the file, opened the file, skipped three lines and read the rest. We can now use

```
tf <- open(tfile)
h <- scan(tf, list(xl = 0, xu = 0, yl = 0, yu = 0, fac = 0),
          n = 5, skip = 2)
pp <- scan(tf, list(x = 0, y = 0))
close(tf)
```

The convention is that if a function finds a connection open it leaves it open, but if it needs to open it, it is also closed.

Connections are explicitly opened by open and closed by close. Their first argument is connection and connections can be generated by any of the functions

```
file("path", open, blocking = T)
pipe("shell command", open)
fifo("path", open="", blocking = F)
textConnection("character vector", open, blocking = T)
stdin(); stdout(); stderr()
```

(Specifying connection as a character string gives an implicit call to file.) These behave as one would expect: in particular a text connection treats each element of a character vector as a line of text (and is read-only). The open argument should either be a logical value (by default the connection is *not* opened) or a character string giving the mode.[18]

One thing that does *not* behave as one would expect is file positioning: S-PLUS 5.x keeps separate positions for reading and writing, initially at the beginning and end of the file, respectively. Function seek allows either position to be changed (if allowed), and will also return the current position (so encompassing the tell POSIX file call).

There are additional input/output functions that are particularly useful with open connections. Functions readLines and writeLines read and write a specified number of text lines; parseSome reads and parses up to one S expression and dataGet and dataPut reads or writes one object in data.dump format.

The function showConnections shows all the currently open connections except the standard ones unless its argument all is true;

```
> showConnections(all=T)
             Class Mode   State      Description
 stdin "terminal" "r"   "Read"   "/dev/tty"
stdout "terminal" "w"   "Write"  "/dev/tty"
stderr "terminal" "w"   "Write"  "/dev/tty"
 Audit "file"     "a"   "Write"  "././.Data/.Audit"
```

---

[18] For example, "r", "ra" (read/append) or "*" (anything allowed by the OS).

When `source` is used, the file is read in as a single expression and then evaluated, so any errors in the file result in nothing being committed (which can result in large memory usage while deferring commitment), and auto-printing will not occur. There is now another possibility, `setReader(file("myfile.q"))`, that starts another S evaluator and so reads the file asynchronously. Unlike `source`, commitments are made immediately and auto-printing does occur.

2.10 Exercises

The exercises in Chapter 4 also exercise the material of this chapter, but are rather harder.

2.1. How would you find the index(es) of specified values within a vector? For example, where is the hill race (in `hills`) with a climb of 2 100 feet?

2.2. The column `ftv` in data frame `birthwt` counts the number of visits. Reduce this to a factor with levels 0, 1 and '2 or more'. [Hint: manipulate the `levels`, or investigate functions `cut` and `merge.levels`.]

2.3. Write a simple function to compute the median absolute deviation (used in robust statistics; see Section 5.5) median$|x - \mu|$ with default μ the sample median. Compare your answer with the S-PLUS function `mad`.

2.4. Suppose `x` is an object with named components and `out` is a character string vector. How would you make a new object obtained from `x` by *excluding* any components whose names are in `out`?

2.5. Given a matrix `X` of distinct rows and a vector `w` of the number of times that each row should occur, reconstruct the original matrix.

2.6. "I calculated a cross-correlation matrix. I want to print only members of this matrix that are larger than 0.90 and I want to include dimnames in the answer."

2.7. "I have a large data frame (5 000 observations) and I would like the cases where a variable indicating ethnic group is in (1,3,4,6,7)."

Chapter 3

Graphics

S-PLUS provides comprehensive graphics facilities, from simple facilities for producing common diagnostic plots by plot(*object*) to fine control over publication-quality graphs. In consequence, the number of graphics parameters is huge. In this chapter, we build up the complexity gradually. Most readers will not need the material in Section 3.4, and indeed the material there is not used elsewhere in this book. However, we *have* needed to make use of it, especially in matching existing graphical styles.

Some graphical ideas are best explored in their statistical context, so that, for example, histograms are covered in Chapter 5, survival curves in Chapter 12, biplots in Chapter 11 and time-series graphics in Chapter 13. Table 3.1 gives an overview of the high-level graphics commands with page references.

There are many books on graphical design. Cleveland (1993) discusses most of the methods of this chapter and the detailed design choices (such as aspect ratios of plots and the presence of grids) that can affect the perception of graphical displays. As these are to some extent a matter of personal preference and this is also a guide to S-PLUS, we have kept to the default choices.

Trellis graphics are a recent addition to S-PLUS with a somewhat different style and philosophy to the basic plotting functions. We describe the basic functions first, then the Trellis functions in Section 3.5. S-PLUS 4.x for Windows (including S-PLUS 2000) has a very different style of object-oriented editable graphics which we cover in Section B.2.

3.1 Graphics devices

Before any plotting commands can be used, a graphics device must be opened to receive graphical output. Most commonly this is a window on the screen of a workstation or a plotter or printer. A list of supported devices on the current hardware with some indication of their capabilities is available from the on-line help system by

```
?Devices
```

Table 3.1: High-level plotting functions. Page references are given to the most complete description in the text. Those marked by † are superseded by Trellis functions.

	Page	
abline	59	Add lines to the current plot in slope-intercept form.
axis	65	Add an axis to the plot.
barplot	58	Bar graphs.
biplot	336	Represent rows and columns of a data matrix.
brush spin	330	Dynamic graphics.
contour †	61	Contour plot. contourplot is preferred.
dotchart		Produce a dot chart.
eqscplot	60	Plot with geometrically equal scales (our library).
faces		Chernoff's faces plot of multivariate data.
frame	63	Advance to next figure region.
hist	118	Histograms. We prefer our function truehist .
hist2d	139	Two-dimensional histogram calculations.
identify locator	65	Interact with an existing plot.
image † image.legend	61	High-density image plot functions. levelplot is preferred.
interaction.plot		Interaction plot for a two-factor experiment.
legend	66	Add a legend to the current plot.
matplot	73	Multiple plots specified by the columns of a matrix.
mtext	65	Add text in the margins.
pairs †	60	All pairwise plots between multiple variables. splom is preferred.
par	67	Set or ask about graphics parameters.
persp † perspp persp.setup	61	Three-dimensional perspective plot functions. Partially superseded by wireframe and cloud.
pie		Produce a pie diagram.
plot		Generic plotting function.
polygon		Add a polygon to the present plot, possibly filled.
points lines	59	Add points or lines to the current plot.
qqplot † qqnorm †	114	Quantile-quantile and normal Q-Q plots. Trellis functions qq and qqmath are preferred.
scatter.smooth	284	Scatterplot with a smooth curve.
segments arrows	73	Draw line segments or arrows on the current plot.
stars		Star plots of multivariate data.
symbols		Draw variable sized symbols on a plot.
text	59	Add text symbols to the current plot.
title	64	Add title(s).

Table 3.2: Some of the graphical devices available for various versions of S-PLUS.

`motif`	UNIX: X11–windows systems.
`openlook`	UNIX: X11–windows systems; S-PLUS 3.x only.
`iris4d`	Silicon Graphics Iris workstations; S-PLUS 3.x only.
`win.graph`	Windows; S-PLUS 3.x.
`graphsheet`	S-PLUS 4.x—see Section B.2.
`postscript`	PostScript printers.[1]
`hplj`	Hewlett-Packard LaserJet printers.
`win.printer`	Use Windows printing.
`pdf.graph`	Adobe's PDF format (not S-PLUS 3.x).

(Note the capital letter.) A number of graphics terminals are supported, but as these are now unlikely to be encountered they are not considered here.

A graphics device is opened by giving the command in Table 3.2, possibly with parameters giving the size and position of the window; for example,

```
motif("-geometry 600x400-0+0")
```

opens a small graphics window initially positioned in the top right-hand corner of the screen. Many screen devices support the argument `ask`, which if true will request permission to clear the screen and start the next plot. (This can be set later by `par(ask=T)` or using `dev.ask`.) S-PLUS 4.x and 5.x will automatically open a graphics device if one is needed, but we often choose to open the device ourselves and so take advantage of the ability to customize it.

Graphics devices may be closed using

```
graphics.off()
```

This will close all the graphics devices currently open and perform any further wrap-up actions that need to be taken, such as sending trailer information to printer devices. Note that quitting with `q()` will automatically perform a `graphics.off` operation.

It is possible to have several graphical devices open at once. By default the most recently opened one is used, but `dev.set` can be used to change the current device (by number). The function `dev.list` lists currently active devices, and `dev.off` closes the current device, or one specified by number. There are also commands `dev.cur`, `dev.next` and `dev.prev` which return the number of the current, next or previous device on the list.

The `motif` and `openlook` devices have a Copy option on their Graph menu which allows a (smaller) copy of the current plot to be copied to a new window, perhaps for comparison with later plots. (The copy window can be dismissed by the Delete or Destroy item from its Graph menu.) There is also a

---

[1] `postscript` exists on both UNIX and Windows systems, but the driver and the function arguments are different.

`dev.copy` function that copies the current plot to the specified device (by default the next device on the list).

Many of the graphics devices on windowing systems have menus of choices, for example, to make hardcopies and to alter the colour scheme in use.

graphsheet devices

There are some special considerations for users of `graphsheet` devices on S-PLUS 4.x. Graphical output is not drawn immediately on a `graphsheet` but delayed until the current S expression has finished or some input is required, for example from the keyboard or by `locator` or `identify` (see page 65). One way to overcome this is to add a call that asks for input at suitable points in your code, and a call to `guiLocator` with an argument of zero is usually the easiest, as this does nothing except flush the graphics queue.

Complicated graphs are not always drawn correctly; pressing function key F9 (redraw) will usually correct this.

Hitting function key F2 with a `graphsheet` in focus zooms it to full screen. Click a mouse button or hit a key to revert.

Graphsheets can have multiple pages. The default is to use these pages for all the plots drawn within an S expression without any input, including code submitted from a scripts window. This is often helpful, but can be changed from the Page Creation of the Options... tab of the graphsheet properties dialog box (brought by right-clicking on the background of a `graphsheet` device).

As `win.graph` exists only for backwards compatibility on S-PLUS 4.x and in fact uses a `graphsheet`, the same considerations apply.

Graphical hardcopy

There are several ways to produce a hardcopy of a (command-line) plot on a suitable printer or plotter. The `motif` and `openlook` devices have a Print item on their Graph menu that will send a full-page copy of the current plot to the default printer. A little more control (the orientation and the print command) is available from the Printing item on the Options or Properties menu.

Users of the Windows version of S-PLUS can use the Print option on their File menu or (in S-PLUS 4.x) the printer icon on the toolbar. This prints the window with focus, so bring the desired graphics window to the top first.

The function `printgraph` (UNIX) will copy the current plot to a PostScript or LASERJET printer, and allows the size and orientation to be selected, as well as paper size and printer resolution where appropriate.

The function `dev.print` will copy the current plot to a printer device (default `postscript` under UNIX and `win.printer` under Windows) and allow its size, orientation, pointsize of text and so on to be set.

Finally, it is normally possible to open an appropriate printer device and repeat the plot commands, although this does preclude interacting with the plot on-screen.

A *warning:* none of these methods appear to work correctly for brush and spin plots. The plots in this book were produced by screen dumps using X11 facilities. Windows users can copy to the clipboard and then paste into other applications.

Hardcopy to a file

It is very useful to be able to copy a plot to a file for inclusion in a paper or book (as here). Since each of the hardcopy methods allows the printer output to be re-directed to a file, there are many possibilities.

Users of S-PLUS 3.x under Windows can use the Print option on the File menu to select printing to a file and may also check the Setup box to include the header in the file. The graphics window can then be printed in the usual way. Under S-PLUS 4.x it is preferable to use the Export Graph... item on the File menu, which can save in a variety of graphics formats including Windows metafile and Encapsulated PostScript (with a preview image if desired).

If the plot file will be rescaled subsequently, the simplest way under UNIX is to edit the print command in the Printing item of a motif or openlook device to be

```
cat > plot_file_name <
```

or to make a Bourne-shell command file rmv by

```
$ cat > rmv
mv $2 $1
^D
$ chmod +x rmv
```

place this in your path and use rmv plot_file_name. (This avoids leaving around temporary files with names like ps.out.0001.ps.) Then click on Print to produce the plot.

PostScript users will probably want Encapsulated PostScript (EPSF) format files. These are produced automatically (on UNIX) by the procedure in the last paragraph, and also by setting both onefile=F and print.it=F as arguments to the postscript device under UNIX. Note that these are *not* EPSI files and do not include a preview image. It would be unusual to want an EPSF file in landscape format; select 'Portrait' on the Printing menu item, or use horizontal=F as argument to the postscript device under UNIX. The default pointsize (14) is set for a full-page landscape plot, and 10 or 11 are often more appropriate for a portrait plot. Set this in the call to postscript or use ps.options (*before* the motif or openlook device is opened, or use ps.options.send to alter the print settings of the current device).

For Windows users the placeable metafile format may be the most useful for plot files, as it is easily incorporated into other Windows applications while retaining full resolution. This is automatically used if the graphics device window is copied to the clipboard, and may also be generated by win.printer.

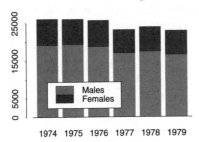
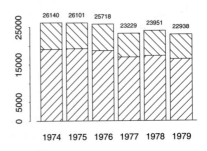

Figure 3.1: Two different styles of barchart showing the annual UK deaths from certain lung diseases. In each case the lower block is for males, the upper block for females.

3.2 Basic plotting functions

The function `plot` is a generic function, which when applied to many S objects will give a plot. (However, in many cases the plot does not seem appropriate to us.) Many of the plots appropriate to univariate data such as boxplots and histograms are considered in Chapter 5.

Barcharts

The function to display barcharts is `barplot`. This has many options (described in the on-line help), but some simple uses are shown in Figure 3.1. (Many of the details are covered in Section 3.3.)

```
lung.deaths <- aggregate(ts.union(mdeaths, fdeaths), 1)
barplot(t(lung.deaths), names=dimnames(lung.deaths)[[1]],
    main="UK deaths from lung disease")
legend(locator(1), c("Males", "Females"), fill=c(2,3))
loc <- barplot(t(lung.deaths), names=dimnames(lung.deaths)[[1]],
    style="old", dbangle=c(45,135), density=10)
total <- apply(lung.deaths, 1, sum)
text(loc, total + par("cxy")[2], total, cex=0.7)
```

Line and scatterplots

The default plot function takes arguments x and y, vectors of the same length, or a matrix with two columns, or a list (or data frame) with components x and y and produces a simple scatterplot. The axes, scales, titles and plotting symbols are all chosen automatically, but can be overridden with additional graphical parameters that can be included as named arguments in the call. The most commonly used ones are:

`type="c"`	Type of plot desired. Values for c are: p for points only (the default), l for lines only, b for both points and lines, (the lines miss the points), s, S for step functions (s specifies the level of the step at the left end, S at the right end), o for overlaid points and lines, h for high density vertical line plotting, and n for no plotting (but axes are still found and set).
`axes=L`	If F all axes are suppressed (default T, axes are automatically constructed).
`xlab="string"` `ylab="string"`	Give labels for the x- and/or y-axes (default: the names, including suffices, of the x- and y-coordinate vectors).
`sub="string"` `main="string"`	sub specifies a title to appear under the x-axis label and main a title for the top of the plot in larger letters (default: both empty).
`xlim=c(lo ,hi)` `ylim=c(lo, hi)`	Approximate minimum and maximum values for x- and/or y-axis settings. These values are normally automatically rounded to make them 'pretty' for axis labelling.

The functions `points`, `lines`, `text` and `abline` can be used to add to a plot, possibly one created with `type="n"`. Brief summaries are:

`points(x,y,...)`	Add points to an existing plot (possibly using a different plotting character). The plotting character is set by `pch=` and the size of the character by `cex=` or `mkh=`.
`lines(x,y,...)`	Add lines to an existing plot. The line type is set by `lty=` and width by `lwd=`. The `type` options may be used.
`text(x,y,labels,...)`	Add text to a plot at points given by x,y. labels is an integer or character vector; `labels[i]` is plotted at point `(x[i],y[i])`. The default is `seq(along=x)`. The character size is set by `cex=`.

`abline(a,b,...)`
`abline(h=c,...)`
`abline(v=c,...)`
`abline(`*lmobject*`,...)`
	Draw a line in intercept and slope form, (a,b), across an existing plot. h=c may be used to specify y-coordinates for the heights of horizontal lines to go across a plot, and v=c similarly for the x-coordinates for vertical lines. The coefficients of a suitable *lmobject* are used.

These are the most commonly used graphics functions; we have shown examples of their use in Chapter 1, and show many more. (There are also functions `arrows` and `symbols` that we do not use in this book.) The plotting characters available for `plot` and `points` can be characters of the form pch="o" or numbered from 0 to 18,[2] which uses the marks shown in Figure 3.2.

---

[2] Up to 27 on a `graphsheet` in S-PLUS 4.x; see the Annotation palette on page 461.

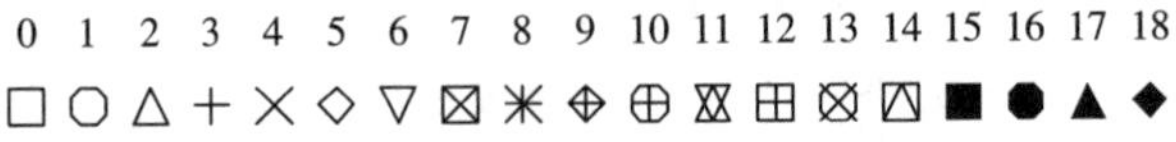

0 1 2 3 4 5 6 7 8 9 10 11 12 13 14 15 16 17 18

□ ○ △ + × ◇ ▽ ⊠ ✳ ⟠ ⊕ ⊠ ⊞ ⊠ ◁ ■ ● ▲ ◆

Figure 3.2: Plotting symbols or marks, specified by pch=n.

Size of text and symbols

Confusingly, the size of plotting characters is selected in one of two very different ways. For plotting characters (by pch="o") or text (by text), the parameter cex (for 'character expansion') is used. This defaults to the global setting (which defaults to 1), and rescales the character by that factor. For a mark set by pch=n, the size is controlled by the mkh parameter which gives the height of the symbol *in inches*. (This will be clear for printers; for screen devices the default device region is about 8in × 6in and this is not changed by re-sizing the window.) However if mkh=0 (the default) the size is then controlled by cex and the default size of each symbol is approximately that of O. Care is needed in changing cex on a call to plot, as this will also change the size of the axis labels. It is better to use, for example,

```
plot(x, y, type="n")              # axes only
points(x, y, pch=4, mkh=0, cex=0.7)  # add the points
```

If cex is used to change the size of all the text on a plot, it will normally be desirable to set mex to the same value to change the interline spacing. An alternative to specifying cex is csi, which specifies the absolute size of a character (in inches). (There is no msi.)

The default text size can be changed for some graphics devices, for example, by argument pointsize for the postscript and win.printer devices.

Equally scaled plots

There are many plots, for example, in multivariate analysis, that represent distances in the plane and for which it is essential to have a scaling of the axes that is geometrically accurate. This can be done in many ways, but most easily by our function eqscplot which behaves as the default plot function but shrinks the scale on one axis until geometrical accuracy is attained.

Warning: when screen devices (except a graphsheet) are resized the S-PLUS process is not informed, so eqscplot can only work for the original window shape.

Multivariate plots

The plots we have seen so far deal with one or two variables. To view more we have several possibilities. A *scatterplot matrix* or *pairs* plot shows a matrix of scatterplots for each pair of variables, as we saw in Figure 1.2, which was produced by splom(~ hills). Enhanced versions of such plots are a *forte* of Trellis graphics, so we do not discuss how to make them in the base graphics system.

Dynamic graphics

The function `brush` allows interaction with the (lower half) of a scatterplot matrix. An example is shown in Figure 1.3 on page 9. As it is much easier to understand these by using them, we suggest you try

```
brush(as.matrix(hills))
```

and experiment.

Points can be highlighted (marked with a symbol) by moving the brush (a rectangular window) over them with button 1 held down. When a point is highlighted, it is shown highlighted in all the displays. Highlighting is removed by brushing with button 2 held down. It is also possible to add or remove points by clicking with button 1 in the scrolling list of row names.

One of four possible (device-dependent) marking symbols can be selected by clicking button 1 on the appropriate one in the display box on the right. The marking is by default persistent, but this can be changed to 'transient' in which only points under the brush are labelled (and button 1 is held down). It is also possible to select marking by row label as well as symbol.

The brush size can be altered under UNIX by picking up a corner of the brush in the `brush size` box with the mouse button 1 and dragging to the required size. Under Windows, move the brush to the background of the main `brush` window, hold down the left mouse button and drag the brush to the required size.

The plot produced by `brush` will also show a three-dimensional plot (unless `spin=F`), and this can be produced on its own by `spin`. Clicking with mouse button 1 will select three of the variables for the x-, y- and z-axes. The plot can be spun in several directions and re-sized by clicking in the appropriate box. The `speed` box contains a vertical line or slider indicating the current position.

Plots from `brush` and `spin` can only be terminated by clicking with the mouse button 1 on the `quit` box or button.

Obviously `brush` and `spin` are only available on suitable screen devices, including `motif`, `openlook`, `graphsheet` and `win.graph`. Hardcopy is only possible by directly printing the window used, not by copying the plot to a printer graphics device.

Plots of surfaces

The functions `contour`, `persp` and `image` allow the display of a function defined on a two-dimensional regular grid. Their Trellis equivalents give more elegant output, so we do not discuss them in detail. The function `contour` allows more control than `contourplot`. We anticipate an example from Chapter 14 of plotting a smooth topographic surface for Figure 3.3, and contrast it with `contourplot`.

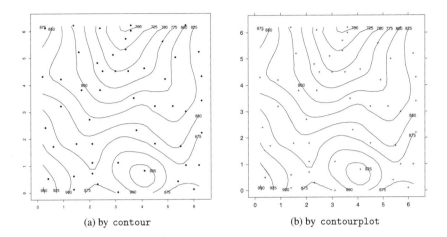

(a) by contour   (b) by contourplot

Figure 3.3: Contour plots of loess smoothing of the topo dataset.

```
topo.loess <- loess(z ~ x * y, topo, degree=2, span = 0.25)
topo.mar <- list(x = seq(0, 6.5, 0.2), y=seq(0, 6.5, 0.2))
topo.lo <- predict(topo.loess, expand.grid(topo.mar))
par(pty="s")        # square plot
contour(topo.mar$x, topo.mar$y, topo.lo, xlab="", ylab="",
    levels = seq(700,1000,25), cex=0.7)
points(topo$x, topo$y)
par(pty="m")
contourplot(z ~ x * y, mat2tr(topo.lo), aspect=1,
    at = seq(700, 1000, 25), xlab="", ylab="",
    panel = function(x, y, subscripts, ...) {
        panel.contourplot(x, y, subscripts, ...)
        panel.xyplot(topo$x,topo$y, cex=0.5)
    }
)
```

This generates values of the surface on a regular 33×33 grid generated by
expand.grid. We provide the functions con2tr and mat2tr to convert objects
designed for input to contour and matrices as produced by predict.loess
into data frames suitable for the Trellis 3D plotting routines.

3.3 Enhancing plots

In this section we cover a number of ways that are commonly used to enhance
plots, without reaching the level of detail of Section 3.4.

Some plots (such as Figure 1.1) are square, whereas others are rectangular.
These are selected by the graphics parameter pty. Setting par(pty="s") selects a square plotting region, whereas par(pty="m") selects a maximally sized
(and therefore usually non-square) region.

Multiple figures on one plot

We have already seen several examples of plotting two or more figures on a single device surface, apart from scatterplot matrices. The graphics parameters `mfrow` and `mfcol` subdivide the plotting region into an array of figure regions. They differ in the order in which the regions are filled. Thus

```
par(mfrow=c(2,3))
par(mfcol=c(2,3))
```

both select a 2×3 array of figures, but with the first they are filled along rows, and with the second along columns. A new figure region is selected for each new plot, and figure regions can be skipped by using `frame`.

All but two of the multi-figure plots in this book were produced with `mfrow`. Most of the side-by-side plots were produced with `par(mfrow=c(2,2))`, but using only the first two figure regions.

The `split.screen` function provides an alternative and more flexible way of generating multiple displays on a graphics device. An initial call such as

```
split.screen(figs=c(3,2))
```

subdivides the current device surface into a 3×2 array of *screens*. The screens created in this example are numbered 1 to 6, *by rows*, and the original device surface is known as screen 0. The current screen is then screen 1 in the upper left corner, and plotting output will fill the screen as it would a figure. Unlike multi-figure displays, the next plot will use the same screen unless another is specified using the `screen` function. For example, the command

```
screen(3)
```

causes screen 3 to become the next current screen.

On screen devices the function `prompt.screen` may be used to define a screen layout interactively. The command

```
split.screen(prompt.screen())
```

allows the user to define a screen layout by clicking mouse button 1 on diagonally opposite corners. In our experience this requires a steady hand, although there is a `delta` argument to `prompt.screen` that can be used to help in aligning screen edges. Alternatively, if the `figs` argument to `split.screen` is specified as an $N \times 4$ matrix, this divides the plot into N screens (possibly overlapping) whose corners are specified by giving (xl, xu, yl, yl) as the row of the matrix (where the whole region is $(0, 1, 0, 1)$).

The `split.screen` function may be used to subdivide the current screen recursively, thus leading to irregular arrangements. In this case the screen numbering sequence continues from where it had reached.

Split-screen mode is terminated by a call to `close.screen(all=T)`; individual screens can be shut by `close.screen(n)`.

The function `subplot` provides a third way to subdivide the device surface. This has call

```
subplot(fun, x, y, size=c(1, 1), vadj=.5, hadj=.5, pars=NULL)
```

which adds the graphics output of `fun` to an existing plot. The size and position can be determined in many ways (see the on-line help); if all but the first argument is missing a call to `locator` is used to ask the user to click on any two opposite corners of the plot region.[3]

Use of the `fig` parameter to `par` provides an even more flexible way to subdivide a plot; see Section 3.4.

With multiple figures it is normally necessary to reduce the size of the text. If either the number of rows or columns set by `mfrow` or `mfcol` is three or more, the text size is halved by setting `cex=0.5` (and `mex=0.5`; see Section 3.4). This may produce characters that are too small and some resetting may be appropriate. (On the other hand, for a 2×2 layout the characters will usually be too large.) For all other methods of subdividing the plot surface the user will have to make an appropriate adjustment to `cex` and `mex` or to the default text size (for example by changing `pointsize` on the `postscript` and `win.printer` devices).

Adding information

The basic plots produced by `plot` often need additional information added to give context, particularly if they are not going to be used with a caption. We have already seen the use of `xlab`, `ylab`, `main` and `sub` with scatterplots. These arguments can all be used with the function `title` to add titles to existing plots. The first argument is `main`, so

```
title("A Useful Plot?")
```

adds a main title to the current plot.

Further points and lines are added by the `points` and `lines` functions. We have seen how plot symbols can be selected with `pch=`. The line type is selected by `lty=`. This is device-specific, but usually includes solid lines (1) and a variety of dotted, dashed and dash-dot lines. Line width is selected by `lwd=`, with standard width being 1, and the effect being device-dependent.

The `type="s"`, `"S"` argument to `lines` allows 'staircase' lines. It is better to use `stepfun` with dashed lines, as the dash sequence is not restarted at each change of angle (at least on the `postscript` device).

Using colour

The colour model of S graphics is quite complex. Colours are referred to as numbers, and set by the parameter `col`. Sometimes only one colour is allowed (e.g., `points`) and sometimes `col` can be a vector giving a colour for each plot item (e.g., `text`). There will always be at least two colours 0 (the background, useful for erasing) and 1. However, how many colours there are and what they appear as is set by the device. Furthermore, there are separate colour groups,

---

[3] Not the figure region; Figure 3.4 on page 66 shows the distinction.

and what they are is device-specific. For example, `motif` devices have separate colour spaces for lines (including symbols), text, polygons (including histograms, bar charts and pie charts) and images, and `graphsheet` devices have two spaces, one for lines and text, the other for polygons and images. Thus the colours can appear completely differently when a graph is copied from device to device, in particular on screen and on a hardcopy. It is usually a good idea to design a colour scheme for each device.

It is necessary to read the device help page thoroughly (and for `postscript` on UNIX, also that for `ps.options.send`).

Identifying points interactively

The function `identify` has a similar calling sequence to `text`. The first two arguments give the x- and y-coordinates of points on a plot and the third argument gives a vector of labels for each point. (The first two arguments may be replaced by a single list argument with two of its components named x and y, or by a two-column matrix.) The labels may be a character string vector or a numeric vector (which is coerced to character). Then clicking with mouse button 1 near a point on the plot causes its label to be plotted; labelling all points or clicking anywhere in the plot with button 2 terminates the process. (The precise position of the click determines the label position, in particular to left or right of the point.) We saw an example in Figure 1.4 on page 10. The function returns a vector of index numbers of the points that were labelled.

In Chapter 1 we used the `locator` function to add new points to a plot. This function is most often used in the form `locator(1)` to return the (x, y) coordinates of a single button click to place a label or legend, but can also be used to return the coordinates of a series of points, terminated by clicking with mouse button 2.

Adding further axes and grids

It is sometimes useful to add further axis scales to a plot, as in Figure 8.1 on page 242 which has scales for both kilograms and pounds. This is done by the function `axis`. There we used

```
attach(wtloss)
oldpar <- par()
# alter margin 4; others are default
par(mar=c(5.1, 4.1, 4.1, 4.1))
plot(Days, Weight, type="p",ylab="Weight (kg)")
Wt.lbs <- pretty(range(Weight*2.205))
axis(side=4, at=Wt.lbs/2.205, lab=Wt.lbs, srt=90)
mtext("Weight (lb)", side=4, line=3)
detach()
par(oldpar)
```

This adds an axis on side 4 (labelled clockwise from the bottom; see Figure 3.4) with labels rotated by 90° (`srt=90`) and then uses `mtext` to add a label 'underneath' that axis. Other parameters are explained in Section 3.4. Please read the

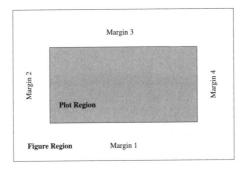

Figure 3.4: Anatomy of a graphics figure.

on-line documentation very carefully to determine which graphics parameters are used in which circumstances. (For example, why did we need `srt=90` for `axis` but not `mtext`?)

Grids can be added by using `axis` with long tick marks, setting parameter `tck=1` (yes, obviously). For example, a dotted grid is created by

```
axis(1, tck=1, lty=2)
axis(2, tck=1, lty=2)
```

and the location of the grid lines can be specified using `at=`.

Adding legends

Legends are added by the function `legend`. Since it can label many types of variation such as line type and width, plot symbol, colour, and fill type, its description is very complex. All calls are of the form

```
legend(x, y, legend, ...)
```

where `x` and `y` give either the upper left corner of the legend box or both upper left and lower right corners. These are often most conveniently specified on-screen by using `locator(1)` or `locator(2)`. Argument `legend` is a character vector giving the labels for each variation. The remaining arguments are vectors of the same length as `legend` giving the appropriate coding for each variation, by `lty=`, `lwd=`, `pch=`, `col=`, `fill=`, `angle=` and `density=`.

By default the legend is contained in a box; the drawing of this box can be suppressed by argument `bty="n"`.

The Trellis function `key` provides a more flexible approach to constructing legends, and can be used with basic plots. (See page 90 for further details.)

Mathematics in labels

Users frequently wish to include the odd subscript, superscript and mathematical symbol in labels. There is no general solution, but for the UNIX `postscript` driver Alan Zaslavsky's package `postscriptfonts` adds these features. It is available from `statlib`: see Appendix C.2.

Under S-PLUS 4.x there is a series of escape codes which can be used to enhance graph labels on graphsheets. These are

`'string'`	*string* in italics
`#string#`	**string** in bold
`x[2]`	x^2, superscript
`x]2[`	x_2, subscript
`\`n	change to font n (the normal font being 0)
`|`n`|`	change to colour n
`~xyz`	character with ASCII code `xyz`

These escape codes can be escaped by preceding them by @. The special characters depend on the font selected: use the **Character Map** accessory to see what is where in each font. In particular, the Greek letters (including variant forms) are in positions `A-Z` and `a-z` of the Symbol font (normally font 1).

Note that these escapes are normally disabled for graphics created by commands from a commands or script window, but can be used by editing text in such graphics.

3.4 Fine control of graphics

The graphics process is controlled by *graphics parameters*, which are set for each graphics device. Each time a new device is opened these parameters for that device are reset to their default values. Graphics parameters may be set, or their current values queried, using the `par` function. If the arguments to `par` are of the `name=value` form the graphics parameter `name` is set to `value`, if possible, and other graphics parameters may be reset to ensure consistency. The value returned is a list giving the previous parameter settings. Instead of supplying the arguments as `name=value` pairs, `par` may also be given a single list argument with named components.

If the arguments to `par` are quoted character strings, `"name"`, the current value of graphics parameter `name` is returned. If more than one quoted string is supplied the value is a list of the requested parameter values, with named components. The call `par()` with no arguments returns a list of all the graphics parameters.

Some of the many graphics parameters are given in Tables 3.3 and 3.4 (on page 70). They have short names, usually of three characters. Those in Table 3.4 can also be supplied as arguments to high-level plot functions, when they apply just to the figure produced by that call. (The layout parameters are ignored by the high-level plot functions.)

Table 3.3: Some graphics layout parameters with example settings.

`din, fin, pin`	Absolute device size, figure size and plot region size in inches. `fin=c(6,4)`
`fig`	Define the figure region as a fraction of the device surface. `fig=c(0,0.5,0,1)`
`font`	Small positive integer determining a text font for characters and hence an interline spacing. `font=3`
`mai, mar`	The four margin sizes, in inches (`mai`), or in text line units (`mar`, that is, *relative* to the current font size). Note that `mar` need not be an integer. `mar=c(3,3,1,1)+0.1`
`mex`	Number of text lines per interline spacing. `mex=0.7`
`mfg`	Define a position within a specified multi-figure display. `mfg=c(2,2,3,2)`
`mfrow, mfcol`	Define a multi-figure display. `mfrow=c(2,2)`
`new`	Logical value indicating whether the current figure has been used. `new=T`
`oma, omi, omd`	Define outer margins in text lines or inches, or by defining the size of the array of figures as a fraction of the device surface. `oma=c(0,0,4,0)`
`plt`	Define the plot region as a fraction of the figure region. `plt=c(0.1,0.9,0.1,0.9)`
`pty`	Plot type, or shape of plotting region, `"s"` or `"m"`
`uin`	Return inches per user coordinate for x and y.
`usr`	Limits for the plot region in user coordinates. `usr=c(0.5, 1.5, 0.75, 10.25)`

The figure region and layout parameters

When a device is opened it makes available a rectangular surface on which one or more plots may appear. Each plot occupies a rectangular section of the device surface called a *figure*. A figure consists of a rectangular *plot region* surrounded by a *margin* on each side. The margins or sides are numbered one to four, clockwise starting from the bottom. The plot region and margins together make up the *figure region*, as in Figure 3.4. The device surface, figure region and plot region have their vertical sides parallel and hence their horizontal sides also parallel.

The size and position of figure and plot regions on a device surface are controlled by *layout parameters*, most of which are listed in Table 3.3. Lengths may be set in either absolute or relative units. Absolute lengths are in *inches*, whereas relative lengths are in *text lines* (so relative to the current font size).

Margin sizes are set using `mar` for text lines or `mai` for inches. These are four-component vectors giving the sizes of the lower, left, upper and right margins in the appropriate units. Changing one causes a consistent change in the other; changing `mex` will change `mai` but not `mar`.

Positions may be specified in relative units using the unit square as a coordinate system for which some enclosing region, such as the device surface or the
figure region, is the unit square. The `fig` parameter is a vector of length four
specifying the current figure as a fraction of the device surface. The first two
components give the lower and upper x-limits and the second two give the y-
limits. Thus to put a point plot in the left-hand side of the display and a Q-Q plot
on the right-hand side we could use:

```
postscript(file="twoplot.ps")    # open a postscript device
par(fig=c(0, 2/3, 0, 1))         # set a figure on the left
plot(x,y)                        # point plot
par(fig=c(2/3, 1, 0, 1))         # set a figure on the right
qqnorm(resid(obj))               # diagnostic plot
dev.off()
```

The left-hand figure occupies $2/3$ of the device surface and the right-hand figure
$1/3$. For regular arrays of figures it is simpler to use `mfrow` or `split.screen`.

Positions in the plot region may also be specified in absolute *user coordinates*. Initially user coordinates and relative coordinates coincide, but any
high-level plotting function changes the user coordinates so that the x- and y-
coordinates range from their minimum to maximum values as given by the plot
axes. The graphics parameter `usr` is a vector of length four giving the lower
and upper x- and y-limits for the user coordinate system. Initially its setting is
`usr=c(0,1,0,1)`. Consider another simple example:

```
> motif()                 # open a device
> par("usr")              # usr coordinates
[1] 0 1 0 1
> x <- 1:20
> y <- x + rnorm(x)       # generate some data
> plot(x, y)              # produce a scatterplot
> par("usr")              # user coordinates now match the plot
[1]  0.2400 20.7600  1.2146 21.9235
```

Any attempt to plot outside the user coordinate limits causes a warning message
unless the general graphics parameter xpd is set to T.

Figure 3.5 shows some of the layout parameters for a multi-figure layout.
Such an array of figures may occupy the entire device surface, or it may have
outer margins, which are useful for annotations that refer to the entire array. Outer
margins are set with the parameter `oma` (in text lines) or `omi` (in inches). Alternatively `omd` may be used to set the region containing the array of figures in a
similar way to which `fig` is used to set one figure. This implicitly determines the
outer margins as the complementary region. In contrast to what happens with the
margin parameters `mar` and `mai`, a change to `mex` will leave the outer margin
size, `omi`, constant but adjust the number of text lines, `oma`.

Text may be put in the outer margins by using `mtext` with parameter
`outer=T`.

Table 3.4: Some of the more commonly used general and high-level graphics parameters with example settings.

Text:

`adj`	Text justification. 0 = left justify, 1 = right justify, 0.5 = centre.
`cex`	Character expansion. `cex=2`
`csi`	Height of font (inches). `csi=0.11`
`font`	Font number: device-dependent.
`srt`	String rotation in degrees. `srt=90`
`cin cxy`	Character width and height in inches and `usr` coordinates (for information, not settable).

Symbols:

`col`	Colour for symbol, line or region. `col=2`
`lty`	Line type: solid, dashed, dotted, etc. `lty=2`
`lwd`	Line width, usually as a multiple of default width. `lwd=2`
`mkh`	Mark height (inches). `mkh=0.05`
`pch`	Plotting character or mark. `pch="*"` or `pch=4` for marks. (See page 60.)

Axes:

`bty`	Box type, as `"o"`, `"l"`, `"7"`, `"c"`, `"n"`.
`exp`	Notation for exponential labels. `exp=1`
`lab`	Tick marks and labels. `lab=c(3,7,4)`
`las`	Label orientation. 0 = parallel to axis, 1 = horizontal, 2 = vertical.
`log`	Control log axis scales. `log="y"`
`mgp`	Axis location. `mgp=c(3,1,0)`
`tck`	Tick mark length as signed fraction of the plot region dimension. `tck=-0.01`
`xaxp yaxp`	Tick mark limits and frequency. `xaxp=c(2, 10, 4)`
`xaxs yaxs`	Style of axis limits. `xaxs="i"`
`xaxt yaxt`	Axis type. `"n"` (null), `"s"` (standard), `"t"` (time) or `"l"` (log)

High Level:

`ask`	Prompt before going on to next plot? `ask=F`
`axes`	Print axes? `axes=F`
`main`	Main title. `main="Figure 1"`
`sub`	Subtitle. `sub="23-Jun-1994"`
`type`	Type of plot. `type="n"`
`xlab ylab`	Axis labels. `ylab="Speed in km/sec"`
`xlim ylim`	Axis limits. `xlim=c(0,25)`
`xpd`	May points or lines go outside the plot region? `xpd=T`

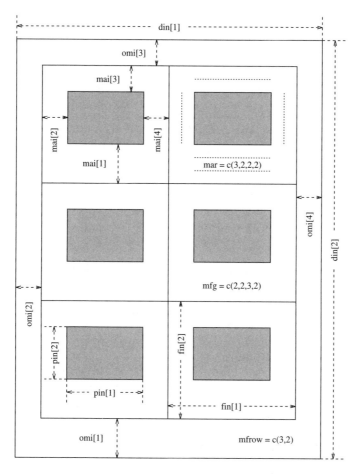

Figure 3.5: An outline of a 3×2 multi-figure display with outer margins showing some graphics parameters. The current figure is at position $(2, 2)$ and the display is being filled by rows. In this figure "fin[1]" is used as a shorthand for par("fin")[1], and so on.

Common axes for figures

There are at least two ways to ensure that several plots share a common axis or axes.

1. Use the same xlim or ylim (or both) setting on each plot and ensure that the parameters governing the way axes are formed, such as lab, las, xaxs and allies, do not change.

2. Set up the desired axis system with the first plot and then set the low-level parameter xaxs="d", yaxs="d" or both as appropriate. This ensures that the axis or axes are not changed by further high-level plot commands on the same device.

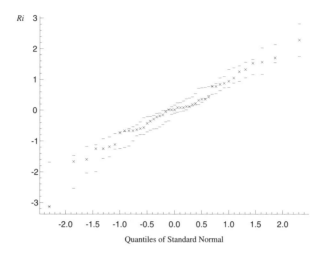

Figure 3.6: The Swiss fertility data. A Q-Q normal plot with envelope for infant mortality.

An example: A Q-Q normal plot with envelope

In Chapter 5 we recommend assessing distributional form by quantile-quantile plots. A simple way to do this is to plot the sorted values against quantile approximations to the expected normal order statistics and draw a line through the 25 and 75 percentiles to guide the eye, performed for the variable Infant.Mortality of the Swiss provinces data (on fertility and socio-economic factors on Swiss provinces in about 1888) by

```
swiss.df <- data.frame(Fertility=swiss.fertility, swiss.x)
attach(swiss.df)
qqnorm(Infant.Mortality)
qqline(Infant.Mortality)
```

The reader should check the result and compare it with the style of Figure 3.6.

Another suggestion to assess departures is to compare the sample Q-Q plot with the envelope obtained from a number of other Q-Q plots from generated normal samples. This is discussed in (Atkinson, 1985, §4.2) and is based on an idea of Ripley (see Ripley, 1981, Chapter 8). The idea is simple. We generate a number of other samples of the same size from a normal distribution and scale all samples to mean 0 and variance 1 to remove dependence on location and scale parameters. Each sample is then sorted. For each order statistic the maximum and minimum values for the generated samples form the upper and lower envelopes. The envelopes are plotted on the Q-Q plot of the scaled original sample and form a guide to what constitutes serious deviations from the expected behaviour under normality. Following Atkinson our calculation uses 19 generated normal samples.

We begin by calculating the envelope and the x-points for the Q-Q plot.

```
samp <- cbind(Infant.Mortality, matrix(rnorm(47*19), 47, 19))
```

```
samp <- apply(scale(samp), 2, sort)
rs <- samp[,1]
xs <- qqnorm(rs, plot=F)$x
env <- t(apply(samp[,-1], 1, range))
```

As an exercise in building a plot with specific requirements we now present
the envelope and Q-Q plot in a style very similar to Atkinson's. To ensure that the
Q-Q plot has a y-axis large enough to take the envelope we could calculate the
y-limits as before, or alternatively use a matrix plot with type=n for the envelope
at this stage. The axes are also suppressed for the present:

```
matplot(xs, cbind(rs,env), type="pnn", pch=4, mkh=0.06, axes=F)
```

The argument setting type="pnn" specifies that the first column (rs) is to produce a point plot and the remaining two (env) no plot at all, but the axes will
allow for them. Setting pch=4 specifies a 'cross' style plotting symbol (see Figure 3.2) similar to Atkinson's, and mkh=0.06 establishes a suitable size for the
plotting symbol.

Atkinson uses small horizontal bars to represent the envelope. We can now
calculate a half length for these bars so that they do not overlap and do not extend
beyond the plot region. Then we can add the envelope bars using segments :

```
xyul <- par("usr")
smidge <- min(diff(c(xyul[1], xs, xyul[2])))/2
segments(xs-smidge, env[,1], xs+smidge, env[,1])
segments(xs-smidge, env[,2], xs+smidge, env[,2])
```

Atkinson's axis style differs from the default S style in several ways. There are
many more tick intervals; the ticks are inside the plot region rather than outside;
there are more labelled ticks, and the labelled ticks are longer than the unlabelled.
From experience ticks along the x-axis at 0.1 intervals with labelled ticks at 0.5
intervals seems about right but this is usually too close on the y-axis. The axes
require four calls to the axis function:

```
xul <- trunc(10*xyul[1:2])/10
axis(1, at=seq(xul[1], xul[2], by=0.1), labels=F, tck=0.01)
xi <- trunc(xyul[1:2])
axis(1, at=seq(xi[1], xi[2], by=0.5), tck=0.02)
yul <- trunc(5*xyul[3:4])/5
axis(2, at=seq(yul[1], yul[2], by=0.2), labels=F, tck=0.01)
yi <- trunc(xyul[3:4])
axis(2, at=yi[1]:yi[2], tck=0.02)
```

Finally we add the L-box, put the x-axis title at the centre and the y-axis title
at the top:

```
box(bty="l")              # lower case "L"
ps.options()$fonts
mtext("Quantiles of Standard Normal", side=1, line=2.5, font=3)
mtext("Ri", side=2, line=2, at=yul[2], font=10)
```

where fonts 3 and 10 are Times-Roman and Times-Italic on the device used
(`postscript` under UNIX), found from the list given by `ps.options`.

The final plot is shown in Figure 3.6.

3.5   Trellis graphics

Trellis graphics were developed at the former AT&T's Bell Laboratories to provide a consistent graphical 'style' and to extend conditioning plots. The style is
a development of that used in Cleveland (1993). We describe the version that is
part of S-PLUS 3.4 and later versions of S-PLUS.

Trellis is very prescriptive, and changing the display style is not always an
easy matter.

It may be helpful to understand that Trellis is written entirely in the S language, as calls to the basic plotting routines. Two consequences are that it can be
slow and memory-intensive, and that it takes over many of the graphics parameters for its own purposes. (Global settings of graphics parameters are usually not
used, the outer margin parameters `omi` being a notable exception.) Computation
of a Trellis plot is done in two passes: once when a Trellis object is produced, and
once when that object is printed (producing the actual plot).

The `trellis` library contains a large number of examples: use

```
?trellis.examples
```

to obtain an up-to-date list. These are all functions that can be called to plot the
example, and listed to see how the effect was achieved.

Trellis graphical devices

The `trellis.device` graphical device is provided by the `trellis` library. It
is perhaps more accurate to call it a meta-device, for it uses one of the underlying graphical devices (currently `motif`, `openlook`, `iris4d`, `postscript`,
`graphsheet`, `win.graph` and `win.printer` where these are available), but
customizes the parameters of the device to use the Trellis style, and in particular
its colour schemes.

Trellis devices by default use colour for screen windows and greylevels for
printer devices. The settings for a particular device can be seen by running the
command[4] `show.settings()` (see Figure 3.7). These settings are not the same
for all colour screens, nor for all printer devices. Trellis colour schemes have a
mid-grey background on colour screens (but not colour printers). If a Trellis plot
is used without a graphics device already in use, a suitable Trellis device is started.

---

[4] Missing from S-PLUS 4.0.

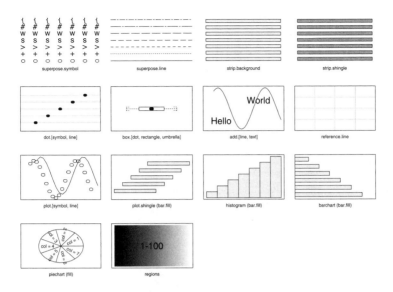

Figure 3.7: The settings of the greylevel PostScript `trellis.device`.

Hardcopy from Trellis plots

The basic paradigm of a Trellis plot is to produce an object that the device 'prints', that is, plots. Thus the simplest way to produce a hardcopy of a Trellis plot is to switch to a printer device, 'print' the object again, and switch back. For example (under UNIX)

```
trellis.device()
p1 <- histogram(geyser$waiting)
p1      # plots it on screen
trellis.device("postscript", file="hist.eps",
    onefile=F, print.it=F)
p1      # print the plot
dev.off()
```

However, it can be difficult to obtain precisely the same layout in this way (since this depends on the aspect ratio and size parameters), and it is impossible to interact with such a graph (for example, by using `identify`). Fortunately, the methods for hardcopy described on page 56 can still be used. It is important to set the options for the `postscript` device to match the colour schemes in use. For example, on UNIX with hardcopy via the `postscript` device we can use

```
ps.options(colors=colorps.trellis[, -1])
```

before the Trellis device is started. Then the `rmv` method and `dev.print` will use the Trellis (printer) colour scheme and produce colour PostScript output. Conversely, if greylevel PostScript output is required (for example, for figures in a book or article) we can use (for a `motif` device)

```
ps.options(colors=bwps.trellis, pointsize=11)
trellis.settings.motif.bw <- trellis.settings.bwps
xcm.trellis.motif.bw <- xcm.trellis.motif.grey
trellis.device(color=F)
```

using xcm.* objects in our library MASS. This sets up a color screen device to mimic the 'black and white' (actually greylevel) PostScript device.

Trellis model formulae

Trellis graphics functions make use of the language for model formulae described in Section 2.5. The Trellis code for handling model formulae to produce a data matrix from a data frame (specified by the data argument) allows the argument subset to select a subset of the rows of the data frame, as one of the first three forms of indexing vector described on page 39. (Character vector indices are not allowed.)

There are a number of inconsistencies in the use of the formula language. There is no na.action argument, and missing values are handled inconsistently; generally rows with NAs are omitted, but splom fails if there are missing values. Surprisingly, splom uses a formula, but does not accept a data argument.

Trellis uses an extension to the model formula language, the operator ' | ' which can be read as 'given'. Thus if a is a factor, lhs ~ rhs | a will produce a plot for each level of a of the subset of the data for which a has that level (so estimating the conditional distribution given a). Conditioning on two or more factors gives a plot for each combination of the factors, and is specified by an interaction, for example, | a*b. For the extension of conditioning to continuous variates via what are known as shingles, see page 87.

Trellis plot objects can be kept, and update can be used to change them, for example, to add a title or change the axis labels.

Basic Trellis plots

As Table 3.5 and Figure 3.7 show, the basic styles of plot exist in Trellis, but with different names and different default styles. Their usage is best seen by considering how to produce some figures in the Trellis style.

Figure 1.2 (page 9) was produced by splom(~ hills). Trellis plots of scatterplot matrices read from bottom to top (as do all multi-panel Trellis displays, like graphs rather than matrices, despite the meaning of the name splom). By default the panels in a splom plot are square.

Figure 3.8 is a Trellis version of Figure 1.4. Note that the y-axis numbering is horizontal by default (equivalent to the option par(las=1)), and that points are plotted by open circles rather than filled circles or stars. It is not possible to add to a Trellis plot (as the user coordinate system is not retained), so the Trellis call has to include all the desired elements. This is done by writing a *panel function*, in this case

Table 3.5: Trellis plotting functions. Page references are given to the most complete description in the text.

	Page	
xyplot	79	Scatterplots.
bwplot	78	Boxplots.
stripplot	84	Display univariate data against a numerical variable.
dotplot	85	ditto in another style,
histogram		'Histogram', actually a frequency plot.
densityplot		Kernel density estimates.
barchart	78	Horizontal bar plots.
piechart		Pie chart.
splom	76	Scatterplot matrices.
contourplot	81	Contour plot of a surface on a regular grid.
levelplot	81	Pseudo-colour plot of a surface on a regular grid.
wireframe	81	Perspective plot of a surface evaluated on a regular grid.
cloud	90	A perspective plot of a cloud of points.
key	90	Add a legend.
color.key	82	Add a color key (as used by levelplot).
trellis.par.get	79	Save Trellis parameters.
trellis.par.set	79	Reset Trellis parameters.
equal.count	87	Compute a shingle.

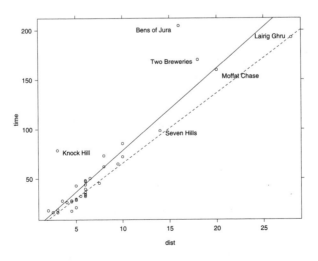

Figure 3.8: A Trellis version of Figure 1.4 (page 10).

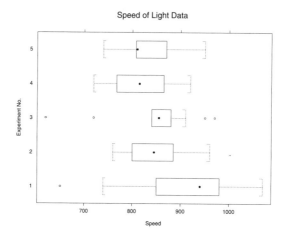

Figure 3.9: A Trellis version of Figure 1.5 (page 11).

```
xyplot(time ~ dist, data = hills,
    panel = function(x, y, ...) {
        panel.xyplot(x, y, ...)
        panel.lmline(x, y, type="l")
        panel.abline(ltsreg(x, y), lty=3)
        identify(x, y, row.names(hills))
    }
)
```

Figure 3.9 is a Trellis version of Figure 1.5. Boxplots are known as box-and-whisker plots, and are displayed horizontally. This figure was produced by

```
bwplot(Expt ~ Speed, data=michelson, ylab="Experiment No.")
title("Speed of Light Data")
```

Note the counter-intuitive way the formula is used. This plot corresponds to a one-way layout splitting Speed by experiment, so it is tempting to use Speed as the response.[5] It may help to remember that the formula is of the y ~ x form for the *x*- and *y*-axes of the plot. (The same ordering is used for all the univariate plot functions.)

Figure 3.10 is a Trellis version of Figure 3.1. This is the first example we have seen of a plot with more than one panel.

```
lung.deaths.df <- data.frame(year = rep(1974:1979, 2),
    deaths = c(lung.deaths[, 1], lung.deaths[ ,2]),
    sex = rep(c("Male", "Female"), rep(6,2)))
barchart(year ~ deaths | sex, lung.deaths.df, xlim=c(0, 20000))
```

Figure 3.11 is an enhanced scatterplot matrix, again using a panel function to add to the basic display. Now we see the power of panel functions, as the basic plot

---

[5] This was the meaning in Trellis 1.x in boxplot.formula and stripplot!

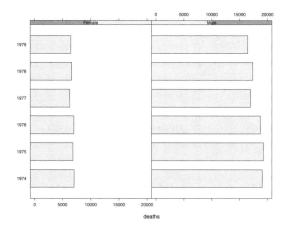

Figure 3.10: A Trellis version of Figure 3.1 (page 58).

commands can easily be applied to multi-panel displays. The `aspect="fill"`
command allows the array of plots to fill the space: by default the panels are
square as in Figure 1.2.

```
splom(~ swiss.df, aspect="fill",
    panel = function(x, y, ...) {
        panel.xyplot(x, y, ...); panel.loess(x, y, ...)
    }
)
```

Most Trellis graphics functions have a `groups` parameter, which we can illustrate on the `stormer` data used in Section 8.4 (see Figure 3.12).

```
sps <- trellis.par.get("superpose.symbol")
sps$pch <- 1:7
trellis.par.set("superpose.symbol", sps)
xyplot(Time ~ Viscosity, stormer, groups = Wt,
    panel = panel.superpose, type = "b",
    key = list(columns = 3,
        text = list(paste(c("Weight:   ", "", ""),
                          unique(stormer$Wt), "gms")),
        points = Rows(sps, 1:3)
        )
)
```

Here we have changed the default plotting symbols (which differ by device) to the
first seven `pch` characters shown in Figure 3.2 on page 60. (We could just use the
argument `pch=1:7` to `xyplot`, but then specifying the key becomes much more
complicated.)

Figure 3.13 shows further Trellis plots of the smooth surface shown in Figure 3.3. Once again panel functions are needed to add the points. The grid has

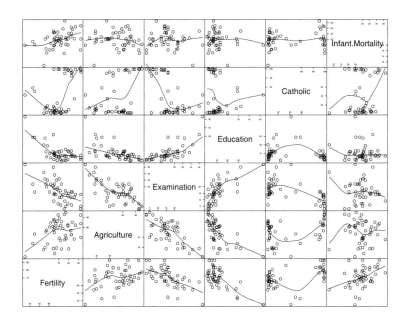

Figure 3.11: A Trellis scatterplot matrix display of the Swiss provinces data.

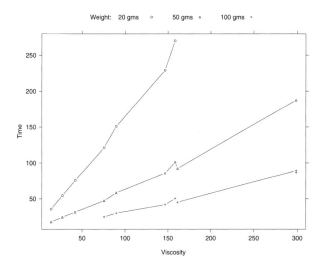

Figure 3.12: A Trellis plot of the `stormer` data.

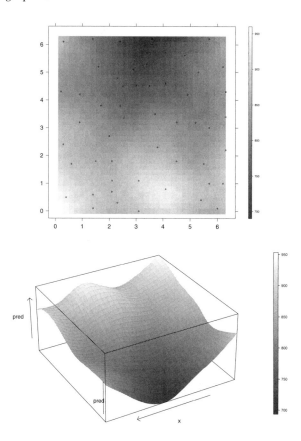

Figure 3.13: Trellis `levelplot` and `wireframe` plots of a `loess` smoothing of the `topo` dataset.

been reduced in extent, since whereas `contourplot` copes with the NAs produced by `predict.loess` for Figure 3.3, prior to S-PLUS 3.4 `levelplot` produced incorrect results (without any warning). The `aspect=1` parameter ensures a square plot. The `drape=T` parameter to `wireframe` is optional, producing the superimposed greylevel (or pseudo-colour) plot.

```
topo.plt <- expand.grid(topo.mar)
topo.plt$pred <- as.vector(predict(topo.loess, topo.plt))
levelplot(pred ~ x * y, topo.plt, aspect=1,
   at = seq(690, 960, 10), xlab="", ylab="",
   panel = function(x, y, subscripts, ...) {
      panel.levelplot(x, y, subscripts, ...)
      panel.xyplot(topo$x,topo$y, cex=0.5, col=1)
   }
)
wireframe(pred ~ x * y, topo.plt, aspect=c(1, 0.5), drape=T,
   screen = list(z = -150, x = -60),
   colorkey=list(space="right", height=0.6))
```

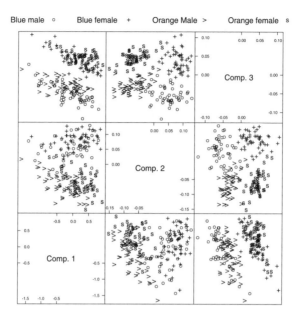

Figure 3.14: A scatterplot matrix of the first three principal components of the `crabs` data.

(This gave a faulty plot and/or key on earlier versions of **S-PLUS**. The arguments given by `colorkey` refer to the `color.key` function.) There is no simple way to add the points to the perspective display.

Trellises of plots

In multivariate analysis we necessarily look at several variables at once, and we explore here several ways to do so. We can produce a scatterplot matrix of the first three principal components of the `crabs` data (Section 11.7) by

```
lcrabs.pc <- predict(princomp(log(crabs[,4:8])))
crabs.grp <- c("B", "b", "O", "o")[rep(1:4, rep(50,4))]
splom(~ lcrabs.pc[, 1:3], groups = crabs.grp,
    panel = panel.superpose,
    key = list(text = list(c("Blue male", "Blue female",
                             "Orange Male", "Orange female")),
        points = Rows(trellis.par.get("superpose.symbol"), 1:4),
        columns = 4)
)
```

A 'black and white' version of this plot is shown in Figure 3.14. On a 'colour' device the groups are distinguished by colour and are all plotted with the same symbol (o).

However, it might be clearer to display these results as a trellis of `splom` plots, by

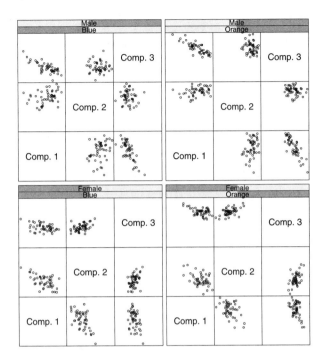

Figure 3.15: A multi-panel version of Figure 3.14.

```
sex <- crabs$sex; levels(sex) <- c("Female", "Male")
sp <- crabs$sp; levels(sp) <- c("Blue", "Orange")
splom(~ lcrabs.pc[, 1:3] | sp*sex, cex=0.5, pscales=0)
```

as shown in Figure 3.15. Notice how this is the easiest method to code. It is at the core of the paradigm of Trellis, which is to display many plots of subsets of the data in some meaningful layout.

Now consider data from a multi-factor study, Quine's data on school absences discussed in Sections 6.6 and 7.4. It will help to set up more informative factor labels, as the factor names are not given (by default) in trellises of plots.

```
Quine <- quine
levels(Quine$Eth) <- c("Aboriginal", "Non-aboriginal")
levels(Quine$Sex) <- c("Female", "Male")
levels(Quine$Age) <- c("primary", "first form",
                       "second form", "third form")
levels(Quine$Lrn) <- c("Average learner", "Slow learner")
bwplot(Age ~ Days | Sex*Lrn*Eth, data=Quine)
```

This gives an array of eight boxplots, which by default takes up two pages. On a screen device there will be no pause between the pages unless the argument ask=T is set for par. It is more convenient to see all the panels on one page, which we can do by asking for a different layout (Figure 3.16).

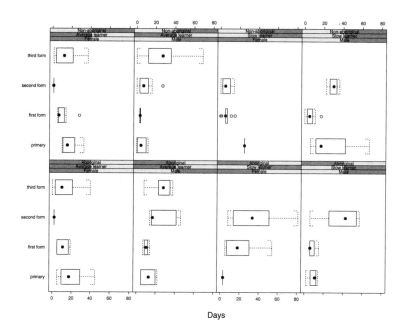

Figure 3.16: A multi-panel boxplot of Quine's school attendance data.

```
bwplot(Age ~ Days | Sex*Lrn*Eth, data=Quine, layout=c(4,2))
```

A `stripplot` allows us to look at the actual data. We jitter the points slightly to avoid overplotting.

```
stripplot(Age ~ Days | Sex*Lrn*Eth, data=Quine,
    jitter = T, layout = c(4,2))

stripplot(Age ~ Days | Eth*Sex, data=Quine,
    groups = Lrn, jitter=T,
    panel = function(x, y, subscripts, jitter.data=F, ...) {
        if(jitter.data)  y <- jitter(y)
        panel.superpose(x, y, subscripts, ...)
    },
    xlab = "Days of absence",
    between = list(y=1), par.strip.text = list(cex=1.2),
    key = list(columns = 2, text = list(levels(Quine$Lrn)),
        points = Rows(trellis.par.get("superpose.symbol"), 1:2)
        ),
    strip = function(...)
            strip.default(..., strip.names=c(T, T), style=1)
)
```

The second form of plot, shown in Figure 3.17, uses different symbols to distinguish one of the factors.

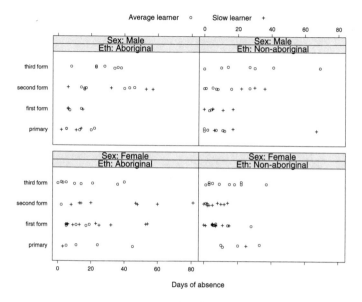

Figure 3.17: A `stripplot` of Quine's school attendance data.

It is possible to include the factor name in the strip labels, by using a custom `strip` function as in Figure 3.17. The parameter `par.strip.text` controls the size, font and colour of the strip labels. The alternative `styles` can be useful with panels labelled by factors; there are currently six.

The Trellis function `dotplot` is very similar to `stripplot`; its panel function includes horizontal lines at each level. `stripplot` uses the styles of `xyplot` whereas `dotplot` has its own set of defaults; for example, the default plotting symbol is a filled rather than open circle.

As a third example, consider our dataset `fgl`. This has 10 measurements on 214 fragments of glass from forensic testing, the measurements being of the refractive index and composition (percent weight of oxides of Na, Mg, Al, Si, K, Ca, Ba and Fe). The fragments have been classified by six sources. We can look at the types for each measurement by

```
fgl0 <- fgl[ ,-10] # omit type.
fgl.df <- data.frame(type = rep(fgl$type, 9),
    y = as.vector(as.matrix(fgl0)),
    meas = factor(rep(1:9, rep(214,9)), labels=names(fgl0)))
stripplot(type ~ y | meas, data=fgl.df, scales=list(x="free"),
    strip=function(...) strip.default(style=1, ...), xlab="")
```

Layout of a trellis

A trellis of plots is generated as a sequence of plots which are then arranged in rows, columns and pages. The sequence is determined by the order in which the

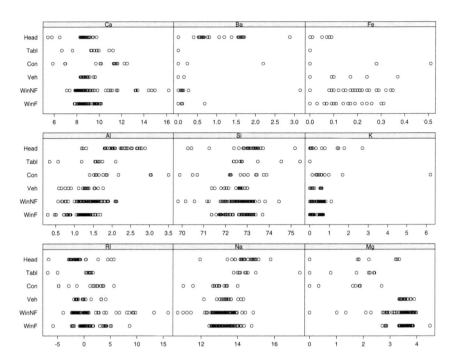

Figure 3.18: Plot by `stripplot` of the forensic glass dataset `fgl`.

conditioning factors are given: the first varying fastest. The order of the levels of the factor is that of its `levels` attribute.

How the sequence of plots is displayed on the page(s) is controlled by an algorithm that tries to optimize the use of the space available, but it can be controlled by the `layout` parameters. A specification `layout = c(c, r, p)` asks for c columns, r rows and p pages. (Note the unusual ordering.) Using $c = 0$ allows the algorithm to choose the number of columns; p is used to produce only the first p pages of a many-page trellis.

If the number of levels of a factor is large and not easily divisible (for example, seven), we may find a better layout by leaving some of the cells of the trellis empty. For example, we might use (extending an example on the next page)

```
Cath <- equal.count(swiss.df$Catholic, number=2, overlap=0)
Agr5 <- equal.count(swiss.df$Agric, number=5, overlap=0.25)
xyplot(Fertility ~ Education | Agr5 * Cath, data=swiss.df,
        layout=c(2,3), skip = c(F,F,F,F,F,T))
```

to use five panels on each page of a 3×2 layout.

The `between` parameter can be used to specify gaps in the trellis layout, as in Figure 3.17. It is a list with `x` and `y` components, numeric vectors which specify the gaps in units of character height. The `page` parameter can be used to invoke a function (with argument n, the page number) to label each page. The default page function does nothing.

It sometimes helps to re-order the levels of a factor to emphasis trends in the data. The `trellis` function `reorder.factor` provides one way to do this. For example, we could use

```
reorder.factor(Quine$Age, Quine$Days, median)
```

to reorder the forms by the median number of days' absence. (This returns an ordered factor.)

Conditioning plots and shingles

The idea of a trellis of plots conditioning on combinations of one or more factors can be extended to conditioning on real-valued variables, in what are known as conditioning plots or *coplots*. Two variables are plotted against each other in a series of plots with the values of further variable(s) restricted to a series of possibly overlapping ranges. This needs an extension of the concept of a factor known as a *shingle*.[6]

Suppose we wished to examine the relationship between `Fertility` and `Education` in the Swiss fertility data as the variable `Catholic` ranges from predominantly non-Catholic to mainly Catholic provinces. We add a smooth fit to each panel.

```
Cath <- equal.count(swiss.df$Catholic, number=6, overlap=0.25)
xyplot(Fertility ~ Education | Cath, data=swiss.df,
   panel = function(x, y) {
      panel.xyplot(x, y); panel.loess(x, y)
   }
)
```

The result is shown in Figure 3.19, with the strips continuing to show the (now overlapping) coverage for each panel. Fertility generally falls as education rises and rises as the proportion of Catholics in the population rises. Note that the level of education is lower in predominantly Catholic provinces.

The function `equal.count` is used to construct a shingle with suitable ranges for the conditioning intervals.

Conditioning plots may also have more than one conditioning variable. Let us condition on Catholic and agriculture simultaneously. Since the dataset is small it seems prudent to limit the number of panels to six in all.

```
Cath <- equal.count(swiss.df$Catholic, number=2, overlap=0)
Agr <- equal.count(swiss.df$Agric, number=3, overlap=0.25)
xyplot(Fertility ~ Education | Cath * Agr, data=swiss.df,
   panel = function(x, y) {
      panel.xyplot(x, y); panel.loess(x, y)
   }
)
```

---

[6] In American usage this is a rectangular wooden tile laid partially overlapping on roofs or walls.

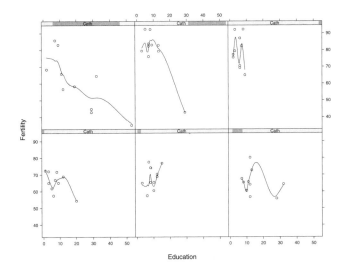

Figure 3.19: A conditioning plot for the Swiss provinces data.

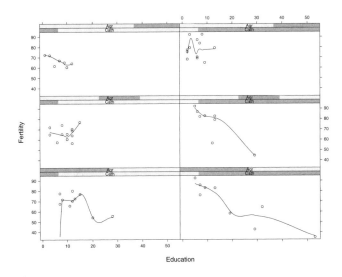

Figure 3.20: Another conditioning plot with two conditioning shingles.

The result is shown in Figure 3.20. In general the fertility rises with the proportion of Catholics and agriculture and falls with education. There is no convincing evidence of substantial interaction.

Shingles have levels, and can be printed and plotted:

```
> Cath <- equal.count(swiss.df$Cath, number=6, overlap=0.25)
> Cath
Data:
```

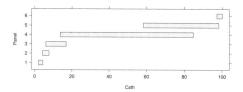

Figure 3.21: A plot of the shingle `Cath`.

```
[1] 10.0  84.8  93.4  33.8   5.2  90.6  92.9  97.2  97.7  91.4
 ....
Intervals:
  min    max count
  2.2    4.5    10
  4.2    7.7    10
  ....
Overlap between adjacent intervals:
[1] 3 2 3 2 3
> levels(Cath)
  min    max
  2.2    4.5
  ....
> plot(Cath, aspect = 0.3)
```

Multiple displays per page

Recall that a Trellis object is plotted by printing it. The method `print.trellis`
has optional arguments `position`, `split` and `more`. The argument `more`
should be set to `T` for all but the last part of a figure. The position of individual
parts on the device surface can be set by either `split` or `position`. A `split`
argument is of the form `c(`x, y, nx, ny`)` for four integers. The second pair give
a division into a $nx \times ny$ layout, just like the `mfrow` and `mfcol` arguments to
`par`. The first pair give the rectangle to be used within that layout, with origin at
the bottom left.

A `position` argument is of the form `c`(*xmin, ymin, xmax, ymax*) giving the
corners of the rectangle within which to plot the object. (This is a different order
from `split.screen`.) The coordinate system for this rectangle is $[0, 1]$ for both
axes, but the limits can be chosen outside this range. It will often be necessary to
experiment with slightly overlapping subregions to avoid whitespace (as shown
in Figure 5.8 on page 135).

The `print.trellis` works by manipulating the graphics parameter `omi`, so
the outer margin settings are preserved. However, none of the basic plot methods
of subdividing the device surface will work, and if a trellis print fails `omi` is not
reset. (Using `par(omi=rep(0,4)`, `new=F)` will reset the usual defaults.)

Fine control

Detailed control of Trellis plots may be accomplished by a series of arguments described in the help page for `trellis.args`, with variants for the `wireframe` and `cloud` perspective plots under `trellis.3d.args`.

We have seen some of the uses of panel functions. Some care is needed with computations inside panel functions that use any data (or user-defined objects or functions) other than their arguments. First, the computations will occur inside a deeply nested set of function calls, so care is needed to ensure that the data are visible. Second, those computations will be done at the time the result is printed (that is, plotted) and so the data need to be in the desired state at plot time.

If non-default panel functions are used, we may want these to help control the coordinate system of the plots, for example, to use a fitted curve to decide the aspect ratio of the panels. This is the function of the `prepanel` argument, and there are prepanel functions corresponding to the `densityplot`, `lmline`, `loess`, `qq`, `qqmath` and `qqmathline` panel functions. These will ensure that the whole of the fitted curve is visible, and they may affect the choice of aspect ratio.

The parameter `aspect` controls the aspect ratio of the panels. A numerical value (most usefully one) sets the ratio, `"fill"` adjusts the aspect ratio to fill the space available and `"xy"` attempts to bank the fitted curves to $\pm45°$. (See Exercise 3.4.)

The `scales` argument determines how the x and y axes are drawn. It is a list of components of `name=value` form, and components x and y may themselves be lists. The default `relation="same"` ensures that the axes on each panel are identical. With `relation="sliced"` the same numbers of data units are used, but the origin may vary by panel, whereas with `relation="free"` the axes are drawn to accommodate just the data for that panel. One can also specify most of the parameters of the `axis` function, and also `log=T` to obtain a $\log_{10}$ scale or even `log=2` for a $\log_2$ scale.

The function `splom` has an argument `varnames` which sets the names of the variables plotted on the diagonal. The argument `pscales` determines how the axes are plotted: set `pscales=0` to omit them.

Keys

The function `key` is a replacement for `legend`, and can also be used as an argument to Trellis functions. If used in this way, the Trellis routines allocate space for the key, and repeat it on each page if the trellis extends to multiple pages.

The call of key specifies the location of the key by the arguments x, y and corner. By default `corner=c(0,1)`, when the coordinate (x, y) specifies the upper left corner of the key. Any other coordinate of the key can be specified by setting `corner`, but the size of the key is computed from its contents. (If the argument `plot=F`, the function returns a two-element vector of the computed width and height, which can be used to allocate space.) When `key` is used as an argument to a Trellis function, the position is normally specified not by x and y but by the argument `space` which defaults to `"top"`.

Most of the remaining arguments to `key` will specify the contents of the key. The (optional) arguments `points`, `lines`, `text` and `rectangles` (for `barchart`) will each specify a column of the key in the order in which they appear. Each argument must be a *list* giving the graphics parameters to be used (and for `text`, the first argument must be the character vector to be plotted). (The function `trellis.par.get` is useful to retrieve the actual settings used for graphics parameters.)

The third group of arguments to `key` fine-tunes its appearance—should it be transparent (`transparent=T`), the presence of a border (specified by giving the border colour as argument `border`), the spacing between columns (`between.columns` in units of character width), the background colour, the font(s) used, the existence of a title and so on. Consult the on-line help for the current details. The argument `columns` specifies the number of columns in the key—we used this in Figures 3.12 and 3.14.

Perspective plots

The argument `aspect` is a vector of two values for the perspective plots, giving the ratio of the y and z sizes to the x size; its effect can be seen in Figure 3.13.

The arguments `distance`, `perspective` and `screen` control the perspective view used. If `perspective=T` (the default), the `distance` argument (default 0.2) controls the extent of the perspective, although not on a physical distance scale as 1 corresponds to viewing from infinity. The `screen` argument (default `list(z = 40, x = -60)`) is a list giving the rotations (in degrees) to be applied to the specified axis in turn. The initial coordinate system has x pointing right, z up and y into the page.

The argument `zoom` (default 1) may be used to scale the final plot, and the argument `par.box` controls how the lines forming the enclosing box are plotted.

3.6   Exercises

3.1.   The data frame `survey` contains the results of a survey of 237 first-year statistics students at Adelaide University. For a graphical summary of all the variables, use `plot(survey)`. Note that this produces a dotchart for factor variables, and a normal scores plot for the numeric variables.

One component of this data frame, `Exer`, is a factor object containing the responses to a question asking how often the students exercised. Produce a barchart of these responses. Use `table` and `pie` or `piechart` to create a pie chart of the responses. Do you like this better than the bar plot? Which is more informative? Which gives a better picture of exercise habits of students? The `pie` function takes an argument `names` which can be used to put labels on each pie slice. Redraw the pie chart with labels. Alternatively, you could add a legend to identify the slices.

You might like to try the same things with the `Smoke` variable, which records responses to the question, "How often do you smoke?" Note that `table` and `levels` ignore missing values; if you wish to include non-respondents in your chart use `summary` to generate the values, and `names` on the summary object to generate the labels.

3.2. Make a plot of petal width versus petal length of the `iris` data for a partially sighted audience, identifying the three species. You will need to double the annotation size, thicken the lines and change the layout to allow larger margins for the larger annotation.

3.3. Plot $\sin(x)$ against x, using 200 values of x between $-\pi$ and π, but do not plot any axes yet (use parameter `axes=F` in the call to `plot`). Add a y-axis passing through the origin using the 'extended' style and horizontal labels. Add an x-axis with tick-marks from $-\pi$ to π in increments of $\pi/4$, twice the usual length.

3.4. Cleveland (1993) recommends that the aspect ratio of line plots is chosen so that lines are 'banked' at $45°$. By this he means that the averaged absolute value of the slope should be around $\pm45°$. Write a function to achieve this for a time-series plot, and try it out on the `sunspots` dataset. (Average along the arc length of the curve. See the function `banking` for Cleveland's solution.)

3.5. The Trellis function `splom` produces a complete matrix of scatterplots, as does the basic plotting functions `pairs`, but in earlier versions of S-PLUS `pairs` only plotted the lower triangle of the matrix. Write a function to emulate the earlier behaviour. [Hint: look at `pairs.default`. The graphics parameter `mfg` may be useful.]

3.6. *Ternary* plots are used for compositional data (Aitchison, 1986) where there are three components whose proportions add to one. These are represented by a point in an equilateral triangle, where the distances to the sides add to a constant.

Write an S function to plot a matrix of compositions on a ternary diagram. Apply this to the dataset `Skye` on the composition of rocks on the Isle of Skye in Scotland. (Our solution can be found on the help page for this dataset, and S-PLUS 2000 has a GUI-graphics example in its `samples` directory.)

Chapter 4

Programming in S

The S language is both an interactive language and a language for adding new functions to the S-PLUS system. It is a complete programming language with control structures, recursion and a useful variety of data types. The S-PLUS environment provides many functions to handle standard operations, but most users need occasionally to write new functions, a topic we discuss in detail in the companion volume. In this chapter we discuss language ideas that will be seen in system functions.

4.1 Control structures

Control structures are the commands that make decisions or execute loops. These statements are most often used inside functions rather than interactively.

Conditional execution of statements

Conditional execution uses either the `if` statement or the `switch` function. The `if` statement has the form

```
if (condition)  true.branch   else   false.branch
```

First the expression `condition` is evaluated. If the result is T (or non-zero) the value of the `if` statement is that of the expression `true.branch`, otherwise that of the expression `false.branch`. The `else` part is optional and omitting it is equivalent to using "`else NULL`". If `condition` has a vector value only the first component is used and a warning is issued. The `if` function can be extended over several lines, and the statements may be compound statements enclosed in braces { }.

Two additional logical operators, && and ||, are useful with `if` statements. Unlike & and |, which operate componentwise on vectors, these operate on scalar logical expressions. With && the right-hand expression is only evaluated if the left-hand one is true, and with || only if it is false. This conditional evaluation property can be used as a safety feature, as in

```
if (is.numeric(x) && min(x) > 0)  sx <- sqrt(x)
else  stop("x must be numeric and all components positive")
```

The expression `min(x) > 0` is invalid for non-numeric x.

The functions `any` and `all` are often useful in defining scalar logicals. They evaluate, respectively, the logical 'or' and 'and' of all components of their argument. For example, the usual test for exact symmetry of a matrix is `all(X == t(X))`; a less strict version would be `all(abs(X-t(X)) < eps)` where `eps` is some appropriate tolerance. Alternatively, we could use `all.equal(X, t(X))`, which uses a tolerance.

The `if` statement should be distinguished from the `ifelse` function, which is its vector counterpart. Its form is

```
ifelse(test, true.value, false.value)
```

where all arguments are vectors, and the recycling rule applies if any are short. All arguments are evaluated and `test` is coerced to logical if necessary. In those component positions where the value is `T` the corresponding component of `true.value` is the result and elsewhere it is that of `false.value`. Since `ifelse` operates on vectors, it is fast and should be used if possible.

Note that `ifelse` can generate `NA` warning messages even in cases where the result contains none. In an assignment such as

```
y.logy <- ifelse(y <= 0, 0, y*log(y))
```

all three arguments to `ifelse` are evaluated, so warning messages will be generated if y has any negative components even though the final result will have no `NA` components. Two ways of producing the desired result for *frequency* vectors y are

```
y.logy <- y * log(y + (y==0))
y.logy <- y * log(pmax(1, y))  # alternative
```

Loops: The `for`, `while` and `repeat` statements

A `for` loop allows a statement to be iterated as a variable assumes values in a specified sequence. The statement has the form

```
for(variable in sequence) statement
```

where `in` is a keyword, `variable` is the loop variable and `sequence` is the vector of values it assumes as the loop proceeds. This is often of the form `1:10` or `seq(along=x)` but it may be a list, in which case `variable` assumes the value of each component in turn. The `statement` part will often be a grouped statement and hence enclosed within braces, `{ }`.

The `while` and `repeat` loops do not make use of a loop variable. Their forms are

```
while (condition) statement
```

and

```
    repeat statement
```

In both cases the commands in the body of the loop are repeated. For a `while`
loop the normal exit occurs when `condition` becomes F; the `repeat` statement
continues indefinitely unless exited by a `break` statement.

The `next` statement within the body of a `for`, `while` or `repeat` loop causes
a jump to the beginning of the next iteration. The `break` statement causes an
immediate exit from the loop.

A single-parameter maximum-likelihood example

For a simple example with a statistical context we estimate the parameter λ of
the zero-truncated Poisson distribution by maximum likelihood. We use artificial
data but the same problem sometimes occurs in practice.

The probability distribution is specified by

$$\Pr(Y = y) = \frac{e^{-\lambda}\lambda^y}{(1 - e^{-\lambda})\,y!} \qquad y = 1, 2, \ldots$$

and corresponds to observing only non-zero values of a Poisson count. The mean
is

$$E(Y) = \frac{\lambda}{1 - e^{-\lambda}}$$

The maximum likelihood estimate $\hat{\lambda}$ is found by equating the sample mean to its
expectation

$$\bar{y} = \frac{\hat{\lambda}}{1 - e^{-\hat{\lambda}}}$$

If this equation is written as $\hat{\lambda} = \bar{y}\,(1 - e^{-\hat{\lambda}})$, Newton's method leads to the
iteration scheme

$$\hat{\lambda}_{m+1} = \hat{\lambda}_m - \frac{\hat{\lambda}_m - \bar{y}\,(1 - e^{-\hat{\lambda}_m})}{1 - \bar{y}\,e^{-\hat{\lambda}_m}}$$

which we now implement. This is a natural situation in which a `while` loop
might be used in **S-PLUS**. First we generate our artificial sample from a distribution with $\lambda = 1$.

```
> yp <- rpois(50, lam=1)    # full Poisson sample of size 50
> table(yp)
  0  1  2 3 5
 21 12 14 2 1
> y <- yp[yp > 0]            # truncate the zeros; n = 29
```

We have a termination condition based both on convergence of the process and an
iteration count limit just for safety. An obvious starting value is $\hat{\lambda}_0 = \bar{y}$.

```
> ybar <- mean(y); ybar
> [1] 1.7586
> lam <- ybar
> it <- 0                        # iteration count
> del <- 1                       # iterative adjustment
> while (abs(del) > 0.0001 && (it <- it + 1) < 10) {
    del <- (lam - ybar*(1 - exp(-lam)))/(1 - ybar*exp(-lam))
    lam <- lam - del
    cat(it, lam, "\n")}
1 1.32394312696735
2 1.26142504977282
3 1.25956434178259
4 1.25956261931933
```

To generate output from a loop in progress an explicit call to a function such as
print or cat has to be used. For tracing output cat is usually convenient since
it can combine several items. Numbers are coerced to character in full precision;
using format(lam) in place of lam is the simplest way to reduce the number
of significant digits to the options default.

4.2 More on character strings

There are several more sophisticated facilities for manipulating character vectors
beyond those we discussed in Section 2.6.

The function abbreviate provides a more general mechanism for generating abbreviations. In this example it gives

```
> as.vector(abbreviate(state.name[44:50]))
[1] "Utah" "Vrmn" "Vrgn" "Wshn" "WsVr" "Wscn" "Wymn"
> as.vector(abbreviate(state.name[44:50], use.classes=F))
[1] "Utah" "Verm" "Virg" "Wash" "WVir" "Wisc" "Wyom"
```

We used as.vector to suppress the names attribute that contains the unabbrevi-
ated names!

The function grep searches for patterns in a vector of character strings, and
returns the indices of the strings in which a match is found. On UNIX systems
it is based on the function egrep. There are important differences between the
pattern matching protocols used by grep on UNIX and Windows so the details in
the on-line help should be noted. On UNIX ' . ' matches any character (use '\.' to
match ' . ') and ' .*' matches zero or more occurrences of any character, that is,
any character string. In Windows versions '*' is a wildcard character matching
zero or more characters and '?' matches precisely one character.

Simpler forms of matching are done by the functions match, pmatch and
charmatch. Each tries to match each element of its first argument against the
elements of its second argument. The function match seeks the first exact match
(equality) whereas the other two look for partial matches (the search string starts
the character string) and match each element in the second argument once only.

All return the value of their `nomatch` argument (which defaults to `NA`) if there is no match. The function `charmatch` returns the index of a unique match, and `0` if there is more than one match. With argument `duplicates.ok=F` (the default), `pmatch` returns `nomatch` for duplicate matches, whereas with `duplicates.ok=T` it returns the index of the first match.

Regular expressions are powerful ways to match character strings familiar to users of such tools as `sed`, `grep`, `awk` and `perl`. They are used in function `regexpr` in S-PLUS 3.4 and later. Function `regexpr` matches one regular expression to a character vector. For example

```
> regexpr("na$", state.name)
 [1] -1 -1  6 -1 -1 -1 -1 -1 -1 -1 -1 -1 -1  6 -1 -1 -1  8 -1
[20] -1 -1 -1 -1 -1 -1  6 -1 -1 -1 -1 -1 -1 13 -1 -1 -1 -1 -1
[39] -1 13 -1 -1 -1 -1 -1 -1 -1 -1 -1
attr(, "match.length"):
 [1] -1 -1  2 -1 -1 -1 -1 -1 -1 -1 -1 -1 -1  2 -1 -1 -1  2 -1
[20] -1 -1 -1 -1 -1 -1  2 -1 -1 -1 -1 -1 -1  2 -1 -1 -1 -1 -1
[39] -1  2 -1 -1 -1 -1 -1 -1 -1 -1 -1
> state.name[regexpr("na$", state.name)> 0]
[1] "Arizona"      "Indiana"         "Louisiana"
[4] "Montana"      "North Carolina" "South Carolina"
```

The functions `regMatch` and `regMatchPos` of S-PLUS 5.x have a very similar role, but encode the answer somewhat differently.

```
> regMatch(state.name, "na$")
 [1] F F T F F F F F F F F F F T F F F T F F F F F F F T F F
[29] F F F F T F F F F F F T F F F F F F F F F
> regMatchPos(state.name, "na$")
integer matrix: 50 rows, 2 columns.
     [,1] [,2]
[1,]  NA   NA
[2,]  NA   NA
[3,]   6    7
[4,]  NA   NA
 ....
```

Both `regMatch` and `regMatchPos` can match multiple regular expressions, recycling arguments as needed. Regular expressions can also be used in specifying index vectors in extracting or replacing subsets of vector-like objects, for example

```
> state.name[regularExpression("na$")]
[1] "Arizona"      "Indiana"         "Louisiana"
[4] "Montana"      "North Carolina" "South Carolina"
```

and as the `first` argument to `substring`, as in

```
> substring(state.name, "^[A-Za-z]+")
 [1] "Alabama"        "Alaska"        "Arizona"
 [4] "Arkansas"       "California"    "Colorado"
 [7] "Connecticut"    "Delaware"      "Florida"
      . . . .
```

which extracts the first word of the name.

4.3 Matrix operations

The standard matrix operators and indexing of matrices have already been covered
in Sections 2.3 and 2.8. In this section we cover more specialized calculations
involving matrices. Many (but not all) of these will work with numerical data
frames, which can always be coerced to matrices by as.matrix(dataframe)
or data.matrix(dataframe) (see page 102).

The function crossprod forms 'crossproducts', meaning that

```
XT.y <- crossprod(X, y)
```

calculates $X^T y$. This matrix could be calculated as t(X) %*% y but using
crossprod is more efficient. If the second argument is omitted it is taken to
be the same as the first. Thus crossprod(X) calculates the matrix $X^T X$.

An important operation on arrays is the *outer product*. If a and b are two
numeric arrays, their outer product is an array whose dimension vector is obtained
by concatenating their two dimension vectors (order is important), and whose data
vector is obtained by forming all possible products of elements of the data vector
of a with those of b. The outer product is formed by the operator %o%:

```
ab <- a %o% b
```

or by the function outer:

```
ab <- outer(a, b, "*")
ab <- outer(a, b)          # as "*" is the default.
```

The multiplication function may be replaced by an arbitrary function of two variables (or its name as a character string). For example if we wished to evaluate the
function

$$f(x, y) = \frac{\cos(y)}{1 + x^2}$$

over a regular grid of values with x- and y-coordinates defined by the S vectors
x and y respectively, we could use

```
f <- function(x, y) cos(y)/(1 + x^2) # define the function
z <- outer(x, y, f)                  # use it.
```

If the function is not required elsewhere we could do the whole operation in one
step using an *anonymous* function as the third argument, as in

```
z <- outer(x, y, function(x, y) cos(y)/(1 + x^2))
```

The function `diag` either creates a diagonal matrix from a vector argument, or extracts as a vector the diagonal of a matrix argument. Used on the assignment side of an expression it allows the diagonal of a matrix to be replaced.[1] For example, to form a covariance matrix in multinomial fitting we could use

```
> p <- dbinom(0:4, size=4, prob=1/3)   # an example prob vector
> CC <- -(p %o% p)
> diag(CC) <- p + diag(CC)
> structure(3^8 * CC, dimnames=list(0:4, 0:4))   # convenience
      0     1     2     3    4
0  1040  -512  -384  -128  -16
1  -512  1568  -768  -256  -32
2  -384  -768  1368  -192  -24
3  -128  -256  -192   584   -8
4   -16   -32   -24    -8   80
```

In addition `diag(n)` for a positive integer n generates an $n \times n$ identity matrix. This is an exception to the behaviour for vector arguments; `diag(x,length(x))` will give a diagonal matrix with diagonal x for a vector of any length, even one.

Functions operating on matrices

The standard operations of linear algebra are either available as functions or can easily be programmed, making S a flexible matrix manipulation language (if rather slower than specialized matrix languages).

The function `solve` inverts matrices and solves systems of linear equations; `solve(A)` inverts A and `solve(A, b)` solves A %*% x = b. (If the system is over-determined, the least-squares fit is found, but matrices of less than full rank give an error.)

The function `chol` returns the Choleski decomposition $A = U^T U$ of a non-negative definite symmetric matrix. (Note that this is not the most common convention, in which the lower-triangular form $A = LL^T$ with $L = U^T$ is used.) Function `backsolve` solves upper triangular systems of matrices, and is often used in conjunction with `chol`. (There is an almost identical function `solve.upper`, but no analogue for lower-triangular matrices.)

Eigenvalues and eigenvectors

The function `eigen` calculates the eigenvalues and eigenvectors of a square matrix. The result is a list of two components, `values` and `vectors`. If we only need the eigenvalues we can use:

```
eigen(Sm, only.values=T)$values
```

---

[1] This is one of the few places where the recycling rule is disabled: the replacement must be a scalar or of the correct length.

Real symmetric matrices have real eigenvalues, and the calculation for
this case can be much simpler and more stable. A further named argument,
symmetric, may be used to specify whether a matrix is (to be regarded as) symmetric. The default value is T if the matrix exactly equals its transpose, otherwise
F.

Singular value decomposition and generalized inverses

An $n \times p$ matrix X has a *singular value decomposition* (SVD) of the form

$$X = U\Lambda V^T$$

where U and V are $n \times \min(n,p)$ and $p \times \min(n,p)$ matrices of orthonormal
columns, and Λ is a diagonal matrix. Conventionally the diagonal elements of
Λ are ordered in decreasing order; the number of non-zero elements is the rank
of X. A proof of its existence can be found in Golub & Van Loan (1989), and a
discussion of the *statistical* value of SVDs in Thisted (1988).

The function svd takes a matrix argument M and calculates the singular value
decomposition. The components of the result are u and v, the orthonormal matrices and d, a vector of singular values. If either U or V is not required its
calculation can be avoided by the argument nu=0 or nv=0.

In some applications a generalized inverse of a matrix X is required. This
is defined as any matrix X^- such that $XX^-X = X$. One choice is known as
the Moore–Penrose or spectral generalized inverse, usually written as X^+ and
defined as

$$X^+ = V\Lambda^- U^T$$

where Λ^- is a diagonal matrix like Λ with entries λ_i^{-1} if $\lambda_i > 0$ or 0 if
$\lambda_i = 0$. Deciding when a singular value is small enough to be considered zero
requires some care and the following function for the generalized inverse uses a
protected reciprocal based on the relative sizes of the singular values.

```
ginv <- function(X, tol = sqrt(.Machine$double.eps))
{
    s <- svd(X)
    nz <- s$d > tol * s$d[1]
    if(any(nz)) s$v[, nz] %*% (t(s$u[, nz])/s$d[nz]) else X * 0
}
```

(An enhanced version[2] is in library MASS.) The generalized inverse may be used
to find a particular solution to a system of consistent linear equations of less than
full rank, or a least-squares solution to an overdetermined system, not necessarily
of maximum rank. Useful references include Rao & Mitra (1971), Pringle &
Rayner (1971) and Dodge (1985).

---

[2] S-PLUS 4.x and later have a similar function ginverse.

The QR decomposition

A faster decomposition to calculate than the SVD is the QR decomposition, defined as

$$M = Q\,R$$

where, if M is $n \times p$, Q is an $n \times n$ matrix of orthonormal columns (that is, an orthogonal matrix) and R is an $n \times p$ matrix with zero elements apart from the first p rows that form an upper triangular matrix. (See Golub & Van Loan, 1989, §5.2). The function qr(M) implements the algorithm detailed in Golub & Van Loan, and hence returns the result in a somewhat inconvenient form to use directly. The result is a list that can be used by other tools. For example,

```
M.qr <- qr(M)            # QR decomposition
Q <- qr.Q(M.qr)          # Extract  a Q (n x p) matrix
R <- qr.R(M.qr)          # Extract an R (p x p) matrix
y.res <- qr.resid(M.qr, y)  # Project onto error space
```

The last command finds the residual vector after projecting the vector y onto the column space of M. Other tools that use the result of qr include qr.fitted for fitted values and qr.coef for regression coefficients.

Note that by default qr.R only extracts the first p rows of the matrix R and qr.Q only the first p columns of Q, which form an orthonormal basis for the column space of M if M is of maximal rank. To find the complete form of Q we need to call qr.Q with an extra argument complete=T. The columns of Q beyond the rth, where $r \leqslant p$ is the rank of M, form an orthonormal basis for the null space or *kernel* of M, which is occasionally useful for computations. A simple function to extract it is

```
Null <- function(M) {
   tmp <- qr(M)
   set <- if(tmp$rank == 0) 1:ncol(M) else -(1:tmp$rank)
   qr.Q(tmp, complete=T)[, set, drop=F]
}
```

Determinant and trace

The only function provided for finding the determinant of a square matrix M is in the Matrix library (see page 102). For older systems lacking the Matrix library there are several ways to write determinant functions. Often it is known in advance that a determinant will be non-negative or that its sign is not needed, in which case methods to calculate the absolute value of the determinant suffice. In this case it may be calculated as the product of the singular values, or slightly faster but possibly less accurately from the QR-decomposition as

```
absdet <- function(M) abs(prod(diag(qr(M)$qr)))
```

If the sign is unknown and important, the determinant may be calculated as the product of the eigenvalues. These will in general be complex and the result may have complex roundoff error even though the exact result is known to be real, so a simple function to perform the calculation is

```
det <- function(M) Re(prod(eigen(M, only.values=T)$values))
```

The following trace function is so simple the only reason for having it might
be to make code using it more readable.

```
tr <- function(M) sum(diag(M))
```

Converting data frames to and from matrices

The coercion function `as.data.frame` takes a matrix argument (of mode numerical, character or logical) and produces a data frame from it using the columns
as variables. If the matrix has a `dimnames` attribute it is retained in the names and
row names of the data frame. If not, names are constructed in a default manner
with the variable (column) names incorporating the original matrix name.

The constructor function `data.frame` may also be used to convert a matrix
to a data frame in a similar way, but in this case by default any column names
that are nonstandard variable names are standardized by changing the offending
characters to periods.

There are two functions that may be used for converting a data frame into a
matrix, and they behave slightly differently.

1. `as.matrix(dataframe)` produces a matrix from the argument data
   frame of mode numeric if all constituent variables in the data frame are
   numeric, and of mode character otherwise. So any factor will cause the
   entire matrix result to be of mode character. Note that it is not possible to
   have mixed mode matrices.

2. `data.matrix(dataframe)` produces a numeric matrix from the argument data frame in all circumstances. Any factors among the variables are
   first coerced to numeric, and the levels information of the factors is retained
   in the attribute `column.levels` of the outcome.

The `Matrix` library

There is a completely distinct set of matrix operations based on the LAPACK
library of FORTRAN linear algebra routines. These are made available by

```
library(Matrix)
```

Note the M. They seem to be little used, but should be explored for specialized
matrix computations. (At the time of writing they were not implemented in S-
PLUS 5.x.)

4.4   Vectorized calculations and loop avoidance functions

Programmers coming to S from other languages are often slow to take advantage
of the power of S to do vectorized calculations, that is, calculations that operate
on entire vectors rather than on individual components in sequence. This often
leads to unnecessary loops. For example, consider calculating the Pearson chi-
squared statistic for testing independence in a two-way contingency table. This is
defined as

$$X_P^2 = \sum_{i=1}^{r} \sum_{j=1}^{s} \frac{(f_{ij} - e_{ij})^2}{e_{ij}}$$

where $e_{ij} = f_{i.}f_{.j}/f_{..}$ are the expected frequencies. Two nested `for` loops
may seem to be necessary, but in fact no explicit loops are needed. Assuming
the frequencies f_{ij} are held as a matrix the most efficient calculation in S uses
matrix operations:

```
fi. <- f %*% rep(1, ncol(f))
f.j <- rep(1, nrow(f)) %*% f
e <- (fi. %*% f.j)/sum(fi.)
X2p <- sum((f-e)^2/e)
```

Explicit loops in S should be regarded as potentially expensive in time and
memory use and ways of avoiding them should be considered. (Note that this will
be impossible with genuinely iterative calculations such as our Newton scheme.)

The functions `apply`, `tapply`, `sapply` and `lapply` offer ways around explicit loops.

The functions `apply` and `sweep`

The function `apply` allows functions to operate on an array using sections suc-
cessively. For example, consider the dataset `iris` which is a $50 \times 4 \times 3$ array of
four observations on 50 specimens of each of three species. Suppose we want the
means for each variable by species; we can use `apply`.

The arguments of `apply` are

1. the name of the array, `X`;

2. an integer vector `MARGIN` giving the indices defining the sections of the
   array to which the function is to be separately applied. It is helpful to note
   that if the function applied has a scalar result, the result of `apply` is an
   array with `dim(X)[MARGIN]` as its dimension vector;

3. the function, or the name of the function `FUN` to be applied separately to
   each section;

4. any additional arguments needed by the function as it is applied to each
   section.

Thus we need to use

```
> apply(iris, c(2,3), mean)
          Setosa Versicolor Virginica
Sepal L.   5.006      5.936     6.588
Sepal W.   3.428      2.770     2.974
Petal L.   1.462      4.260     5.552
Petal W.   0.246      1.326     2.026
> apply(iris, c(2,3), mean, trim=0.1)
          Setosa Versicolor Virginica
Sepal L. 5.0025     5.9375    6.5725
Sepal W. 3.4150     2.7800    2.9625
Petal L. 1.4600     4.2925    5.5100
Petal W. 0.2375     1.3250    2.0325
```

where we also show how arguments can be passed to the function, in this case to give a trimmed mean. If we want the overall means we can use

```
> apply(iris, 2, mean)
 Sepal L. Sepal W. Petal L. Petal W.
   5.8433   3.0573    3.758   1.1993
```

Note how dimensions have been dropped to give a vector. If the result of `FUN` is itself a vector of length d, say, then the result of `apply` is an array with dimension vector `c(d, dim(X)[MARGIN])`, with single-element dimensions dropped. Also note that matrix results are reduced to vectors; if we ask for the covariance matrix for each species,

```
ir.var <- apply(iris, 3, var)
```

we get a 16×3 matrix. We can add back the correct dimensions, but in so doing we lose the `dimnames`. We can restore both by

```
ir.var <- array(ir.var, dim = dim(iris)[c(2,2,3)],
                 dimnames = dimnames(iris)[c(2,2,3)])
```

The function `apply` is often useful to replace explicit loops. Earlier versions of the function used loops internally, but since S-PLUS 3.2 these have been re-placed by a call to an internal function. Note too that for *linear* computations it is rather inefficient. We can form the means by matrix multiplication:

```
> matrix(rep(1/50, 50) %*% matrix(iris, nrow = 50), nrow = 4,
         dimnames = dimnames(iris)[-1])
          Setosa Versicolor Virginica
Sepal L.   5.006      5.936     6.588
Sepal W.   3.428      2.770     2.974
Petal L.   1.462      4.260     5.552
Petal W.   0.246      1.326     2.026
```

which will be very much faster on larger examples, but is much less transparent.

The function `aperm` is often useful with array/matrix arithmetic of this sort. It permutes the indices, so that `aperm(iris, c(2,3,1))` is a $4 \times 3 \times 50$ array. Note that the matrix transpose operation is a special case.

Functions sweep *and* scale

The functions apply and sweep are often used together. For example, having found the means of the iris data, we may want to remove them by subtraction or perhaps division. We can use sweep in each case:

```
ir.means <- apply(iris, c(2,3), mean)
sweep(iris, c(2,3), ir.means)
log(sweep(iris, c(2,3), ir.means, "/"))
```

Of course, we could have subtracted the log means in the second case.

The function scale takes a matrix argument and operates on each column in turn. By default it mean-corrects the column and rescales to unit variance. If the argument center is F no centring is done, and if it is a vector of length the number of columns, that vector is subtracted from each row. Similarly, the *argument* scale controls the scaling of each column. Thus for a matrix A, scale(A, T, F) is equivalent to sweep(A, apply(A,2,mean), "-"), but more efficient.

Functions operating on factors and lists

Factors are used to define groups in vectors that have the same length. Each level of the factor defines a group, possibly empty.

In this section we use the quine data frame from the MASS library giving data from a survey of school children. The study giving rise to the data set is described more fully on page 180; in brief the data frame has four factors, Sex, Eth (ethnicity—two levels), Age (four levels) and Lrn (Learner group—two levels) and a quantitative response, Days, the number of days the child in the sample was away from school in a year.

The function table

It is often necessary to tabulate factors to find frequencies. The main function for this purpose is table which returns a frequency cross-tabulation as an array. For example:

```
> attach(quine)
> table(Age)
 F0 F1 F2 F3
 27 46 40 33
> table(Sex, Age)
   F0 F1 F2 F3
 F 10 32 19 19
 M 17 14 21 14
```

Note that the factor levels become the appropriate names or dimnames attribute for the frequency array. If the arguments given to table are not factors they are effectively coerced to factors.

The function crosstabs may also be used. It takes a formula and data frame as its first two arguments, so the data frame need not be attached. A call to crosstabs for the same table is

```
> tab <- crosstabs(~Sex + Age, quine)
> print.default(tab)
   F0 F1 F2 F3
F  10 32 19 19
M  17 14 21 14
....
```

The print method for objects of class crosstabs gives extra information of no interest to us here, but the object behaves in calculations as a frequency table.

Ragged arrays and the function tapply

The combination of a vector and a labelling factor or factors is an example of what is called a *ragged array*, since the group sizes can be irregular. (When the group sizes are all equal the indexing may be done implicitly and more efficiently using arrays, as was done with the iris data.)

To calculate the average number of days absent for each age group we can use the ragged array function tapply.

```
> tapply(Days, Age, mean)
     F0     F1    F2     F3
 14.852 11.152 21.05 19.606
```

The first argument is the vector for which functions on the groups are required, the second argument, INDICES, is the factor defining the groups and the third argument, FUN, is the function to be evaluated on each group. If the function requires more arguments they may be included as additional arguments to the function call, as in

```
> tapply(Days, Age, mean, trim = 0.1)
     F0     F1     F2    F3
 12.565 9.0789 18.406 18.37
```

for 10% trimmed means.

If the second argument is a list of factors the function is applied to each group of the cross-classification. Thus to find the average days absent for age by sex classes we could use

```
> tapply(Days, list(Sex, Age), mean)
       F0     F1     F2     F3
F  18.700 12.969 18.421 14.000
M  12.588  7.000 23.429 27.214
```

To find the standard errors of these we could use an anonymous function as the third argument, as in

```
> tapply(Days, list(Sex, Age),
         function(x) sqrt(var(x)/length(x)))
       F0     F1     F2     F3
F  4.2086 2.3299 5.3000 2.9409
M  3.7682 1.4181 3.7661 4.5696
```

As with table coercion to factor takes place where necessary.

Operating along structures: The functions `lapply,` `sapply` *and* `split`

The functions `lapply` and `sapply` are the analogues of `tapply` and `apply` for operations on the individual components of lists or vectors. The two functions are called with exactly the same arguments but whereas `sapply` will simplify the result to a vector or an array if possible, `lapply` will always return a list.

The function `split` takes as arguments a data vector and a vector defining groups within it. The value is a list of vectors, one component for each group.

For ragged arrays defined by a single factor the calls `tapply(vec, fac, fun)` and `sapply(split(vec, fac), fun)` often give the same result. (The only time they will differ is when `sapply` simplifies a list to an array and `tapply` leaves it as a list.) Even though `tapply` actually uses `lapply` with `split` for the critical computations, calling them directly can sometimes be much faster. For example, consider the problem of finding the *convolution* of two numeric vectors. If they have components a_i and b_j, respectively, their convolution has components $c_k = \sum_{i+j=k} a_i b_j$. This is the same problem as finding the coefficient vector for the product of two polynomials. A simple function to do this is

```
> conv1 <- function(a, b) {
    ab <- outer(a, b)
    unlist(lapply(split(ab, row(ab) + col(ab)), sum))
}
```

A second version, `conv2`, replaces the last line with

```
tapply(ab, row(ab) + col(ab), sum)
```

We can compare times (on a PC):

```
> a <- runif(1000)
> b <- runif(100)
> dos.time(conv1(a, b))
[1] 2.98
> dos.time(conv2(a, b))
[1] 4.38
```

Notice that with `sapply`, `lapply` and `split` if the relevant argument is not a factor or list of factors, once again the appropriate coercions to factor take place.

The first argument to `lapply` or `sapply` may be any vector, in which case the function is applied to each component. For example, one way to discover which single-letter names are currently visible on the search path is

```
> Letters <- c(LETTERS, letters)
> Letters[sapply(Letters, function(xx) exists(xx))]
[1] "C" "D" "F" "I" "T" "c" "q" "s" "t"
```

This example can be used to expose some anomalies. First, if we had used x as an argument instead of xx it would have shown up as an object in the result although the object is not globally visible.

```
> Letters[sapply(Letters, function(x) exists(x))]
 [1] "C" "D" "F" "I" "T" "c" "q" "s" "t" "x"
```

Second, if we had used the function `exists` by name, another phantom variable, X, would be found, again because of the temporary visibility of internal coding.

```
> Letters[sapply(Letters, exists)]
 [1] "C" "D" "F" "I" "T" "X" "c" "q" "s" "t"
```

These problems occur because `exists` is a generic function and the relevant method, `exists.default`, contains a reference to the function `.Internal`. In general any function with an `.Internal` reference may give problems if used as an argument to `lapply` or `sapply`, particularly as a simple name. For example, `dim` is such a function

```
> dim
function(x)
.Internal(dim(x), "S_extract", T, 10)
> sapply(list(quine, quine), dim)  # does NOT work!
[[1]]:
NULL
[[2]]:
NULL
> sapply(list(quine, quine), function(x) dim(x))
      [,1] [,2]
[1,]  146  146
[2,]    5    5
```

The second call should be equivalent to the first, but in this case the second gives correct results and the first does not.

In S-PLUS 5.x there is a new member of the `apply` family, `rapply`, which is like `lapply` but applied recursively, that is, to all lists contained within the supplied list, and to lists within those lists and so on.

Frequency tables as data frames

Returning to the `quine` data frame, we may find which of the variables are factors using

```
> sapply(quine, is.factor)
 Eth Sex Age Lrn Days
   T   T   T   T    F
```

since a data frame is a list.

Consider the general problem of taking a set of n factors and constructing the complete n-way frequency table as a data frame, that is, as a frequency vector and a set of n classifying factors. The four factors in the `quine` data serve as an example, but we work in a way that generalizes immediately to any number of factors.

First we remove any non-factor components from the data frame.

```
quineF0 <- quine[sapply(quine, is.factor)]
```

To calculate the frequencies we could `attach` the data frame `quineF0` and call
`table` with all factors, but this would be tedious if the number of factors were
large, and more importantly this method could not be used in a general function.

The function `do.call` takes two arguments: the name of a function (as a
character string) and a list. The result is a call to that function with the list supplying the arguments. List names become argument names. Hence we may find
the frequency table using

```
tab <- do.call("table", quineF0)
```

The result is a multi-way array of frequencies. We next find the classifying factors, which correspond to the indices of this array. One way to do this is to use a
`for` loop and `slice.index(tab, k)` to generate the factor levels, but a more
convenient way is to use `expand.grid`. This function takes any number of vectors and generates a data frame consisting of all possible combinations of values,
in the correct order to match the elements of a multi-way array with the lengths
of the vectors as the index sizes. Argument names become component names in
the data frame. Alternatively `expand.grid` may take a single list whose components are used as individual vector arguments.

Hence to find the index vectors for our data frame we may use

```
QuineF <- expand.grid(lapply(quineF0, levels))
```

If `expand.grid` did not accept the single list form of argument we could have
used

```
QuineF <- do.call("expand.grid", lapply(quineF0, levels))
```

Finally we put together the frequency vector and classifying factors.

```
> QuineF$Freq <- as.vector(tab)
> QuineF
   Eth Sex Age Lrn Freq
1   A   F  F0  AL    4
2   N   F  F0  AL    4
3   A   M  F0  AL    5
....
30  N   F  F3  SL    0
31  A   M  F3  SL    0
32  N   M  F3  SL    0
```

We use `as.vector` to remove all attributes since some might conflict with the
data frame properties.

Functions operating on data frames

Three functions, `merge`, `by` and `aggregate`, operate on data frames in a way that mimics common database operations in other languages. Both `by` and `aggregate` are closely related to `tapply`. Using `aggregate` is equivalent to using `tapply` on each column of a data frame. Instead of an `INDICES` argument, the grouping factors are specified by an argument called `by`, which is most usefully a named list. For example

```
> aggregate(crabs[, 4:8], list(sp=crabs$sp, sex=crabs$sex),
        median)
  sp sex    FL    RW    CL    CW    BD
1  B   F 13.15 12.20 27.90 32.35 11.60
2  O   F 18.00 14.65 34.70 39.55 15.65
3  B   M 15.10 11.70 32.45 37.10 13.60
4  O   M 16.70 12.10 33.35 36.30 15.00
```

It is important to ensure that the function used is applicable to each column, which is why we must omit the character columns here.

The function `by` takes a data frame and splits it by the second argument, `INDICES`, passing each data frame in turn to its `FUN` argument. This is equivalent to using `tapply` on the columns in parallel rather than one at a time.

```
> by(crabs[,4:8], list(crabs$sp, crabs$sex), summary)
crabs$sp:B
crabs$sex:F
        FL                  RW               CL               CW
 Min.   : 7.2    Min.   : 6.5    Min.   :14.7    Min.   :17.1
 1st Qu.:11.5    1st Qu.:10.6    1st Qu.:23.9    1st Qu.:27.9
 Median :13.1    Median :12.2    Median :27.9    Median :32.4
 Mean   :13.3    Mean   :12.1    Mean   :28.1    Mean   :32.6
 3rd Qu.:15.3    3rd Qu.:13.9    3rd Qu.:32.8    3rd Qu.:37.8
 Max.   :19.2    Max.   :16.9    Max.   :40.9    Max.   :47.9
    ....
```

The function `merge` provides a *join* of two data frames as databases. That is, it combines pairs of rows which have common values in specified columns to a row with all the information contained in either data frame. See the on-line help for full details.

As an example, consider combining the `Animals` and `mammals` data frames, selecting the entries with the same row names in each. We could do this using

```
> merge(Animals, mammals, by="row.names")
        Row.names   body.x brain.x   body.y brain.y
1 African elephant 6654.000  5712.0 6654.000  5712.0
2   Asian elephant 2547.000  4603.0 2547.000  4603.0
 ....
20   Rhesus monkey    6.800   179.0    6.800   179.0
21           Sheep   55.500   175.0   55.500   175.0
```

Note how the original row names are discarded.

4.5 Introduction to object orientation

The primary purpose of the S programming environment is to construct and manipulate objects. These objects may be fairly simple, such as numeric vectors, factors, arrays or data frames, or reasonably complex such as an object conveying the results of a model-fitting process. The manipulations fall naturally into broad categories such as plotting, printing, summarizing and so forth. They may also involve several objects such as performing an arithmetic operation on two objects to construct a third.

Since S is intended to be an extensible environment new kinds of object are designed by users to fill new needs, but it will usually be convenient to manipulate them using familiar functions such as `plot`, `print` and `summary`. For the new kind of object the standard manipulations will usually have an obvious purpose, even though the precise action required differs at least in detail from any previous action in this category.

Consider the `summary` function. If a user designs a new kind of object called, say, a "`newfit`" object, it will often be useful to make available a method of summarizing such objects so that the important features are easy to appreciate and the less important details are suppressed. One way of doing this is to write a new function, say `summary.newfit`, which could be used to perform the particular kind of printing action needed. If `myobj` is a particular `newfit` object it could then be summarized using

```
> summary.newfit(myobj)
```

We could also write functions with names `plot.newfit`, `residuals.newfit`, `coefficients.newfit` for the particular actions appropriate for `newfit` objects for plotting, summarizing, extracting residuals, extracting coefficients and so on. It would be most convenient if the user could just type `summary(myobj)`, and this is one important idea behind *object-oriented programming*. To make it work we need to have a standard method by which the evaluator may recognise the different classes of object being presented to it. In S this is done by giving the object a `class`. There is a `class` replacement function available to set the class, which would normally be done when the object was created. For example,

```
> class(myobj) <- "newfit"
```

Functions like `summary`, `plot` and `residuals` are called *generic* functions. They have the property of adapting their action to match the class of object presented to them. The specific implementations such as `summary.lm` are known as *method* functions. It may not be completely clear if a function *is* generic, but S-PLUS 5.x has the test `isGeneric`.

At this point we need to emphasize that the two S engines, S3 in S-PLUS 3.x and 4.x and S4 in S-PLUS 5.x, handle classes and generic functions internally quite differently, and the details are given in the companion volume. However, there is fairly extensive backwards compatibility, and most of S-PLUS 5.x uses the compatibility features. You may notice that many classes in S-PLUS 5.x are set by `oldClass` not `class` when listing functions in S-PLUS 5.x.

Method dispatch

Generic functions use the class of their first argument (S-PLUS 3.x and 4.x) or
first few arguments (S-PLUS 5.x) to decide which method function to use. In the
simple cases the generic function will consist of a call to UseMethod as in[3]

```
> summary
function(object, ...) UseMethod("summary")
```

Then a call to methods will list all the method functions currently available, for
example

```
> methods(summary)
$SHOME/splus/.Functions $SHOME/splus/.Functions
"summary.aov"           "summary.aovlist"
$SHOME/splus/.Functions $SHOME/splus/.Functions
"summary.coxph"         "summary.cts"
$SHOME/stat/.Functions  $SHOME/splus/.Functions
"summary.data.frame"    "summary.default"
$SHOME/splus/.Functions $SHOME/splus/.Functions
"summary.factanal"      "summary.factor"
  ....
```

in S-PLUS 3.4. The method for the closest matching class is used; the details of
'closest matching' are complex and given in Volume 2. However, the underlying
idea is simple. Objects from linear models have class "lm", those from generalized linear models class "glm" and those from analysis of variance models class
"aov". Each of the last two classes is defined to *inherit* from class "lm", so if
there is no specific method function for the class, that for "lm" will be tried. This
can greatly reduce the number of method functions needed. Furthermore, method
functions can build on those from simpler classes: for example predict.glm
works by manipulating the results of a call to predict.lm (not in all cases successfully!).

---

[3] Beware: in S-PLUS 5.x printing the function need not give you the generic function; this is true
of plot, for example.

Chapter 5

Univariate Statistics

In this chapter we cover a number of topics from classical univariate statistics plus some modern versions.

5.1 Probability distributions

In this section we confine attention to *univariate* distributions, that is, the distributions of random variables X taking values in the real line $\mathbb{R}$.

The standard distributions used in statistics are defined by probability density functions (for continuous distributions) or probability functions $P(X = n)$ (for discrete distributions). We refer to both as *densities* since the probability functions can be viewed mathematically as densities (with respect to counting measure). The *cumulative distribution function* or CDF is $F(x) = P(X \leqslant x)$ which is expressed in terms of the density by a sum for discrete distributions and as an integral for continuous distributions. The *quantile function* $Q(u) = F^{-1}(u)$ is the inverse of the CDF where this exists. Thus the quantile function gives the percentage points of the distribution. (For discrete distributions the quantile is the smallest integer m such that $F(m) \geqslant u$.)

S-PLUS has built-in functions to compute the density, cumulative distribution function and quantile function for many standard distributions. In many cases the functions can not be written in terms of standard mathematical functions, and the built-in functions are very accurate approximations. (But this is also true for the numerical calculation of cosines, for example.)

The first letter of the name of the S function indicates the function so, for example, dnorm, pnorm, qnorm are respectively, the density, CDF and quantile functions for the normal distribution. The rest of the function name is an abbreviation of the distribution name. The distributions available are listed in Table 5.1.

The first argument of the function is always the observation value q (for quantile) for the densities and CDF functions, and the probability p for quantile functions. Additional arguments specify the parameters, with defaults for 'standard' versions of the distributions where appropriate. Precise descriptions of the parameters are given in the on-line help pages.

Table 5.1: S function names and parameters for standard probability distributions.

Distribution	S name	Parameters
beta	beta	shape1, shape2
binomial	binom	size, prob
Cauchy	cauchy	location, scale
chi-squared	chisq	df
exponential	exp	rate
F	f	df1,df2
gamma	gamma	shape, rate
geometric	geom	prob
hypergeometric	hyper	m, n, k
log-normal	lnorm	meanlog, sdlog
logistic	logis	location, scale
negative binomial	nbinom	size, prob
normal	norm	mean, sd
normal range	nrange	size
Poisson	pois	lambda
stable	stab	index, skewness
T	t	df
uniform	unif	min, max
Weibull	weibull	shape, scale
Wilcoxon	wilcox	m, n
multivariate normal	mvnorm	mean, cov ...

These functions can be used to replace statistical tables. For example, the 5% critical value for a (two-sided) t test on 11 degrees of freedom is given by `qt(0.975, 11)`, and the p-value associated with a Poisson(25)-distributed count of 32 is given by (by convention) `1 - ppois(31, 25)`. The functions can be given a vector first argument to calculate several p-values or quantiles.

Q-Q Plots

One of the best ways to compare the distribution of a sample `x` with a distribution is to use a Q-Q plot. The normal probability plot is the best-known example. For a sample `x` the quantile function is the inverse of the empirical CDF; that is,

$$\text{quantile}\,(p) = \min\,\{z \mid \text{proportion } p \text{ of the data } \leqslant z\,\}$$

The function `quantile` calculates the quantiles of a single set of data. The function `qqplot(x, y, ...)` plots the quantile functions of two samples `x` and `y` against each other, and so compares their distributions. The function `qqnorm(x)` replaces one of the samples by the quantiles of a standard normal distribution. This idea can be applied quite generally. For example, to test a sample against a t_9 distribution we might use

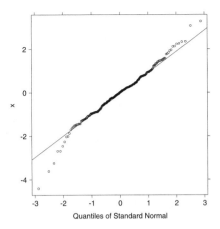

Figure 5.1: Normal probability plot of 250 simulated points from the t_9 distribution, plotted using Trellis.

```
plot( qt(ppoints(x),9), sort(x) )
```

where the function `ppoints` computes the appropriate set of probabilities for the plot. These values are $(i - 1/2)/n$ for $n \geqslant 11$ and $(i - 3/8)/(n + 1/4)$ for $n \leqslant 10$, and are generated in increasing order.

The function `qqline` helps assess how straight a `qqnorm` plot is by plotting a straight line through the upper and lower quartiles. To illustrate this we generate 250 points from a t distribution and compare them to a normal distribution (Figure 5.1). This can be done using the base graphics or with Trellis:

```
x <- rt(250, 9)
qqnorm(x); qqline(x)
# OR
qqmath(~ x, distribution=qnorm, aspect="xy",
    prepanel = prepanel.qqmathline,
    panel = function(x, y, ...) {
        panel.qqmathline(y, distribution=qnorm, ...)
        panel.qqmath(x, y, ...)
    },
    xlab= "Quantiles of Standard Normal"
)
```

The greater spread of the extreme quantiles for the data is indicative of a long-tailed distribution. The `prepanel` argument is used to ensure that the y axis covers both the line and the points, and the `aspect` argument banks the plot to $45°$.

This method is most often applied to residuals from a fitted model. Some people prefer the roles of the axes to be reversed, so that the data go on the x-axis and the theoretical values on the y-axis; this is achieved for `qqnorm` by the argument `datax=T`.

5.2   Generating random data

There are S functions to generate independent random samples from all the probability distributions listed in Table 5.1. These have prefix r and first argument n, the size of the sample required. For most of the functions the parameters can be specified as vectors, allowing the samples to be non-identically distributed. For example, we can generate 100 samples from the contaminated normal distribution in which a sample is from $N(0, 1)$ with probability 0.95 and otherwise from $N(0, 9)$, by

```
contam <- rnorm( 100, 0, (1 + 2*rbinom(100, 1, 0.05)) )
```

The function sample resamples from a data vector, with or without replacement. It has a number of quite different forms. Here n is an integer, x is a data vector and p is a probability distribution on 1, ..., length(x):

sample(n)	select a random permutation from $1, \dots, n$
sample(x)	randomly permute x
sample(x, replace=T)	a bootstrap sample
sample(x, n)	sample n items from x without replacement
sample(x, n, replace=T)	sample n items from x with replacement
sample(x, n, replace=T, prob=p)	probability sample of n items from x.

The last of these provides a way to sample from an arbitrary (finite) discrete distribution; set x to the vector of values and prob to the corresponding probabilities.

The numbers produced by these functions are of course *pseudo*-random rather than genuinely random. The state of the generator is controlled by a set of integers stored in the S object .Random.seed. Whenever sample or an 'r' function is called, .Random.seed is read in, used to initialize the generator, and then its current value is written back out at the end of the function call. If there is no .Random.seed in the current working database, one is created with default values. To re-run a simulation the initial value of .Random.seed needs to be recorded. For some purposes it may be more convenient to use the function set.seed with argument between 1 and 1 000, which selects one of 1 000 preselected seeds. (In fact there are appear to be 1 024 seeds, differing only in .Random.seed[4:6], and the argument of set.seed is used modulo 1024.)

Details of the pseudo-random number generator

Some users will want to understand the internals of the pseudo-random number generator; others should skip the rest of this section. Background knowledge on pseudo-random number generators is given by Ripley (1987, Chapter 2). We describe the generator that was current at the time of writing and which had been in use for several years (with a slight change in 1994); it is of course not guaranteed to remain unchanged.

The current generator is based on George Marsaglia's "Super-Duper" package from about 1973. The generator produces a 32-bit integer whose top 31 bits are divided by 2^{31} to produce a real number in $[0, 1)$. The 32-bit integer is produced by a bitwise exclusive-or of two 32-bit integers produced by separate generators. The C code for the S random number generator is (effectively, since S-PLUS 3.2)

```
unsigned long congrval, tausval; # assume 32-bit
static double xuni()
{
  unsigned long n, lambda = 69069, res;
  do {
    congrval = congrval * lambda;
    tausval ^= tausval >> 15;
    tausval ^= tausval << 17;
    n = tausval ^ congrval;
    res = (n>>1) & 017777777777;
  } while(res == 0);
  return (res / 2147483648.);
}
```

The integer `congrval` follows the congruential generator

$$X_{i+1} = 69069X_i \bmod 2^{32}$$

as unsigned integer overflows are (silently) reduced modulo 2^{32}; that is, the overflowing bits are discarded. As this is a multiplicative generator (no additive constant), its period is 2^{30} and the bottom bit must always be odd (including for the seed).

The integer `tausval` follows a 32-bit Tausworthe generator (of period 4 292 868 097 $< 2^{32} - 1$):

$$b_i = b_{i-p} \operatorname{xor} b_{i-(p-q)}, \qquad p = 32, \quad q = 15$$

This follows from a slight modification of Algorithm 2.1 of Ripley (1987, p. 28). (In fact, the period quoted is true only for the vast majority of starting values; for the remaining 2 099 198 non-zero initial values there are shorter cycles.)

For most starting seeds the period of the generator is $2^{30} \times$ 4 292 868 097 $\approx 4.6 \times 10^{18}$, that is quite sufficient for calculations that can be done in a reasonable time in S. The current values of `congrval` and `tausval` are encoded in the vector .Random.seed, a vector of 12 integers in the range $0, \ldots, 63$. If x represents .Random.seed, we have

$$\texttt{congrval} = \sum_{i=1}^{6} x_i\, 2^{6(i-1)} \quad \text{and} \quad \texttt{tausval} = \sum_{i=1}^{6} x_{i+6}\, 2^{6(i-1)}$$

5.3 Data summaries

Standard univariate summaries such as mean, median and var are available. The summary function returns the mean, quartiles and the number of missing values, if non-zero.

The `var` function will take a data matrix and give the variance-covariance matrix, and `cor` computes the correlations, either from two vectors or a data matrix. The function `cov.wt` returns the means and variance[1] matrix and optionally the correlation matrix of a data matrix. As its name implies, the rows of the data matrix can be weighted in these summaries.

The function `quantile` computes quantiles at specified probabilities, by default $(0, 0.25, 0.5, 0.75, 1)$ giving a "five number summary" of the data vector. This function linearly interpolates, so if $x_{(1)}, \ldots, x_{(n)}$ is the ordered sample,

$$\texttt{quantile(x, p)} = \left[1 - (p(n-1) - \lfloor p(n-1) \rfloor)\right] x_{1+\lfloor p(n-1) \rfloor}$$
$$+ \left[p(n-1) - \lfloor p(n-1) \rfloor\right] x_{2+\lfloor p(n-1) \rfloor}$$

where $\lfloor\ \rfloor$ denotes the 'floor' or integer part of. (This differs from the definitions of a quantile given earlier.) There are also standard functions `max`, `min` and `range`.

The functions `mean` and `cor` will compute trimmed summaries using the argument `trim`. More sophisticated robust summaries are discussed in Section 5.5.

These functions differ in the way they handle missing values. Functions `mean`, `median`, `max`, `min`, `range` and `quantile` have an argument `na.rm` which defaults to false, but can be used to remove missing values. The functions `var` and `cor` fail with missing values in **S-PLUS 3.x**, but later versions allow several options for the handling of missing values.

Histograms and stem-and-leaf plots

The standard histogram function is `hist(x, ...)` which plots a conventional histogram. More control is available via the extra arguments; `probability=T` gives a plot of unit total area rather than of cell counts, and `style="old"` gives the more conventional representation by outlines of bars rather than grey bars. There is also a Trellis function `histogram`, but neither is totally satisfactory.

The argument `nclass` of `hist` suggests the number of bins, and `breaks` specifies the breakpoints between bins. The default for `nclass` is $\log_2 n + 1$. This is known as Sturges' formula. One problem is that `nclass` is only a suggestion, and it is often exceeded. Another is that the definition of the bins that is of the form $[x_0, x_1], (x_1, x_2], \ldots$, not the convention most people prefer.[2]

Sturges' formula corresponds to a bin width of $\text{range}(x)/(\log_2 n + 1)$, and was based on a histogram of a normal distribution (Scott, 1992, p. 48). Note that outliers may inflate the range dramatically and so increase the bin width in the centre of the distribution. Two rules based on compromises between the bias and variance of the histogram for a reference normal distribution are to choose bin width as

$$h = 3.5\hat{\sigma}n^{-1/3} \tag{5.1}$$
$$h = 2Rn^{-1/3} \tag{5.2}$$

---

[1] `cov.wt` does not adjust its divisor for the estimation of the mean, using divisor n even when unweighted.

[2] The special treatment of x_0 is governed by the switch `include.lowest` which defaults to T.

due to Scott (1979) and Freedman & Diaconis (1981), respectively. Here $\hat{\sigma}$ is the estimated standard deviation and R the inter-quartile range. The Freedman–Diaconis formula is immune to outliers, and chooses rather smaller bins than the Scott formula. We program these to suggest the number of classes and leave `hist` to choose 'pretty' breakpoints.

```
nclass.scott <- function(x)
{
    h <- 3.5 * sqrt(var(x)) * length(x)^(-1/3)
    ceiling(diff(range(x))/h)
}
nclass.FD <- function(x)
{
    r <- as.vector(quantile(x, c(0.25, 0.75)))
    h <- 2 * (r[2] - r[1]) * length(x)^(-1/3)
    ceiling(diff(range(x))/h)
}
hist.scott <-
  function(x, prob = T, xlab = deparse(substitute(x)), ...)
      invisible(hist(x, nclass.scott(x), prob = prob,
                     xlab = xlab, ...))
hist.FD <-
  function(x, prob = T, xlab = deparse(substitute(x)), ...)
      invisible(hist(x, nclass.FD(x), prob = prob,
                     xlab = xlab, ...))
```

These are not always satisfactory, as Figure 5.2 shows. (The suggested numbers of bins are 8, 5, 25, 7, 42 and 35.) Column `duration` of our data frame `geyser` gives the duration (in minutes) of 299 eruptions of the Old Faithful geyser in the Yellowstone National Park (from Azzalini & Bowman, 1990); `chem` is discussed later in this section and `tperm` in Section 5.7.

The beauty of S is that it is easy to write one's own function to plot a histogram. Our function is called `truehist` (in library MASS). The primary control is by specifying the bin width h, but suggesting the number of bins by argument `nbins` will give a 'pretty' choice of h. Function `truehist` is used in Figures 5.5 and later. There is also a Trellis version called `histplot` for use in conditioning plots.

A *stem-and-leaf* plot is an enhanced histogram. The data are divided into bins, but the 'height' is replaced by the next digits in order. We apply this to `swiss.fertility`, the standardized fertility measure for each of 47 French-speaking provinces of Switzerland at about 1888.

```
> stem(swiss.fertility)

N = 47    Median = 70.4
Quartiles = 64.4, 79.3

Decimal point is 1 place to the right of the colon
```

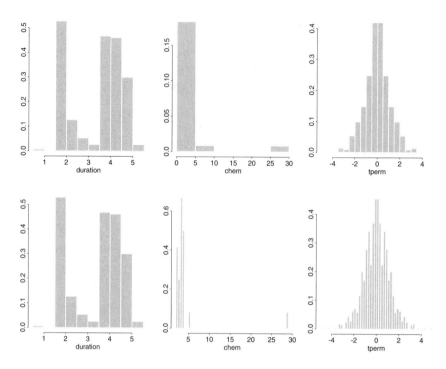

Figure 5.2: Histograms with bin widths chosen by the Scott rule (5.1) (top row) and the Freedman–Diaconis rule (5.2) (bottom row) for datasets `geyser$duration`, `chem` and `tperm`.

```
3 : 5
4 : 35
5 : 46778
6 : 024455555678899
7 : 00222345677899
8 : 0233467
9 : 222
```

Apart from giving a visual picture of the data, this gives more detail. The actual data, in sorted order, are 35, 43, 45, 54, ... and this can be read from the plot. Sometimes the pattern of numbers (all odd? many 0s and 5s?) gives clues. Quantiles can be computed (roughly) from the plot. If there are outliers, they are marked separately:

```
> stem(chem)

N = 24    Median = 3.385
Quartiles = 2.75, 3.7

Decimal point is at the colon
```

```
  2 : 22445789
  3 : 00144445677778
  4 :
  5 : 3

High: 28.95
```

(This dataset on measurements of copper in flour from the Analytical Methods Committee (1989a) is discussed further in Section 5.5.) Sometimes the result is less successful, and manual override is needed:

```
> stem(abbey)

N = 31    Median = 11
Quartiles = 8, 16

Decimal point is at the colon

   5 : 2
   6 : 59
   7 : 0004
   8 : 00005
   ....
  26 :
  27 :
  28 : 0

High:    34   125

> stem(abbey, scale=-1)

N = 31    Median = 11
Quartiles = 8, 16

Decimal point is 1 place to the right of the colon

   0 : 56777778888899
   1 : 011224444
   1 : 6778
   2 : 4
   2 : 8

High:    34   125
```

Here the `scale` argument sets the backbone to be 10s rather than units. The `nl` argument controls the number of rows per backbone unit as 2, 5 or 10. The details of the design of stem-and-leaf plots are discussed by Mosteller & Tukey (1977), Velleman & Hoaglin (1981) and Hoaglin, Mosteller & Tukey (1983).

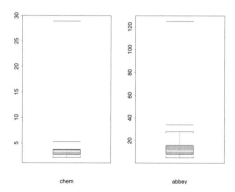

Figure 5.3: Boxplots for the chem and abbey data.

Boxplots

A *boxplot* is a way to look at the overall shape of a set of data. The central box shows the data between the 'hinges' (roughly quartiles), with the median represented by a line. 'Whiskers' go out to the extremes of the data, and very extreme points are shown by themselves.

```
par(mfrow=c(1,2))
boxplot(chem, sub="chem", range=0.5)
boxplot(abbey, sub="abbey")
par(mfrow=c(1,1))
bwplot(type ~ y | meas, data=fgl.df, scales=list(x="free"),
    strip=function(...) strip.default(..., style=1), xlab="")
```

There is a bewildering variety of optional parameters to boxplot documented in the on-line help page. Note how these plots are dominated by the outliers. It is possible to plot boxplots for groups side by side (see Figure 13.16 on page 423) but the Trellis function bwplot (see Figure 5.4 and Figure 3.9 on page 78) will probably be preferred.

5.4 Classical univariate statistics

S-PLUS has a section on classical statistics. The same functions are used to perform tests and to calculate confidence intervals.

Table 5.2 shows the amounts of shoe wear in an experiment reported by Box, Hunter & Hunter (1978). There were two materials (*A* and *B*) that were randomly assigned to the left and right shoes of 10 boys. We use these data to illustrate one-sample and paired and unpaired two-sample tests. (The rather voluminous output has been edited.)

First we test for a mean of 10 and give a confidence interval for the mean, both for material *A*.

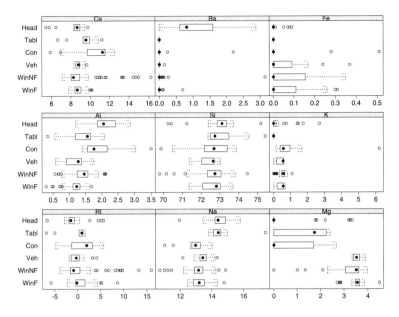

Figure 5.4: Boxplots by type for the `fgl` dataset.

Table 5.2: Data on shoe wear from Box, Hunter & Hunter (1978).

boy	A	B
1	13.2 (L)	14.0 (R)
2	8.2 (L)	8.8 (R)
3	10.9 (R)	11.2 (L)
4	14.3 (L)	14.2 (R)
5	10.7 (R)	11.8 (L)
6	6.6 (L)	6.4 (R)
7	9.5 (L)	9.8 (R)
8	10.8 (L)	11.3 (R)
9	8.8 (R)	9.3 (L)
10	13.3 (L)	13.6 (R)

```
> attach(shoes)
> t.test(A, mu=10)

        One-sample t-Test

data:  A
t = 0.8127, df = 9, p-value = 0.4373
alternative hypothesis: true mean is not equal to 10
95 percent confidence interval:
  8.8764 12.3836
```

```
sample estimates:
 mean of x
     10.63

> t.test(A)$conf.int
[1]   8.8764 12.3836
attr(, "conf.level"):
[1] 0.95

> wilcox.test(A, mu = 10)

        Exact Wilcoxon signed-rank test

data:   A
signed-rank statistic V = 34, n = 10, p-value = 0.5566
alternative hypothesis: true mu is not equal to 10
```

Next we consider two-sample paired and unpaired tests, the latter assuming equal variances (the default) or not. Note that we are using this example for illustrative purposes; only the paired analyses are really appropriate.

```
> var.test(A, B)

        F test for variance equality

data:   A and B
F = 0.9474, num df = 9, denom df = 9, p-value = 0.9372
95 percent confidence interval:
 0.23532 3.81420
sample estimates:
 variance of x variance of y
         6.009          6.3427

> t.test(A, B)

        Standard Two-Sample t-Test

data:   A and B
t = -0.3689, df = 18, p-value = 0.7165
95 percent confidence interval:
 -2.7449   1.9249
sample estimates:
 mean of x mean of y
     10.63      11.04

> t.test(A, B, var.equal=F)

        Welch Modified Two-Sample t-Test

data:   A and B
```

```
t = -0.3689, df = 17.987, p-value = 0.7165
95 percent confidence interval:
 -2.745  1.925
    ....

> wilcox.test(A, B)

         Wilcoxon rank-sum test

data:  A and B
rank-sum normal statistic with correction Z = -0.5293,
   p-value = 0.5966

> t.test(A, B, paired=T)

         Paired t-Test

data:  A and B
t = -3.3489, df = 9, p-value = 0.0085
95 percent confidence interval:
 -0.68695 -0.13305
sample estimates:
 mean of x - y
        -0.41

> wilcox.test(A, B, paired=T)

         Wilcoxon signed-rank test

data:  A and B
signed-rank normal statistic with correction Z = -2.4495,
   p-value = 0.0143
```

The sample size is rather small, and one might wonder about the validity of the *t*-distribution. An alternative for a randomized experiment such as this is to base inference on the permutation distribution of d = B-A. Figure 5.5 shows that the agreement is very good. The computation of the permutations is discussed in Section 5.7.

The full list of classical tests is:

binom.test	chisq.test	cor.test	fisher.test
friedman.test	kruskal.test	mantelhaen.test	mcnemar.test
prop.test	t.test	var.test	wilcox.test
chisq.gof	ks.gof		

Many of these have alternative methods—for cor.test there are methods "pearson", "kendall" and "spearman". We have already seen one- and two-sample versions of t.test and wilcox.test, and var.test which compares the variances of two samples. The function cor.test tests for non-zero correlation between two samples, either classically or via ranks.

Empirical and Hypothesized t CDFs

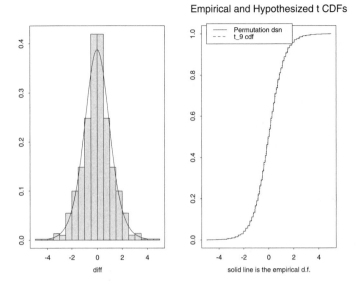

Figure 5.5: Histogram and empirical CDF of the permutation distribution of the paired *t*–test in the shoes example. The density and CDF of t_9 are shown overlaid.

Functions `chisq.gof` and `ks.gof` compute chi-square and Kolmogorov–Smirnov tests of goodness-of-fit. Function `cdf.compare` plots comparisons of cumulative distributions such as the right-hand panel of Figure 5.5.

```
par(mfrow=c(1,2))
truehist(tperm, xlab="diff")
x <- seq(-4,4, 0.1)
lines(x, dt(x,9))
cdf.compare(tperm, distribution="t", df=9)
legend(-5, 1.05, c("Permutation dsn","t_9 cdf"), lty=c(1,3))
```

5.5 Robust summaries

Outliers are sample values that cause surprise in relation to the majority of the sample. This is not a pejorative term; outliers may be correct, but they should always be checked for transcription errors. They can play havoc with standard statistical methods, and many *robust* and *resistant* methods have been developed since 1960 to be less sensitive to outliers.

The sample mean $\bar{y}$ can be upset completely by a single outlier; if any data value $y_i \to \pm\infty$, then $\bar{y} \to \pm\infty$. This contrasts with the sample median, which is little affected by moving any single value to $\pm\infty$. We say that the median is *resistant* to *gross errors* whereas the mean is not. In fact the median will tolerate up to 50% gross errors before it can be made arbitrarily large; we say its *break-down point* is 50% whereas that for the mean is 0%. Although the mean is the optimal estimator of the location of the normal distribution, it can be substantially sub-optimal for distributions close to the normal. Robust methods aim to have high efficiency in a neighbourhood of the assumed statistical model.

There are now a number of books on robust statistics. Huber (1981) is rather theoretical, Hampel *et al.* (1986) and Staudte & Sheather (1990) less so. Rousseeuw & Leroy (1987) is principally concerned with regression, but is very practical. Robust and resistant methods have long been one of the strengths of S and S-PLUS.

Why will it not suffice to screen data and remove outliers? There are several aspects to consider.

1. Users, even expert statisticians, do not always screen the data.
2. The sharp decision to keep or reject an observation is wasteful. We can do better by down-weighting dubious observations than by rejecting them, although we may wish to reject completely wrong observations.
3. It can be difficult or even impossible to spot outliers in multivariate or highly structured data.
4. Rejecting outliers affects the distribution theory, which ought to be adjusted. In particular, variances will be underestimated from the 'cleaned' data.

For a fixed underlying distribution, we define the *relative efficiency* of an estimator $\tilde{\theta}$ relative to another estimator $\hat{\theta}$ by

$$RE(\tilde{\theta}; \hat{\theta}) = \frac{\text{variance of } \hat{\theta}}{\text{variance of } \tilde{\theta}}$$

since $\hat{\theta}$ needs only RE times as many observations as $\tilde{\theta}$ for the same precision, approximately. The asymptotic relative efficiency (ARE) is the limit of the RE as the sample size $n \to \infty$. (It may be defined more widely via asymptotic variances.) If $\hat{\theta}$ is not mentioned, it is assumed to be the optimal estimator. There is a difficulty with biased estimators whose variance can be small or zero. One solution is to use the mean-square error, another to rescale by $\theta/E(\hat{\theta})$. Iglewicz (1983) suggests using $\text{var}(\log \hat{\theta})$ (which is scale-free) for estimators of scale.

We can apply the concept of ARE to the mean and median. At the normal distribution $ARE(\text{median}; \text{mean}) = 2/\pi \approx 64\%$. For longer-tailed distributions the median does better; for the t distribution with five degrees of freedom (which is often a better model of error distributions than the normal) $ARE(\text{median}; \text{mean}) \approx 96\%$.

The following example from Tukey (1960) is more dramatic. Suppose we have n observations $Y_i \sim N(\mu, \sigma^2), i = 1, \ldots, n$ and we want to estimate σ^2. Consider $\hat{\sigma}^2 = s^2$ and $\tilde{\sigma}^2 = d^2 \pi/2$ where

$$d = \frac{1}{n} \sum_i |Y_i - \overline{Y}|$$

and the constant is chosen since for the normal $d \to \sqrt{2/\pi}\,\sigma$. The $ARE(\tilde{\sigma}^2; s^2)$ $= 0.876$. Now suppose that each Y_i is from $N(\mu, \sigma^2)$ with probability $1 - \epsilon$ and from $N(\mu, 9\sigma^2)$ with probability ϵ. (Note that both the overall variance and the variance of the uncontaminated observations are proportional to σ^2.) We have

ϵ (%)	$ARE(\tilde{\sigma}^2; s^2)$
0	0.876
0.1	0.948
0.2	1.016
1	1.44
5	2.04

Since the mixture distribution with $\epsilon = 1\%$ is indistinguishable from normality for all practical purposes, the optimality of s^2 is very fragile. We say it lacks *robustness of efficiency*.

There are better estimators of σ than $d\sqrt{\pi/2}$ (which has breakdown point 0%). Two alternatives are proportional to

$$IQR = X_{(3n/4)} - X_{(n/4)}$$
$$MAD = \underset{i}{\text{median}}\,\{|Y_i - \underset{j}{\text{median}}\,(Y_j)|\}$$

(Order statistics are linearly interpolated where necessary.) At the normal,

$$MAD \to \text{median}\,\{|Y - \mu|\} \approx 0.6745\sigma$$
$$IQR \to \sigma\left[\Phi^{-1}(0.75) - \Phi^{-1}(0.25)\right] \approx 1.35\sigma$$

(We refer to $MAD/0.6745$ as the MAD estimator, calculated by function `mad`.) Both are not very efficient but are very resistant to outliers in the data. The MAD estimator has ARE 37% at the normal (Staudte & Sheather, 1990, p. 123).

Consider n independent observations Y_i from a location family with pdf $f(y - \mu)$ for a function f symmetric about zero, so it is clear that μ is the centre (median, mean if it exists) of the distribution of Y_i. We also think of the distribution as being not too far from the normal. There are a number of obvious estimators of μ, including the sample mean, the sample median, and the MLE.

The *trimmed mean* is the mean of the central $1 - 2\alpha$ part of the distribution, so αn observations are removed from each end. This is implemented by the function `mean` with the argument `trim` specifying α. Obviously, `trim=0` gives the mean and `trim=0.5` gives the median (although it is easier to use the function `median`). (If αn is not an integer, the integer part is used.)

Most of the location estimators we consider are *M-estimators*. The name derives from 'MLE-like' estimators. If we have density f, we can define $\rho = -\log f$. Then the MLE would solve

$$\min_{\mu} \sum_i -\log f(y_i - \mu) = \min_{\mu} \sum_i \rho(y_i - \mu)$$

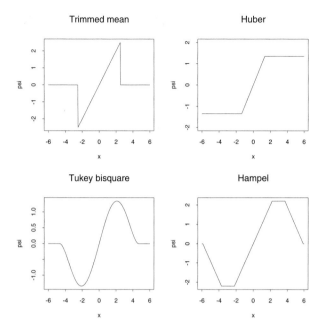

Figure 5.6: The ψ-functions for four common M-estimators.

and this makes sense for functions ρ not corresponding to pdfs. Let $\psi = \rho'$ if this exists. Then we will have $\sum_i \psi(y_i - \hat{\mu}) = 0$ or $\sum_i w_i (y_i - \hat{\mu}) = 0$ where $w_i = \psi(y_i - \hat{\mu})/(y_i - \hat{\mu})$. This suggests an iterative method of solution, updating the weights at each iteration.

Examples of M-estimators

The mean corresponds to $\rho(x) = x^2$, and the median to $\rho(x) = |x|$. (For even n any median will solve the problem.) The function

$$\psi(x) = \begin{cases} x & |x| < c \\ 0 & \text{otherwise} \end{cases}$$

corresponds to *metric trimming* and large outliers have no influence at all. The function

$$\psi(x) = \begin{cases} -c & x < -c \\ x & |x| < c \\ c & x > c \end{cases}$$

is known as *metric Winsorizing*[3] and brings in extreme observations to $\mu \pm c$. The corresponding $-\log f$ is

$$\rho(x) = \begin{cases} x^2 & \text{if } |x| < c \\ c(2|x| - c) & \text{otherwise} \end{cases}$$

---

[3] A term attributed by Dixon (1960) to Charles P. Winsor.

and corresponds to a density with a Gaussian centre and double-exponential tails. This estimator is due to Huber. Note that its limit as $c \to 0$ is the median, and as $c \to \infty$ the limit is the mean. The value $c = 1.345$ gives 95% efficiency at the normal.

Tukey's *biweight* has

$$\psi(t) = t\left[1 - \left(\frac{t}{R}\right)^2\right]_+^2.$$

where $[\,]_+$ denotes the positive part of. This implements 'soft' trimming. The value $R = 4.685$ gives 95% efficiency at the normal.

Hampel's ψ has several linear pieces,

$$\psi(x) = \text{sgn}(x) \begin{cases} |x| & 0 < |x| < a \\ a & a < |x| < b \\ a(c - |x|)/(c - b) & b < |x| < c \\ 0 & c < |x| \end{cases}$$

for example, with $a = 2.2s, b = 3.7s, c = 5.9s$. Figure 5.6 illustrates these functions.

There is a scaling problem with the last four choices, since they depend on a scale factor (c, R or s). We can apply the estimator to rescaled results, that is,

$$\min_{\mu} \sum_i \rho\left(\frac{y_i - \mu}{s}\right)$$

for a scale factor s, for example the MAD estimator. Alternatively, we can estimate s in a similar way. The MLE for density $s^{-1}f((x - \mu)/s)$ gives rise to the equation

$$\sum_i \psi\left(\frac{y_i - \mu}{s}\right)\left(\frac{y_i - \mu}{s}\right) = n$$

which is not resistant (and is biased at the normal). We modify this to

$$\sum_i \chi\left(\frac{y_i - \mu}{s}\right) = (n - 1)\gamma$$

for bounded χ, where γ is chosen for consistency at the normal distribution, so $\gamma = E\chi(N)$. The main example is "Huber's proposal 2" with

$$\chi(x) = \psi(x)^2 = \min(|x|, c)^2 \qquad (5.3)$$

In very small samples we need to take account of the variability of $\hat{\mu}$ in performing the Winsorizing.

If the location μ is known we can apply these estimators with $n - 1$ replaced by n to estimate the scale s alone.

The main S-PLUS function for location M-estimation is `location.m` which has parameters including

```
location.m(x, location, scale, weights, na.rm=F,
    psi.fun="bisquare", parameters)
```

The initial values specified by `location` defaults to the low median, and `scale` is by default fixed as MAD (with the high median). The `psi.fun` can be Tukey's bi-square, `"huber"` or a user-supplied function. The `parameters` are passed to the `psi.fun` and default to 5 for bi-square and 1.45 for Huber, for efficiency at the normal of about 96%. Hampel's ψ function can be implemented by the example (using C) given on the help page.

For scale estimation there are the functions `mad` described previously and `scale.tau` which implements Huber's scale based on (5.3) with parameter 1.95 chosen for 80% efficiency at the normal. We supply functions `huber` and `hubers` for the Huber M-estimator with MAD and "proposal 2" scale respectively, with default $c = 1.5$.

Examples

We give two datasets taken from analytical chemistry (Abbey, 1988; Analytical Methods Committee, 1989a,b). The dataset `abbey` contains 31 determinations of nickel content ($\mu g\ g^{-1}$) in SY-3, a Canadian syenite rock, and `chem` contains 24 determinations of copper ($\mu g\ g^{-1}$) in wholemeal flour. These data are part of a larger study that suggests $\mu = 3.68$.

```
> sort(chem)
 [1]   2.20  2.20  2.40  2.40  2.50  2.70  2.80  2.90  3.03
[10]   3.03  3.10  3.37  3.40  3.40  3.40  3.50  3.60  3.70
[19]   3.70  3.70  3.70  3.77  5.28 28.95
> mean(chem)
[1] 4.2804
> median(chem)
[1] 3.385
> location.m(chem)
[1] 3.1452
    ....
> location.m(chem, psi.fun="huber")
[1] 3.2132
    ....
> mad(chem)
[1] 0.52632
> scale.tau(chem)
[1] 0.639
> scale.tau(chem, center=3.68)
[1] 0.91578
> unlist(huber(chem))
      mu       s
 3.2067 0.52632
> unlist(hubers(chem))
      mu       s
 3.2055 0.67365
```

The sample is clearly highly asymmetric with one value that appears to be out by a factor of 10. It was checked and reported as correct by the laboratory. With such a distribution the various estimators are estimating different aspects of the distribution and so are not comparable. Only for symmetric distributions do all the location estimators estimate the same quantity, and although the true distribution is unknown here, it is unlikely to be symmetric.

```
> sort(abbey)
 [1]    5.2    6.5    6.9    7.0    7.0    7.0    7.4    8.0    8.0
[10]    8.0    8.0    8.5    9.0    9.0   10.0   11.0   11.0   12.0
[19]   12.0   13.7   14.0   14.0   14.0   16.0   17.0   17.0   18.0
[28]   24.0   28.0   34.0  125.0
> mean(abbey)
[1] 16.006
> median(abbey)
[1] 11
> location.m(abbey)
[1] 10.804
> location.m(abbey, psi.fun="huber")
[1] 11.517
> unlist(hubers(abbey))
      mu      s
 11.732 5.2585
> unlist(hubers(abbey, k=2))
      mu      s
 12.351 6.1052
> unlist(hubers(abbey, k=1))
      mu      s
 11.365 5.5673
```

Note how reducing the constant k (representing c) reduces the estimate of location, as this sample (like many in analytical chemistry) has a long right tail.

5.6 Density estimation

The non-parametric estimation of probability density functions is a large topic; several books have been devoted to it, notably Silverman (1986), Scott (1992), Wand & Jones (1995) and Simonoff (1996). Bowman & Azzalini (1997) concentrate on providing an introduction to kernel-based methods, with an easy-to-use S-PLUS library sm.[4] This has the unusual ability to compute and plot kernel density estimates of three-dimensional and spherical data.

The histogram with probability=T is of course an estimator of the density function. The histogram depends on the starting point of the grid of bins. The effect can be surprisingly large; see Figure 5.7. The figure also shows that by averaging the histograms we can obtain a much clearer view of the distribution.

---

[4] Available from http://www.stats.gla.ac.uk/~adrian/sm and http://www.stat. unipd.it/dip/homes/azzalini/SW/Splus/sm.

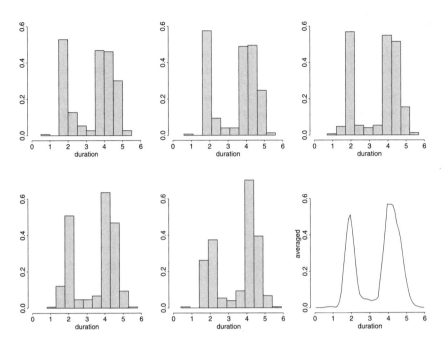

Figure 5.7: Five shifted histograms with bin width 0.5 and the frequency polygon of their average, for the Old Faithful geyser `duration` data.

```
attach(geyser)
par(mfrow=c(2,3))
truehist(duration, h=0.5, x0=0.0, xlim=c(0, 6), ymax=0.7)
truehist(duration, h=0.5, x0=0.1, xlim=c(0, 6), ymax=0.7)
truehist(duration, h=0.5, x0=0.2, xlim=c(0, 6), ymax=0.7)
truehist(duration, h=0.5, x0=0.3, xlim=c(0, 6), ymax=0.7)
truehist(duration, h=0.5, x0=0.4, xlim=c(0, 6), ymax=0.7)

breaks <- seq(0, 5.9, 0.1)
counts <- numeric(length(breaks))
for(i in (0:4)) counts[i+(1:55)] <- counts[i+(1:55)] +
    rep(hist(duration, breaks=0.1*i + seq(0, 5.5, 0.5),
    prob=T, plot=F)$counts, rep(5,11))
plot(breaks+0.05, counts/5, type="l", xlab="duration",
    ylab="averaged", bty="n", xlim=c(0, 6), ylim=c(0, 0.7))
```

This idea of an *average shifted histogram* or *ASH* density estimate is a useful one and is discussed in detail in Scott (1992). Suppose we average over m shifts by $\delta = h/m$. Another way to look at the result is to take a histogram with bin width δ, and to compute the histograms of bin width h by aggregating counts in m adjacent bins. Specifically, if the counts are ν_k, the ASH for a point x in small bin k is

$$\hat{f}(x) = \frac{1}{nh} \sum_{i=1-m}^{m-1} \left[1 - \frac{|i|}{m} \right] \nu_{k+i} = \frac{1}{nh} \sum_i \left[1 - \frac{|i\delta|}{h} \right]_+ \nu_{k+i}$$

so is a weighted average of the bin counts with a triangular weighting function. Now suppose we let $m \to \infty$. The counts in bins reduce to zero or one, so the limit is

$$\hat{f}(x) = \frac{1}{nh} \sum_{j=1}^n \left[1 - \frac{|x - x_j|}{h} \right]_+$$

This is a *kernel density* estimate with a triangular kernel. We prefer a smoother kernel of the form

$$\hat{f}(x) = \frac{1}{b} \sum_{j=1}^n K\left(\frac{x - x_j}{b} \right) \tag{5.4}$$

for a sample $x_1, \ldots, x_n$, a fixed kernel $K()$ and a bandwidth b; the kernel is normally chosen to be a probability density function.

The only built-in S function specifically for density estimation is `density`. The default kernel is the normal (argument `window="g"` for Gaussian), with alternatives `"rectangular"`, `"triangular"` and `"cosine"` (the latter being $(1 + \cos \pi x)/2$ over $[-1, 1]$). The bandwidth is the length of the non-zero section for the alternatives, and four times the standard deviation for the normal. (Note that these definitions are twice and four times those most commonly used.)

The choice of bandwidth is a compromise between smoothing enough to remove insignificant bumps and not smoothing too much to smear out real peaks. Mathematically there is a compromise between the bias of $\hat{f}(x)$, which increases as b is increased, and the variance which decreases. Theory (see (5.6) on page 137) suggests that the bandwidth should be proportional to $n^{-1/5}$, but the constant of proportionality depends on the unknown density.

We can illustrate this effect with the permutation distribution of Figure 5.5, which is stored in vector `tperm`. Under-smoothing (`width=0.2`) shows spurious bumps (Figure 5.8); over-smoothing (`width=1.5`) has no bumps in the tails but reduces the height of the main peak. By default the density is computed at 50 equally-spaced points, which is too few for this example.

```
tperm <- test.t.2(B-A)   # see Section 5.7
par(mfrow=c(2,2))
x <- seq(-5, 5, 0.1)
plot(c(-5, 5), c(0, 0.45), type="n", bty="l",
    sub="default", xlab="", ylab="")
lines(x, dt(x,9), lty=2); rug(jitter(tperm))
lines(density(tperm, n=200, from=-5, to=5))
plot(c(-5, 5), c(0, 0.45), type="n", bty="l",
    sub="width=0.2", xlab="", ylab="")
lines(x, dt(x,9), lty=2); rug(jitter(tperm))
lines(density(tperm, n=200, width=0.2, from=-5, to=5))
```

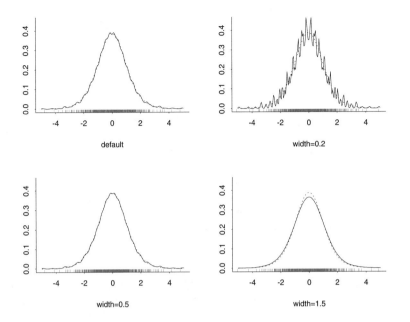

Figure 5.8: Density plots for the permutation distribution for the t statistic for the difference in shoe materials. In each case the reference t_9 density is shown dotted.

```
plot(c(-5, 5), c(0, 0.45), type="n", bty="l",
    sub="width=0.5", xlab="", ylab="")
lines(x, dt(x,9), lty=2); rug(jitter(tperm))
lines(density(tperm, n=200, width=0.5, from=-5, to=5))
plot(c(-5, 5), c(0, 0.45), type="n", bty="l",
    sub="width=1.5", xlab="", ylab="")
lines(x, dt(x,9), lty=2); rug(jitter(tperm))
lines(density(tperm, n=200, width=1.5, from=-5, to=5))
```

Function `rug` plots the data as ticks and `jitter` randomly perturbs them to eliminate ties.

The default width b is given[5] by a rule of thumb

$$\frac{\text{range}(x)}{2(1 + \log_2 n)}$$

which happens to work quite well in this example (with $b \approx 0.45$). (This seems to be based on Sturges' rule and a rough equivalence with a histogram density estimate of bin width equal to the bandwidth.) There are better-supported rules of thumb such as

$$\hat{b} = 1.06 \min(\hat{\sigma}, R/1.34)n^{-1/5} \tag{5.5}$$

for the IQR R and the Gaussian kernel of bandwidth the standard deviation, to be quadrupled for use with `density` Silverman (1986, pp. 45–47).

---

[5] Up to S-PLUS 4.5 and 5.0.

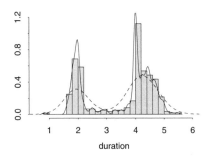 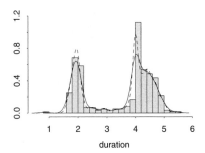

Figure 5.9: Density plots for the Old Faithful duration data. Superimposed on a histogram are the kernel density estimates with a Gaussian kernel and default bandwidth (left, solid), normal reference bandwidth (left, dashed) and the Sheather–Jones 'direct plug-in' (right, solid) and 'solve-the-equation' (right, dashed) bandwidth estimates with a Gaussian kernel.

We return to the geyser duration data. This is an example in which the default bandwidth, 0.25, is too small, about 17% of the value suggested by (5.5), which is too large (Figure 5.9). (Scott, 1992, §6.5.1.2) suggests that the value

$$b_{OS} = 1.144\hat{\sigma}n^{-1/5}$$

provides an upper bound on the (Gaussian) bandwidths one would want to consider, again to be quadrupled for use in density. The normal reference (5.5) is only slightly below this upper bound, and will be too large for densities with multiple modes.

```
attach(geyser)
truehist(duration, nbins=15, xlim=c(0.5,6), ymax=1.2)
lines(density(duration, n=200))
bandwidth.nrd <- function(x)
{
    r <- quantile(x, c(0.25, 0.75))
    h <- (r[2] - r[1])/1.34
    4 * 1.06 * min(sqrt(var(x)), h) * length(x)^(-1/5)
}
bandwidth.nrd(duration)
[1] 1.5565
lines(density(duration, width=1.5565, n=200), lty=3)
bandwidth.nrd(tperm)
[1] 1.1536
```

Note that the data in this example are (negatively) serially correlated, so the theory for independent data must be viewed as only a guide.

Bandwidth selection

Ways to find automatically compromise value(s) of b are discussed by Härdle (1991) and Scott (1992). These are based on estimating the *mean integrated square error*

$$MISE = E \int |\hat{f}(x;b) - f(x)|^2 \, \mathrm{d}x = \int E|\hat{f}(x;b) - f(x)|^2 \, \mathrm{d}x$$

and choosing the smallest value as a function of b. We can expand $MISE$ as

$$MISE = E \int \hat{f}(x;b)^2 \, \mathrm{d}x - 2E\hat{f}(X;b) + \int f(x)^2 \, \mathrm{d}x$$

where the third term is constant and so can be dropped.

The estimates are based on *cross-validation*, that is using the remaining observations to fit a density at each observation in turn. (See for example Stone, 1974.) Let the fitted value be $\hat{f}(x_i;b)_{-i}$. Then the *unbiased* (or least-squares) *cross-validation* estimates $E\hat{f}(X;b)$ by the mean of $\hat{f}(x_i;b)_{-i}$ to give

$$UCV(b) = \int \hat{f}(x;b)^2 \, \mathrm{d}x - \frac{2}{n} \sum_{i=1}^{n} \hat{f}(x_i;b)_{-i}$$

The minimizer $\hat{b}$ is found to be accurate but rather variable. A more involved approximation gives the *biased cross-validation* criterion $BCV(b)$ which is minimized over $b \leqslant b_{OS}$, and is usually less variable. Both $UCV(b)$ and $BCV(b)$ can have multiple local minima.

Scott (1992, §6.5.13) provides formulae for the Gaussian kernel which we can evaluate via the functions ucv and bcv of our library MASS. (These contain typographic errors that we have corrected.) As the formulae involve calculations using the distances between all pairs of points, they cannot be used for very large datasets (say thousands of observations). Our functions round the data to the nearest point on a grid so the formulae are summed over the resulting much smaller set of distances between pairs.

Ideas in bandwidth selection have developed rapidly since 1990, and UCV and BCV are now regarded as 'first generation' rules (Jones, Marron & Sheather, 1996). Another approach is to make an asymptotic expansion of $MISE$ of the form

$$MISE = \tfrac{1}{nb} \int K^2 + \tfrac{1}{4}b^4 \int (f'')^2 \{ \int x^2 K \}^2 + O(1/nb + b^4)$$

so if we neglect the remainder, the optimal bandwidth would be

$$b_{AMISE} = \left[\frac{\int K^2}{n \int (f'')^2 \{ \int x^2 K \}^2} \right]^{1/5} \tag{5.6}$$

The 'direct plug-in' estimators use (5.6), which involves the integral $\int (f'')^2$. This in turn is estimated using the second derivative of a kernel estimator with a different bandwidth, chosen by repeating the process, this time using a reference bandwidth. The 'solve-the-equation' estimators solve (5.6) when the bandwidth for estimating $\int (f'')^2$ is taken as function of b (in practice proportional to $b^{5/7}$). Details are given by Sheather & Jones (1991) and Wand & Jones (1995, §3.6); this is implemented for the Gaussian kernel in our function width.SJ.

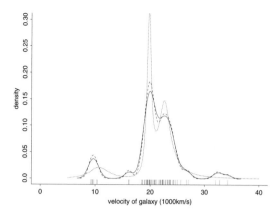

Figure 5.10: Density estimates for the 82 points of the `galaxies` data. The solid and dashed lines are Gaussian kernel density estimates with bandwidths chosen by two variants of the Sheather–Jones method. The dotted line is a logspline estimate.

```
> c(width.SJ(duration, method="dpi"), width.SJ(duration))
[1] 0.57479 0.35966
truehist(duration, nbins=15, xlim=c(0.5,6), ymax=1.2)
lines(density(duration, width=0.57, n=200), lty=1)
lines(density(duration, width=0.36, n=200), lty=3)
> c( ucv(tperm), bcv(tperm), width.SJ(tperm) )
[1] 1.2958 1.0750 1.1200
Warning messages:
  minimum occurred at one end of the range in: ucv(tperm)
```

There have been a number of comparative studies of bandwidth selection rules (including Park & Turlach, 1992, and Cao, Cuevas & González-Manteiga, 1994) and a review by Jones, Marron & Sheather (1996). 'Second generation' rules such as `width.SJ` seem preferred and indeed to be close to optimal.

As another example, consider our dataset `galaxies` from Roeder (1990), which show evidence of at least four peaks (Figure 5.10).

```
gal <- galaxies/1000
c(width.SJ(gal, method="dpi"), width.SJ(gal))
[1] 3.2562 2.5655
plot(x=c(0, 40), y=c(0, 0.3), type="n", bty="l",
     xlab="velocity of galaxy (1000km/s)", ylab="density")
rug(gal)
lines(density(gal, width=3.25, n=200), lty=1)
lines(density(gal, width=2.56, n=200), lty=3)
median(gal)
[1] 20.834
```

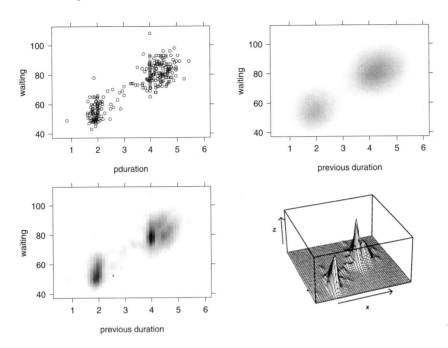

Figure 5.11: Scatter plot and two-dimensional density plots of the bivariate Old Faithful geyser data. Note the effects of the observations of duration rounded to two or four minutes.

End effects

Most density estimators will not work well when the density is non-zero at an end of its support, such as the exponential and half-normal densities. (They are designed for continuous densities and this is discontinuity.) One trick is to reflect the density and sample about the endpoint, say, a. Thus we compute the density for the sample c(x, 2a-x), and take double its density on $[a, \infty)$ (or $(-\infty, a]$ for an upper endpoint). This will impose a zero derivative on the estimated density at a, but the end effect will be much less severe. For details and further tricks see Silverman (1986, §3.10). The alternative is to modify the kernel near an endpoint (Wand & Jones, 1995, §2.1), but we know of no S implementation of such *boundary kernels*.

Two-dimensional data

It is often useful to look at densities in two dimensions. Visualizing in more dimensions is difficult, although Scott (1992) provides some examples of visualization of three-dimensional densities.

The dataset on the Old Faithful geyser has two components, the duration duration which we have studied, and waiting, the waiting time in minutes until the next eruption. There is also evidence of non-independence of the durations. S provides a function hist2d for two-dimensional histograms, but its output is too rough to be useful. We apply two-dimensional kernel analysis directly; this is most straightforward for the normal kernel aligned with axes, that is, with variance $\mathrm{diag}(h_x^2, h_y^2)$. Then the kernel estimate is

$$f(x,y) = \frac{\sum_s \phi\big((x-x_s)/h_x\big)\phi\big((y-y_s)/h_y\big)}{n\,h_x h_y}$$

which can be evaluated on a grid as XY^T where $X_{is} = \phi\big((gx_i - x_s)/h_x\big)$ and (gx_i) are the grid points, and similarly for Y. Our function kde2d implements this; the results are shown in Figures 5.11 and 5.12. We reverse the default greylevel postscript colour scheme to make black indicate high (rather than low) levels. Users of **S-PLUS 4.x** can use the 3D plots in the GUI graphics to explore the fitted density surface.

```
geyser2 <- data.frame(as.data.frame(geyser)[-1,],
  pduration=geyser$duration[-299])
attach(geyser2)
xyplot(waiting ~ pduration, xlim=c(0.5,6), ylim=c(40,110))
f1 <- kde2d(pduration, waiting, n=50, lims=c(0.5,6,40,110))
levelplot(z ~ x*y, con2tr(f1),
  xlab="previous duration", ylab="waiting",
  at = seq(0, 0.07, 0.001), colorkey=F,
  col.regions = rev(trellis.par.get("regions")$col))
f2 <- kde2d(pduration, waiting, n=50, lims=c(0.5,6,40,110),
  h = c(width.SJ(duration), width.SJ(waiting)) )
levelplot(z ~ x*y, con2tr(f2),
  xlab="previous duration", ylab="waiting",
  at = seq(0, 0.07, 0.001), colorkey=F,
  col.regions = rev(trellis.par.get("regions")$col))
wireframe(z ~ x*y, con2tr(f2),
  aspect = c(1, 0.5), screen=list(z=20, x=-60), zoom=1.2)

xyplot(duration ~ pduration, xlim=c(0.5, 6),
    ylim=c(0.5, 6),xlab="previous duration", ylab="duration")
f1 <- kde2d(pduration, duration,
  h=rep(1.5, 2), n=50, lims=c(0.5,6,0.5,6))
contourplot(z ~ x * y, con2tr(f1) ,xlab="previous duration",
    ylab="duration", at = c(0.05, 0.1, 0.2, 0.4) )
f1 <- kde2d(pduration, duration,
  h=rep(0.6, 2), n=50, lims=c(0.5,6,0.5,6))
contourplot(z ~ x * y, con2tr(f1) ,xlab="previous duration",
    ylab="duration", at = c(0.05, 0.1, 0.2, 0.4) )
f1 <- kde2d(pduration, duration,
  h=rep(0.4, 2), n=50, lims=c(0.5,6,0.5,6))
contourplot(z ~ x * y, con2tr(f1) ,xlab="previous duration",
    ylab="duration", at = c(0.05, 0.1, 0.2, 0.4) )
```

An alternative approach using binning and the 2D fast Fourier transform is taken by Wand's library KernSmooth available from statlib.

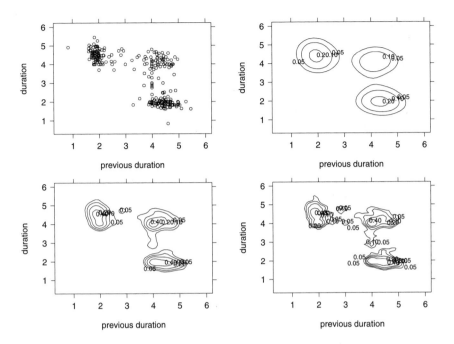

Figure 5.12: Scatter plot and two-dimensional density plots of previous and current duration for the Old Faithful geyser data. The contour plots show the effect of varying the amount of smoothing.

Alternative approaches

There are several proposals (Simonoff, 1996, pp. 67–70, 90–92) to use a univariate density estimator of the form

$$f(y) = \exp g(y; \theta) \tag{5.7}$$

for a parametric family $g(\cdot; \theta)$ of smooth functions, most often splines. The fit criterion is maximum likelihood, possibly with a smoothness penalty. The advantages of (5.7) are that it automatically provides a non-negative density estimate, and that it may be more natural to consider 'smoothness' on a relative rather than absolute scale. The library `logspline` by Charles Kooperberg implements one variant on this theme by Kooperberg & Stone (1992). This is discussed in our on-line complements;[6] see Figures 5.10 and 5.13 for a preview.

Kernel density estimation can be seen as fitting a locally constant function to the data; other approaches to density estimation use local polynomials, which have the advantage of being much less sensitive to end effects. Implementations in Wand's library `KernSmooth` and Loader's library `locfit` are discussed in our on-line complements.

---

[6] See page 467 for where to obtain these.

5.7   Bootstrap and permutation methods

Several modern methods of what is often called *computer-intensive statistics* make use of extensive repeated calculations to explore the sampling distribution of a parameter estimator $\hat{\theta}$. Suppose we have a random sample $x_1, \ldots, x_n$ drawn independently from one member of a parametric family $\{F_\theta \mid \theta \in \Theta\}$ of distributions. Suppose further that $\hat{\theta} = T(x)$ is a *symmetric* function of the sample, that is, does not depend on the sample order.

The *bootstrap* procedure (Efron & Tibshirani, 1993; Davison & Hinkley, 1997) is to take m samples from x *with replacement* and to calculate $\hat{\theta}^*$ for these samples, where conventionally the asterisk is used to denote a bootstrap resample. Note that the new samples consist of an integer number of copies of each of the original data points, and so will normally have ties. Efron's idea was to assess the variability of $\hat{\theta}$ about the unknown true θ by the variability of $\hat{\theta}^*$ about $\hat{\theta}$. In particular, the bias of $\hat{\theta}$ may be estimated by the mean of $\hat{\theta}^* - \hat{\theta}$.

As an example, suppose that we needed to know the median m of the `galaxies` data. The obvious estimator is the sample median, which is $20\,833$ km/s. How accurate is this estimator? The large-sample theory says that the median is asymptotically normal with mean m and variance $1/4n\, f(m)^2$. But this depends on the unknown density at the median. We can use our best density estimators to estimate $f(m)$, but as we have seen we can find considerable bias and variability if we are unlucky enough to encounter a peak (as in a unimodal symmetric distribution). Let us try the bootstrap:

```
density(gal, n=1, from=20.833, to=20.834, width=3.2562)$y
[1] 0.13546
density(gal, n=1, from=20.833, to=20.834, width=2.5655)$y
[1] 0.13009
1/(2*sqrt(length(gal))*0.13)
[1] 0.42474
set.seed(101); m <- 1000
res <- numeric(m)
for (i in 1:m) res[i] <- median(sample(gal, replace=T))
mean(res - median(gal))
[1] 0.032258
sqrt(var(res))
[1] 0.50883
```

which took about 2.3 seconds and confirms the adequacy of the large-sample mean and variance for our example. In this example the bootstrap resampling can be avoided, for the bootstrap distribution of the median can be found analytically (Efron, 1982, Chapter 10; Staudte & Sheather, 1990, p. 84), at least for odd n.

The bootstrap distribution of $\hat{\theta}_i$ about $\hat{\theta}$ is far from normal (Figure 5.13).

```
truehist(res, h=0.1)
lines(density(res, width=width.SJ(res, method="dpi"), n=200))
```

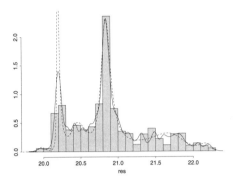

Figure 5.13: Histogram of the bootstrap distribution for the median of the `galaxies` data, with a kernel density estimate (solid) and a logspline density estimate (dashed).

```
quantile(res, p=c(0.025, 0.975))
    2.5%  97.5%
  20.175 22.053
```

In larger problems it is important to do bootstrap calculations efficiently, and there are two suites of S functions to do so. S-PLUS 4.x and 5.x have bootstrap, and library[7] boot has boot which is very comprehensive.

```
> library(boot)
> set.seed(101)
> gal.boot <- boot(gal, function(x,i) median(x[i]), R=1000)
> gal.boot

Bootstrap Statistics :
      original     bias    std. error
t1*    20.834 0.038747       0.52269

boot.ci(gal.boot, conf=c(0.90, 0.95),
            type=c("norm","basic","perc","bca"))

Intervals :
Level       Normal                Basic
90%   (19.94, 21.65 )    (19.78, 21.48 )
95%   (19.77, 21.82 )    (19.59, 21.50 )

Level       Percentile            BCa
90%   (20.19, 21.89 )    (20.18, 21.87 )
95%   (20.17, 22.07 )    (20.14, 21.96 )

> plot(gal.boot)
```

The bootstrap suite of functions has fewer options:

---

[7] Associated with Davison & Hinkley (1997), written by Angelo Canty.

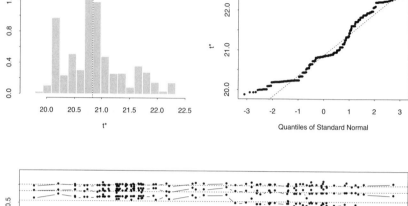

Figure 5.14: Plot for bootstrapped median from the `galaxies` data. The top row shows a histogram (with dashed line the observed value) and a Q-Q plot of the bootstrapped samples. The bottom plot is from `jack.after.boot` and displays the effects of individual observations (here negligible).

```
> gal.bt <- bootstrap(gal, median, seed=101, B=1000)
> summary(gal.bt)
      ....
Summary Statistics:
        Observed    Bias   Mean      SE
median      20.83 0.03226 20.87 0.5088

Empirical Percentiles:
          2.5%     5%   95% 97.5%
median 20.18 20.19 21.87 22.05

BCa Confidence Limits:
          2.5%     5%   95% 97.5%
median 20.17 20.19 21.81 21.97
> limits.emp(gal.bt)
           2.5%      5%     95%   97.5%
median 20.175 20.188 21.867 22.053
> limits.bca(gal.bt)
```

```
            2.5%     5%      95%  97.5%
     median 20.174 20.186 21.814 21.965
     > plot(gal.bt)
     > qqnorm(gal.bt)
```

One approach to a confidence interval for the parameter θ is to use the quantiles of the bootstrap distributions; this is termed the *percentile confidence interval* and was the original approach suggested by Efron. The bootstrap distribution here is quite asymmetric, and the intervals based on normality are not adequate. The 'basic' intervals are based on the idea that the distribution of $\hat{\theta}_i - \hat{\theta}$ mimics that of $\hat{\theta} - \theta$. If this were so, we would get a $1 - \alpha$ confidence interval as

$$1 - \alpha = P(L \leqslant \hat{\theta} - \theta \leqslant U) \approx P(L \leqslant \hat{\theta}_i - \hat{\theta} \leqslant U)$$

so the interval is $(\hat{\theta} - U, \hat{\theta} - L)$ where $L + \hat{\theta}$ and $U + \hat{\theta}$ are the $\alpha/2$ and $1 - \alpha/2$ points of the bootstrap distribution, say $k_{\alpha/2}$ and $k_{1-\alpha/2}$. Then the basic bootstrap interval is

$$(\hat{\theta} - U, \hat{\theta} - L) = \left(\hat{\theta} - [k_{1-\alpha/2} - \hat{\theta}], \hat{\theta} - [k_{\alpha/2} - \hat{\theta}]\right) = (2\hat{\theta} - k_{1-\alpha/2}, 2\hat{\theta} - k_{\alpha/2})$$

which is the percentile interval reflected about the estimate $\hat{\theta}$. In asymmetric problems the basic and percentile intervals will differ considerably (as here), and the basic intervals seem more rational.

The BC_a intervals are an attempt to shift and scale the percentile intervals to compensate for their biases, apparently unsuccessfully in this example. The idea is that if for some unknown increasing transformation g we had $g(\hat{\theta}) - g(\theta) \sim F_0$ for a symmetric distribution F_0, the percentile intervals would be exact. Suppose rather that if $\phi = g(\theta)$,

$$g(\hat{\theta}) - g(\theta) \sim N\left(w\,\sigma(\phi), \sigma^2(\phi)\right) \qquad \text{with } \sigma(\phi) = 1 + a\,\phi$$

Standard calculations (Davison & Hinkley, 1997, p. 204) show that the α confidence limit is given by the $\hat{\alpha}$ percentile of the bootstrap distribution, where

$$\hat{\alpha} = \Phi\left(w + \frac{w + z_\alpha}{1 - a(w + z_\alpha)}\right)$$

and a and w are estimated from the bootstrap samples.

The median (and other sample quantiles) is appreciably affected by discreteness, and it may well be better to sample from a density estimate rather than from the empirical distribution. This is known as the *smoothed bootstrap*. We can do that neatly with boot, using a Gaussian bandwidth of 2 and so standard deviation 0.5.

```
     sim.gen  <- function(data, mle) {
       n <- length(data)
       data[sample(n, replace=T)]  + mle*rnorm(n)
     }
```

```
gal.boot2 <- boot(gal, median, R=1000,
   sim="parametric", ran.gen=sim.gen, mle=0.5)
boot.ci(gal.boot2, conf=c(0.90, 0.95),
          type=c("norm","basic","perc"))
Intervals :
Level       Normal            Basic            Percentile
90%   (20.03, 21.46 )   (19.95, 21.36 )   (20.30, 21.71 )
95%   (19.89, 21.59 )   (19.79, 21.44 )   (20.22, 21.87 )
```

The constants a and w in the BC_a interval cannot be estimated by boot in this case. The smoothed bootstrap slightly inflates the sample variance (here by 1.2%) and we could rescale the sample if this was appropriate.

For a smaller but simpler example, we return to the differences in shoe wear between materials. There is a fifth type of confidence interval that boot.ci can calculate, which needs a variance v^* estimate of the statistic $\hat{\theta}^*$ from each bootstrap sample. Then the confidence interval can be based on the basic confidence intervals for the *studentized* statistics $(\hat{\theta}^* - \hat{\theta})/\sqrt{v^*}$. Theory suggests that the studentized confidence interval may be the most reliable of all the methods we have discussed.

```
> attach(shoes)
> t.test(B - A)
95 percent confidence interval:
 0.1330461 0.6869539
> shoes.boot <- boot(B-A, function(x,i) mean(x[i]), R=1000)
> boot.ci(shoes.boot, type = c("norm", "basic", "perc", "bca"))

Intervals :
Level       Normal                Basic
95%   ( 0.1861,  0.6437 )   ( 0.1800,  0.6497 )

Level       Percentile            BCa
95%   ( 0.1703,  0.6400 )   ( 0.2100,  0.6518 )

mean.fun <- function(d, i) {
  n <- length(i)
  c(mean(d[i]), (n-1)*var(d[i])/n^2)
}
> shoes.boot2 <- boot(B-A, mean.fun, R=1000)
> boot.ci(shoes.boot2, type = "stud")

Intervals :
Level       Studentized
95%   ( 0.1384,  0.7179 )
```

We have only scratched the surface of bootstrap methods: Davison & Hinkley (1997) provide an excellent practically-oriented account, S software and practical exercises.

Permutation tests

Inference for designed experiments is often based on the distribution over the random choices made during the experimental design, on the belief that this randomness alone will give distributions which can be approximated well by those derived from normal-theory methods. There is considerable empirical evidence that this is so, but with modern computing power we can check it for our own experiment, by selecting a large number of re-labellings of our data and computing the test statistics for the re-labelled experiments.

Consider again the shoe-wear data of Section 5.4. The most obvious way to explore the permutation distribution of the t-test of d = B-A is to select random permutations. The supplied function t.test computes much more than we need. In this case it is simple to write a replacement function to do exactly what we need:

```
attach(shoes)
d <- B - A
ttest <- function(x) mean(x)/sqrt(var(x)/length(x))
n <- 1000
res <- numeric(n)
for(i in 1:n) res[i] <- ttest(x <- d*sign(runif(10)-0.5))
```

which takes around seven seconds on our PC.

As the permutation distribution has only $2^{10} = 1\,024$ points we can explore it directly for Figure 5.5:

```
perm.t.test <- function(d) {
    binary.v <- function(x, digits) {
        if(missing(digits)) {
            mx <- max(x)
            digits <- if(mx > 0) 1 + floor(log(mx,2)) else 1
        }
        ans <- 0:(digits - 1); lx <- length(x)
        x <- rep(x, rep(digits, lx))
        x <- (x %/% 2^ans) %% 2
        dim(x) <- c(digits, lx)
        x
    }
    digits <- length(d)
    n <- 2^digits; x <- d * 2 * (binary.v(1:n, digits) - 0.5)
    mx <- matrix(1/digits, 1, digits) %*% x
    s <- matrix(1/(digits - 1), 1, digits)
    vx <- s %*% (x - matrix(mx, digits, n, byrow=T))^2
    as.vector(mx/sqrt(vx/digits))
}
tperm <- perm.t.test(B - A)
```

As the t-test is the ratio of linear operations, we can do many at once via matrix operations, taking only 0.3 sec! The speed-up in this example is unusually large,[8] but it is always worth looking for simplifying ideas that allow a large number of calculations to be done by one function.

---

[8] We are grateful to Bill Dunlap of MathSoft DAPD for his suggestions for this problem.

5.8   Exercises

5.1.   Rice (1995, p. 390) gives the following data (Natrella, 1963) on the latent heat of the fusion of ice (*cal/gm*):

```
Method A: 79.98 80.04 80.02 80.04 80.03 80.03 80.04 79.97
          80.05 80.03 80.02 80.00 80.02
Method B: 80.02 79.94 79.98 79.97 79.97 80.03 79.95 79.97
```

(a) Assuming normality, test the hypothesis of equal means, both with and without making the assumption of equal variances.

Compare the result with a Wilcoxon/Mann–Whitney nonparametric two-sample test.

(b) Inspect the data graphically in various ways, for example, boxplots, Q-Q plots and histograms.

(c) Fit a one-way analysis of variance and compare it with your *t*-test. (Look ahead to the next chapter, or investigate function oneway.)

5.2.   Write functions to produce Q-Q plots for a gamma and a Weibull distribution. Note that unlike the normal Q-Q plot, the shape parameters may need to be estimated.

5.3.   The ideas used in bandwidth selection for kernel density estimation that are implemented in width.SJ can also be applied to the choice of bin width in a histogram (Wand, 1997). Implement such a bin-width estimator in S.

5.4.   Experiment with dataset galaxies. How many modes do you think there are in the underlying density?

Chapter 6

Linear Statistical Models

Linear models form the core of classical statistics and are still the basis of much of statistical practice; many modern modelling and analytical techniques build on the methodology developed for linear models.

In S most modelling exercises are conducted in a fairly standard way. The data set is usually held in a single *data frame* object. A primary model is fitted using a *model fitting function*, for which a *formula* specifying the form of the model and the data frame specifying the variables to be used are the basic arguments. The resulting *fitted model object* can be interrogated, analysed and even modified in various ways using generic functions. The important point to note is that the fitted model object carries with it the information that the fitting process has revealed.

Although most modelling exercises conform to this rough paradigm some features of linear models are special. The *formula* for a linear model specifies the response variable and the explanatory variables (or factors) used to model the mean response by a version of the Wilkinson–Rogers notation (Wilkinson & Rogers, 1973) for specifying models which we discuss in Section 6.2.

We begin with an example to give a feel for the process and to present some of the details.

6.1 An analysis of covariance example

The data frame `whiteside` contains a data set collected in the 1960s by Mr Derek Whiteside of the UK Building Research Station and reported in the collection of small data sets edited by Hand *et al.* (1993, No. 88, p. 69). Whiteside recorded the weekly gas consumption and (purportedly) average external temperature at his own house in south-east England for 26 weeks before, and 30 weeks after cavity-wall insulation was installed. The object of the exercise was to assess the effect of the insulation on gas consumption.

The variables in data frame `whiteside` are `Insul`, a factor with levels `Before` and `After`, `Temp`, for the weekly average external temperature in degrees Celsius and `Gas`, the weekly gas consumption in 1 000 cubic feet units. We begin by plotting the data in two panels showing separate least-squares lines.

149

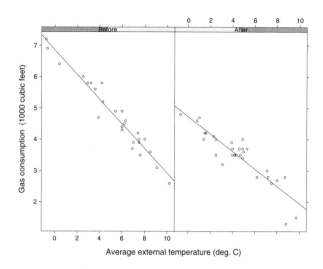

Figure 6.1: Whiteside's data showing the effect of insulation on household gas consumption.

```
xyplot(Gas ~ Temp | Insul, whiteside, panel =
  function(x, y, ...) {
    panel.xyplot(x, y, ...)
    panel.lmline(x, y, ...)
  }, xlab = "Average external temperature (deg. C)",
  ylab = "Gas consumption  (1000 cubic feet)")
```

The result is shown in Figure 6.1. Within the range of temperatures given a straight line model appears to be adequate. The plot shows that insulation reduces the gas consumption for equal external temperatures, but it also appears to affect the slope, that is, the rate at which gas consumption increases as external temperature falls.

To explore these issues quantitatively we will need to fit linear models, the primary function for which is lm. The main arguments to lm are

```
lm(formula, data, weights, subset, na.action)
```

where

        formula is the model formula (the only required argument),

        data is an optional data frame,

        weights is a vector of positive weights, if non-uniform weights are needed,

        subset is an index vector specifying a subset of the data to be used (by default all items are used),

        na.action is a function specifying how missing values are to be handled (by default, missing values are not allowed).

If the argument `data` is specified it gives a data frame from which variables are selected ahead of the search path. Working with data frames and using this argument is strongly recommended.

It should be noted that setting `na.action=na.omit` will allow models to be fitted omitting cases that have missing components on a required variable. If any cases are omitted the fitted values and residual vector will no longer match the original observation vector in length.

Formulae have been discussed in outline in Section 2.5 on page 33. For `lm` the right-hand side specifies the explanatory variables. Operators on the right-hand side of linear model formulae have the special meaning of the Wilkinson–Rogers notation and not their arithmetical meaning, as we show.

To fit the separate regressions of gas consumption on temperature as shown in Figure 6.1 we may use

```
gasB <- lm(Gas ~ Temp, whiteside, subset = Insul=="Before")
gasA <- update(gasB, subset = Insul=="After")
```

The first line fits a simple linear regression for the 'before' temperatures. The right-hand side of the formula needs only to specify the variable `Temp` since an intercept term (corresponding to a column of unities of the model matrix) is always implicitly included. It may be explicitly included using `1 + Temp`, where the `+` operator implies *inclusion* of a term in the model, not addition.

The function `update` is a convenient way to modify a fitted model. Its first argument is a fitted model object that results from one of the model fitting functions such as `lm`. The remaining arguments of `update` specify the desired changes to arguments of the call that generated the object. In this case we simply wish to switch subsets from `Insul=="Before"` to `Insul=="After"`; the formula and data frame remain the same. Notice that variables used in the `subset` argument may also come from the data frame and need not be visible on the (global) search path.

Fitted model objects are *lists* of an appropriate *class*, in this case class `"lm"`. Generic functions to perform further operations on the object include

> `print` for a simple display,
>
> `summary` for a conventional regression analysis output,
>
> `coef` (or `coefficients`) for extracting the regression coefficients,
>
> `resid` (or `residuals`) for residuals,
>
> `fitted` (or `fitted.values`) for fitted values,
>
> `deviance` for the residual sum of squares,
>
> `anova` for a sequential analysis of variance table, or a comparison of several hierarchical models,
>
> `predict` for predicting means for new data, optionally with standard errors, and
>
> `plot` for diagnostic plots.

Many of these method functions are very simple, merely extracting a component of the fitted model object. The only component likely to be accessed for which no extractor function is supplied is df.residual, the residual degrees of freedom.

The output from summary is self-explanatory. Edited results for our fitted models are

```
> summary(gasB)
    ....
Coefficients:
              Value Std. Error t value Pr(>|t|)
(Intercept)   6.854    0.118    57.876    0.000
       Temp  -0.393    0.020   -20.078    0.000

Residual standard error: 0.281 on 24 degrees of freedom

> summary(gasA)
    ....
Coefficients:
              Value Std. Error t value Pr(>|t|)
(Intercept)   4.724    0.130    36.410    0.000
       Temp  -0.278    0.025   -11.036    0.000

Residual standard error: 0.355 on 28 degrees of freedom
```

The difference in residual variances is relatively small, but the formal textbook F-test for equality of variances could easily be done. The sample variances could be extracted in at least two ways, for example

```
varB <- deviance(gasB)/gasB$df.resid    # direct calculation
varB <- summary(gasB)$sigma^2           # alternative
```

It is known this F-test is highly non-robust to non-normality (see, for example, Hampel *et al.* (1986, pp. 55, 188)) so its usefulness here would be doubtful.

To fit both regression models in the same "lm" model object we may use

```
> gasBA <- lm(Gas ~ Insul/Temp - 1, whiteside)
> summary(gasBA)
    ....
Coefficients:
                  Value Std. Error t value Pr(>|t|)
   InsulBefore    6.854    0.136    50.409    0.000
    InsulAfter    4.724    0.118    40.000    0.000
InsulBeforeTemp  -0.393    0.022   -17.487    0.000
 InsulAfterTemp  -0.278    0.023   -12.124    0.000

Residual standard error: 0.323 on 52 degrees of freedom
    ....
```

Notice that the estimates are the same but the standard errors are different because they are now based on the pooled estimate of variance.

Terms of the form a/x, where a is a factor, are best thought of as "separate regression models of type 1 + x within the levels of a." In this case an intercept is not needed, since it is replaced by two separate intercepts for the two levels of insulation, and the formula term - 1 removes it.

We can check for curvature in the mean function by fitting separate quadratic rather than linear regressions in the two groups. This may be done as

```
> gasQ <- lm(Gas ~ Insul/(Temp + I(Temp^2)) - 1, whiteside)
> summary(gasQ)$coef
                       Value Std. Error  t value Pr(>|t|)
        InsulBefore   6.7592     0.1508  44.8263   0.0000
         InsulAfter   4.4964     0.1607  27.9855   0.0000
    InsulBeforeTemp  -0.3177     0.0630  -5.0450   0.0000
     InsulAfterTemp  -0.1379     0.0731  -1.8876   0.0649
InsulBeforeI(Temp^2) -0.0085     0.0066  -1.2789   0.2068
 InsulAfterI(Temp^2) -0.0150     0.0074  -2.0114   0.0497
```

The 'identity' function I(...) is used almost exclusively in this context. It evaluates its argument with operators having their arithmetical meaning and returns the result. Hence it allows arithmetic operators to be used in linear model formulae, although if any function call is used in a formula their arguments are evaluated in this way.

The separate regression coefficients show that a second degree term is possibly needed for the After group only, but the evidence is not overwhelming.[1] We retain the separate linear regressions model on the grounds of simplicity.

An even simpler model that might be considered is one with parallel regressions. We can fit this model and test it within the separate regression model using

```
> gasPR <- lm(Gas ~ Insul+Temp, whiteside)
> anova(gasPR, gasBA)
Analysis of Variance Table
    ....
                         Terms Resid. Df    RSS  Test Df
1                 Insul + Temp       53 6.7704
2 Insul + Temp %in% Insul - 1       52 5.4252 1 vs. 2  1

  Sum of Sq F Value      Pr(F)
1
2    1.3451  12.893 0.00073069
```

When anova is used with two or more nested models it gives an analysis of variance table for those models. In this case it shows that separate slopes are indeed necessary. Note the unusual layout of the analysis of variance table. Here we could conduct this test in a simpler and more informative way. We now fit the same model using a different parametrization:

---

[1] Notice that when the quadratic terms are present first degree coefficients mean 'the slope of the curve at temperature zero', so a non-significant value does not mean that the linear term is not needed. Removing the non-significant linear term for the 'after' group, for example, would be unjustified.

```
> options(contrasts = c("contr.treatment", "contr.poly"))
> gasBA1 <- lm(Gas ~ Insul*Temp, whiteside)
> summary(gasBA1)$coef
                Value Std. Error  t value    Pr(>|t|)
(Intercept)  6.85383   0.135964  50.4091 0.0000e+00
      Insul -2.12998   0.180092 -11.8272 2.2204e-16
       Temp -0.39324   0.022487 -17.4874 0.0000e+00
 Insul:Temp  0.11530   0.032112   3.5907 7.3069e-04
```

The call to `options` is explained more fully in Section 6.2; for now we note that it affects the way regression models are parametrized when factors are used. The formula `Insul*Temp` expands to `1 + Insul + Temp + Insul:Temp` and the corresponding coefficients are, in order, the intercept for the 'before' group, the *difference* in intercepts, the slope for the 'before' group and the *difference* in slopes. Since this last term is significant we conclude that the two separate slopes are required in the model. Indeed note that the F-statistic in the analysis of variance table is the square of the final t-statistic and that the tail areas are identical.

6.2 Model formulae and model matrices

This section contains some rather technical material and might be skimmed at first reading.

A linear model is specified by the response vector y and by the matrix of explanatory variables, or *model matrix*, X. The model formula conveys both pieces of information, the left-hand side providing the response and the right-hand side instructions on how to generate the model matrix according to a particular convention.

A multiple regression with three quantitative determining variables might be specified as `y ~ x1+x2+x3`. This would correspond to a model with a familiar algebraic specification

$$y_i = \beta_0 + \beta_1 x_{i1} + \beta_2 x_{i2} + \beta_3 x_{i3} + \epsilon_i, \qquad i = 1, 2, \ldots, n$$

The model matrix has the partitioned form

$$X = \begin{bmatrix} 1 & x_1 & x_2 & x_3 \end{bmatrix}$$

The intercept term (β_0 corresponding to the leading column of ones in X) is implicitly present; its presence may be confirmed by giving a formula such as `y ~ 1 + x1 + x2 + x3`, but wherever the 1 occurs in the formula the column of ones will always be the first column of the model matrix. It may be omitted and a regression through the origin fitted by giving a `- 1` term in the formula, as in `y ~ x1 + x2 + x3 - 1`.

Factor terms in a model formula are used to specify classifications leading to what are often called analysis of variance models. Suppose `a` is a factor. An analysis of variance model for the one-way layout defined by a might be written in the algebraic form

$$y_{ij} = \mu + \alpha_j + \epsilon_{ij} \qquad i = 1, 2, \ldots, n_j; \quad j = 1, 2, \ldots, k$$

where there are k classes and the n_j is the size of the jth. Let $n = \sum_j n_j$. This specification is over-parametrized, but we could write the model matrix in the form

$$X = \begin{bmatrix} \mathbf{1} & X_{\mathrm{a}} \end{bmatrix}$$

where X_{a} is an $n \times k$ binary incidence (or 'dummy variable') matrix where each row has a single unity in the column of the class to which it belongs.

The redundancy comes from the fact that the columns of X_{a} add to $\mathbf{1}$, making X of rank k rather than $k + 1$. One way to resolve the redundancy is to remove the column of ones. This amounts to setting $\mu = 0$, leading to an algebraic specification of the form

$$y_{ij} = \alpha_j + \epsilon_{ij} \qquad i = 1, 2, \ldots, n_j; \ j = 1, 2, \ldots, k$$

so the α_j parameters are the class means. This formulation may be specified by y ~ a - 1.

If we do not break the redundancy by removing the intercept term it must be done some other way, since otherwise the parameters are not identifiable. The way this is done in S is most easily described in terms of the model matrix. The model matrix generated has the form

$$X^{\star} = \begin{bmatrix} \mathbf{1} & X_{\mathrm{a}}C_{\mathrm{a}} \end{bmatrix}$$

where C_{a}, the *contrast matrix* for a, is a $k \times (k - 1)$ matrix chosen so that $X^{\star}$ has rank k, the number of columns. A necessary (and usually sufficient) condition for this to be the case is that the square matrix $\begin{bmatrix} \mathbf{1} & C_{\mathrm{a}} \end{bmatrix}$ be non-singular.

The reduced model matrix $X^{\star}$ in turn defines a linear model, but the parameters are often not directly interpretable and an algebraic formulation of the precise model may be difficult to write down. Nevertheless, the relationship between the newly defined and original (redundant) parameters is clearly given by

$$\alpha = C_{\mathrm{a}}\alpha^{\star} \tag{6.1}$$

where α are the original α parameters and $\alpha^{\star}$ are the new.

If c_{a} is a non-zero vector such that $c_{\mathrm{a}}^T C_{\mathrm{a}} = \mathbf{0}$ it can be seen immediately that using $\alpha^{\star}$ as parameters amounts to estimating the original parameters, α subject to the *identification constraint* $c_{\mathrm{a}}^T \alpha = 0$ which is usually sufficient to make them unique. Such a vector (or matrix) c_{a} is called an *annihilator* of C_{a} or a basis for the orthogonal complement of the range of C_{a}.

If we fit the one-way layout model using the formula

y ~ a

the coefficients we obtain will be estimates of μ and α^*. The corresponding
constrained estimates of the α may be obtained by multiplying by the contrasts
matrix or by using the function dummy.coef. Consider an artificial example.

```
> dat <- data.frame(a = factor(rep(1:3, 3)),
                    y = rnorm(9, rep(2:4, 3), 0.1))
> obj <- lm(y ~ a, dat)
> alf.star <- coef(obj)
> alf.star
  (Intercept)        a1        a2
       2.9719 0.51452 0.49808
> Ca <- contrasts(dat$a)        # contrast matrix for 'a'
> alf <- drop(Ca %*% alf.star[-1])
> alf
        1        2        3
  -1.0126 0.016443 0.99615
> dummy.coef(obj)
$"(Intercept)":
[1] 2.9719

$a:
        1        2        3
  -1.0126 0.016443 0.99615
```

Notice that the estimates of α sum to zero because the contrast matrix used here
implies the identification constraint $\mathbf{1}^T\alpha = 0$.

Contrast matrices

By default S uses so-called *Helmert* contrast matrices for unordered factors and
orthogonal polynomial contrast matrices for ordered factors. The forms of these
can be deduced from the following artificial example.

```
> N <- factor(Nlevs <- c(0,1,2,4))
> contrasts(N)
   [,1] [,2] [,3]
0   -1   -1   -1
1    1   -1   -1
2    0    2   -1
4    0    0    3
> contrasts(ordered(N))
        .L      .Q        .C
0 -0.67082   0.5 -0.22361
1 -0.22361  -0.5  0.67082
2  0.22361  -0.5 -0.67082
4  0.67082   0.5  0.22361
```

For the poly contrasts it can be seen that the corresponding parameters α^* can
be interpreted as the coefficients in an orthogonal polynomial model of degree
$r - 1$, *provided* the ordered levels are equally spaced (which is not the case for

the example) *and* the class sizes are equal. The $\alpha^\star$ parameters corresponding to the Helmert contrasts also have an easy interpretation, as we see in the following. Since both the Helmert and polynomial contrast matrices satisfy $\mathbf{1}^T C = \mathbf{0}$ the implied constraint on α will be $\mathbf{1}^T \alpha = 0$ in both cases.

The default contrast matrices can be changed by resetting the `contrasts` option. This is a character vector of length two giving the names of the functions that generate the contrast matrices for unordered and ordered factors respectively. For example,

```
options(contrasts=c("contr.treatment", "contr.poly"))
```

sets the default contrast matrix function for factors to `contr.treatment` and for ordered factors to `contr.poly` (the original default). Four supplied contrast functions are as follows.

`contr.helmert`   for the Helmert contrasts.

`contr.treatment`   for contrasts such that each coefficient represents a comparison of that level with level 1 (omitting level 1 itself). This corresponds to the constraint $\alpha_1 = 0$.

`contr.sum`   where the coefficients are constrained to add to zero; that is, in this case the components of $\alpha^\star$ are the same as the first $r - 1$ components of α, with the latter constrained to add to zero.

`contr.poly`   for the equally spaced, equally replicated orthogonal polynomial contrasts.

Others can be written using these as templates (as we do with our function `contr.sdif`). We recommend the use of the treatment contrasts for unbalanced layouts, including generalized linear models and survival models, because the unconstrained coefficients obtained directly from the fit are then easy to interpret.

Notice that the `helmert`, `sum` and `poly` contrasts ensure the rank condition on C is met by choosing C so that the columns of $[\mathbf{1}\,C]$ are mutually orthogonal whereas the `treatment` contrasts choose C so that $[\mathbf{1}\,C]$ is in echelon form.

Contrast matrices for particular factors may also be set as an attribute of the factor itself. This can be done either by the `contrasts` replacement function or by using the function `C` which takes three arguments, the factor, the matrix from which contrasts are to be taken (or the abbreviated name of a function that will generate such a matrix) and the number of contrasts. On some occasions a p-level factor may be given a contrast matrix with fewer than $p - 1$ columns, in which case it contributes fewer than $p-1$ degrees of freedom to the model, or the unreduced parameters α have additional constraints placed on them apart from the one needed for identification. An alternative method is to use the replacement form with a specific number of contrasts as the second argument. For example, suppose we wish to create a factor `N2` that would generate orthogonal linear and quadratic polynomial terms, only. Two equivalent ways of doing this would be

```
> N2 <- N
> contrasts(N2, 2) <- poly(Nlevs, 2)
> N2 <- C(N, poly(Nlevs, 2), 2)          # alternative
> contrasts(N2)
            1          2
0  -0.591608   0.56408
1  -0.253546  -0.32233
2   0.084515  -0.64466
4   0.760639   0.40291
```

In this case the constraints imposed on the α parameters are not merely for identification but actually change the model subspace.

Parameter interpretation

The `poly` contrast matrices lead to $\alpha^\star$ parameters that are sometimes interpretable as coefficients in an orthogonal polynomial regression. The `treatment`
contrasts set $\alpha_1 = 0$ and choose the remaining αs as the $\alpha^\star$s. Other cases are
often not so direct, but an interpretation is possible.

To interpret the $\alpha^\star$ parameters in general, consider the relationship (6.1).
Since the contrast matrix C is of full column rank it has a unique left inverse
C^+, so we can reverse this relationship to give

$$\alpha^\star = C^+\alpha \quad \text{where} \quad C^+ = (C^TC)^{-1}C^T \tag{6.2}$$

The pattern in the matrix C^+ then provides an interpretation of each unconstrained parameter as a linear function of the (usually) readily appreciated constrained parameters. For example, consider the Helmert contrasts for $r = 4$. To
exhibit the pattern in C^+ more clearly we use the function `fractions` from the
MASS library for rational approximation and display.

```
> fractions(ginv(contr.helmert(n = 4)))
             1       2      3      4
[1,]      -1/2    1/2      0      0
[2,]      -1/6   -1/6    1/3      0
[3,]     -1/12  -1/12  -1/12    1/4
```

Hence $\alpha_1^\star = \frac{1}{2}(\alpha_2 - \alpha_1)$, $\alpha_2^\star = \frac{1}{3}\{\alpha_3 - \frac{1}{2}(\alpha_1 + \alpha_2)\}$ and in general $\alpha_j^\star$ is
a comparison of α_{j+1} with the average of all preceding αs, divided by $j + 1$.
This is a comparison of the (unweighted) mean of class $j + 1$ with that of the
preceding classes.

It can sometimes be important to use contrast matrices that give a simple interpretation to the fitted coefficients. This can be done by noting that $(C^+)^+ = C$.
For example, suppose we wished to choose contrasts so that the $\alpha_j^\star = \alpha_{j+1} - \alpha_j$,
that is, the successive differences of class effects. For $r = 5$, say, the C^+ matrix
is then given by

```
> Cp <- diag(-1, 4, 5);  Cp[row(Cp) == col(Cp) - 1] <- 1
> Cp
      [,1] [,2] [,3] [,4] [,5]
[1,]   -1    1    0    0    0
[2,]    0   -1    1    0    0
[3,]    0    0   -1    1    0
[4,]    0    0    0   -1    1
```

Hence the contrast matrix to obtain these linear functions as the estimated coefficients is

```
> fractions(ginv(Cp))
      [,1] [,2] [,3] [,4]
[1,] -4/5 -3/5 -2/5 -1/5
[2,]  1/5 -3/5 -2/5 -1/5
[3,]  1/5  2/5 -2/5 -1/5
[4,]  1/5  2/5  3/5 -1/5
[5,]  1/5  2/5  3/5  4/5
```

Note that again the columns have zero sums, so the implied constraint is that the effects add to zero. (If it were not obvious we could find the induced constraint using our function `Null` to find a basis for the null space of the contrast matrix.)

The pattern is obvious from this example and a contrast matrix function for the general case can now be written. To be usable as a component of the `contrasts` option such a function has to conform with a fairly strict convention, but the key computational steps are

```
    . . . .
      contr <- col(matrix(nrow = n, ncol = n - 1))
      upper.tri <- !lower.tri(contr)
      contr[upper.tri] <- contr[upper.tri] - n
      contr/n
    . . . .
```

The complete function is supplied as `contr.sdif` in library `MASS`.

Higher-way layouts

Two- and higher-way layouts may be specified by two or more factors and formula operators. The way the model matrix is generated is then an extension of the conventions for a one-way layout.

If a and b are r- and s-level factors, respectively, the model formula y ~ a+b specifies an additive model for the two-way layout. Using the redundant specification the algebraic formulation would be

$$y_{ijk} = \mu + \alpha_i + \beta_j + \epsilon_{ijk}$$

and the model matrix would be

$$X = \begin{bmatrix} \mathbf{1} & X_a & X_b \end{bmatrix}$$

The reduced model matrix then has the form

$$X^\star = \begin{bmatrix} 1 & X_a C_a & X_b C_b \end{bmatrix}$$

However if the intercept term is explicitly removed using, say, y ~ a + b - 1, the reduced form is

$$X^\star = \begin{bmatrix} X_a & X_b C_b \end{bmatrix}$$

Note that this is asymmetric in a and b and order-dependent.

A two-way non-additive model has a redundant specification of the form

$$y_{ijk} = \mu + \alpha_i + \beta_j + \gamma_{ij} + \epsilon_{ijk}$$

The model matrix can be written as

$$X = \begin{bmatrix} 1 & X_a & X_b & X_a{:}X_b \end{bmatrix}$$

where we use the notation $A{:}B$ to denote the matrix obtained by taking each column of A and multiplying it elementwise by each column of B. In the example $X_a{:}X_b$ generates an incidence matrix for the sub-classes defined jointly by a and b. Such a model may be specified by the formula

y ~ a + b + a:b

or equivalently by y ~ a*b. The reduced form of the model matrix is then

$$X^\star = \begin{bmatrix} 1 & X_a C_a & X_b C_b & (X_a C_a){:}(X_b C_b) \end{bmatrix}$$

It may be seen that $(X_a C_a){:}(X_b C_b) = (X_a{:}X_b)(C_b \otimes C_a)$, where $\otimes$ denotes the Kronecker product, so the relationship between the γ parameters and the corresponding $\gamma^\star$s is

$$\gamma = (C_b \otimes C_a)\gamma^\star$$

The identification constraints can be most easily specified by writing γ as an $r \times s$ matrix. If this is done the relationship has the form $\gamma = C_a \gamma^\star C_b^T$ and the constraints have the form $c_a^T \gamma = 0^T$ and $\gamma c_b = 0$, separately.

If the intercept term is removed, however, such as using y ~ a*b -1 the form is different, namely,

$$X^\star = \begin{bmatrix} X_a & X_b C_b & X_a^{(-r)}{:}(X_b C_b) \end{bmatrix}$$

where (somewhat confusingly) $X_a^{(-r)}$ is a matrix obtained by removing the *last* column of X_a. Furthermore, if a model specified as y ~ - 1 + a + a:b the model matrix generated is $\begin{bmatrix} X_a & X_a{:}(X_b C_b) \end{bmatrix}$. In general addition of a term a:b extends the previously constructed design matrix to a complete non-additive model in some non-redundant way (unless the design is deficient, of course).

Even though a*b expands to a + b + a:b it should be noted that a + a:b is not always the same as a*b - b or even a + b - b + a:b. When used in model-fitting functions the last two formulae construct the design matrix for a*b and only then remove any columns corresponding to the b term. (The result is not a statistically meaningful model.) Model matrices are constructed within the fitting functions by arranging the positive terms in order of their complexity, sequentially adding columns to the model matrix according to the redundancy resolution rules and then removing any generated columns corresponding to negative terms. The exception to this rule is the intercept term which is always removed initially. (With update, however, the formula is expanded and all negative terms are removed *before* the model matrix is constructed.)

The model a + a:b generates the same matrix as a/b, which expands to a + b %in% a. There is no compelling reason for the additional operator, %in%, but it does serve to emphasize that the slash operator should be thought of as specifying separate submodels of the form 1 + b for each level of a. The operator behaves like the colon formula operator when the second main effect term is not given, but is conventionally reserved for nested models.

Star products of more than two terms, such as a*b*c, may be thought of as expanding (1 + a):(1 + b):(1 + c) according to ordinary algebraic rules and may be used to define higher-way non-additive layouts. There is also a power operator, ^, for generating models up to a specified degree of interaction term. For example (a+b+c)^3 generates the same model as a*b*c but (a+b+c)^2 has the highest order interaction absent.

Combinations of factors and non-factors with formula operators are useful in an obvious way. We have seen already that a/x - 1 generates separate simple linear regressions on x within the levels of a. The same model may be specified as a + a:x - 1, whereas a*x generates an equivalent model using a different resolution of the redundancy. It should be noted that (x + y + z)^3 does *not* generate a general third-degree polynomial regression in the three variables, as might be expected. This is because terms of the form x:x are regarded as the same as x, not as I(x^2). However, a single term such as x^2 is silently promoted to I(x^2) and interpreted as a power.

6.3 Regression diagnostics

The message in the Whiteside example is relatively easy to discover and we did not have to work hard to find an adequate linear model. There is an extensive literature (for example Atkinson, 1985) on examining the fit of linear models to consider whether one or more points are not fitted as well as they should be or have undue influence on the fitting of the model. This can be contrasted with the robust regression methods we discuss in Section 6.5, which automatically take account of anomalous points.

The basic tool for examining the fit is the residuals, and we have already looked for patterns in residuals and assessed the normality of their distribution. The residuals are not independent (they sum to zero if an intercept is present) and they do not have the same variance. Indeed, their variance-covariance matrix is

$$\text{var}\,(e) = \sigma^2[I - H] \tag{6.3}$$

where $H = X(X^T X)^{-1} X^T$ is the orthogonal projector matrix onto the model space, or *hat* matrix. If a diagonal entry h_{ii} of H is large, changing y_i will move the fitted surface appreciably towards the altered value. For this reason h_{ii} is said to measure the *leverage* of the observation y_i. The trace of H is p, the dimension of the model space, so 'large' is taken to be greater than two or three times the average, p/n.

Having large leverage has two consequences for the corresponding residual. First, its variance will be lower than average from (6.3). We can compensate for this by rescaling the residuals to have unit variance. The *standardized residuals* are

$$e_i' = \frac{e_i}{s\sqrt{1 - h_{ii}}}$$

where as usual we have estimated σ^2 by s^2, the residual mean square. Second, if one error is very large, the variance estimate s^2 will be too large, and this deflates all the standardized residuals. Let us consider fitting the model omitting observation i. We then get a prediction for the omitted observation, $\hat{y}_{(i)}$, and an estimate of the error variance, $s_{(i)}^2$, from the reduced sample. The *studentized residuals* are

$$e_i^* = \frac{y_i - \hat{y}_{(i)}}{\sqrt{\text{var}\left(y_i - \hat{y}_{(i)}\right)}}$$

but with σ replaced by $s_{(i)}$. Fortunately, it is not necessary to re-fit the model each time an observation is omitted, since it can be shown that

$$e_i^* = e_i' \left/ \left[\frac{n - p - e_i'^2}{n - p - 1}\right]^{1/2}\right.$$

Notice that this implies that the standardized residuals, e_i, must be bounded by $\pm\sqrt{n - p}$.

The terminology used here is not universally adopted; in particular studentized residuals are sometimes called *jackknifed* residuals.

It is usually better to compare studentized residuals rather than residuals; in particular we recommend that they be used for normal probability plots.

There are no system functions to compute studentized or standardized residuals, but we have provided functions `studres` and `stdres`. There is a function `hat`, but this expects the model matrix as its argument. (There is a useful function, `lm.influence`, for most of the fundamental calculations. The diagonal of the hat matrix can be obtained by `lm.influence(lmobject)$hat`.)

Scottish hill races

As an example of regression diagnostics, let us return to the data on 35 Scottish hill races in our data frame `hills` considered in Chapter 1. The data come from Atkinson (1986) and are discussed further in Atkinson (1988) and Staudte & Sheather (1990). The columns are the overall race distance, the total height climbed and the record time. In Chapter 1 we considered a regression of `time` on `dist`. We can now include `climb`:

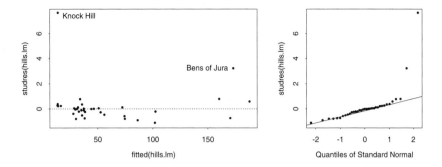

Figure 6.2: Diagnostic plots for Scottish hills data, unweighted model.

```
> hills.lm <- lm(time ~ dist + climb, hills)
> hills.lm
      ....
Coefficients:
 (Intercept)  dist    climb
      -8.992 6.218 0.011048

Degrees of freedom: 35 total; 32 residual
Residual standard error: 14.676
> frame(); par(fig=c(0, 0.6, 0, 0.55))
> plot(fitted(hills.lm), studres(hills.lm))
> abline(h=0, lty=2)
> identify(fitted(hills.lm), studres(hills.lm),
     row.names(hills))
> par(fig=c(0.6, 1, 0, 0.55), pty="s")
> qqnorm(studres(hills.lm))
> qqline(studres(hills.lm))
> hills.hat <- lm.influence(hills.lm)$hat
> cbind(hills, lev=hills.hat)[hills.hat > 3/35, ]
               dist climb    time     lev
 Bens of Jura    16  7500 204.617 0.42043
 Lairig Ghru     28  2100 192.667 0.68982
   Ben Nevis     10  4400  85.583 0.12158
Two Breweries    18  5200 170.250 0.17158
 Moffat Chase    20  5000 159.833 0.19099
```

so two points have very high leverage, two points have large residuals, and Bens of Jura is in both sets. (See Figure 6.2.)

If we look at Knock Hill we see that the prediction is over an hour less than the reported record:

```
> cbind(hills, pred=predict(hills.lm))["Knock Hill", ]
           dist climb time    pred
Knock Hill    3   350 78.65 13.529
```

and Atkinson (1988) suggests that the record is one hour out. We drop this observation to be safe:

```
> hills1.lm <- lm(time ~ dist + climb, hills[-18, ])
> hills1.lm
Call:
lm(formula = time ~ dist + climb, data = hills[-18,  ])

Coefficients:
 (Intercept)    dist     climb
      -13.53  6.3646  0.011855

Degrees of freedom: 34 total; 31 residual
Residual standard error: 8.8035
```

Since Knock Hill did not have a high leverage, as expected deleting it did not change the fitted model greatly. On the other hand, Bens of Jura had both a high leverage and a large residual and so does affect the fit:

```
> lm(time ~ dist + climb, hills[-c(7,18), ])
    ....
Coefficients:
 (Intercept)    dist     climb
     -10.362  6.6921  0.0080468

Degrees of freedom: 33 total; 30 residual
Residual standard error: 6.0538
```

If we consider this example carefully we find a number of unsatisfactory features. First, the prediction is negative for short races. Extrapolation is often unsafe, but on physical grounds we would expect the model to be a good fit with a zero intercept; indeed hill-walkers use a prediction of this sort (3 miles/hour plus 20 minutes per 1 000 feet). We can see from the summary that the intercept is significantly negative:

```
> summary(hills1.lm)
    ....
Coefficients:
               Value Std. Error t value Pr(>|t|)
(Intercept) -13.530     2.649    -5.108    0.000
       dist   6.365     0.361    17.624    0.000
      climb   0.012     0.001     9.600    0.000
    ....
```

Furthermore, we would not expect the predictions of times that range from 15 minutes to over 3 hours to be equally accurate, but rather that the accuracy be roughly proportional to the time. This suggests a log transform, but that would be hard to interpret. Rather we weight the fit using distance as a surrogate for time. We want weights inversely proportional to the variance:

```
> summary(lm(time ~ dist + climb, hills[-18, ],
    weight=1/dist^2))
    ....
```

```
Coefficients:
              Value  Std. Error  t value  Pr(>|t|)
(Intercept)  -5.809     2.034     -2.855    0.008
       dist   5.821     0.536     10.858    0.000
      climb   0.009     0.002      5.873    0.000

Residual standard error: 1.16 on 31 degrees of freedom
```

The intercept is still significantly non-zero. If we are prepared to set it to zero on physical grounds, we can achieve the same effect by dividing the prediction equation by distance, and regressing inverse speed (time/distance) on gradient (climb/distance):

```
> lm(time ~ -1 + dist + climb, hills[-18, ], weight=1/dist^2)
Coefficients:
 dist      climb
 4.9 0.0084718

Degrees of freedom: 34 total; 32 residual
Residual standard error (on weighted scale): 1.2786
> hills <- hills    # make a local copy (needed in 5.x)
> hills$ispeed <- hills$time/hills$dist
> hills$grad <- hills$climb/hills$dist
> hills2.lm <- lm(ispeed ~ grad, hills[-18, ])
> hills2.lm
Coefficients:
 (Intercept)       grad
         4.9 0.0084718

Degrees of freedom: 34 total; 32 residual
Residual standard error: 1.2786
> frame(); par(fig=c(0, 0.6, 0, 0.55))
> plot(hills$grad[-18], studres(hills2.lm), xlab="grad")
> abline(h=0, lty=2)
> identify(hills$grad[-18], studres(hills2.lm),
     row.names(hills)[-18])
> par(fig=c(0.6, 1, 0, 0.55), pty="s")
> qqnorm(studres(hills2.lm))
> qqline(studres(hills2.lm))
> hills2.hat <- lm.influence(hills2.lm)$hat
> cbind(hills[-18,], lev=hills2.hat)[hills2.hat > 1.8*2/34, ]
              dist climb    time  ispeed    grad     lev
Bens of Jura    16  7500 204.617 12.7886  468.75 0.11354
  Creag Dubh     4  2000  26.217  6.5542  500.00 0.13915
```

The two highest-leverage cases are now the steepest two races, and are outliers pulling in opposite directions. We could consider elaborating the model, but this would be to fit only one or two exceptional points; for most of the data we have the formula of 5 minutes/mile plus 8 minutes per 1 000 feet. We return to this example on page 172 where robust fits do support a zero intercept.

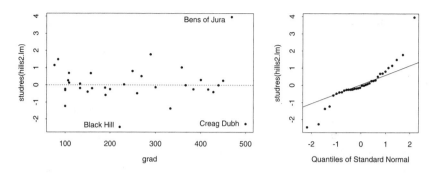

Figure 6.3: Diagnostic plots for Scottish hills data, weighted model.

6.4   Safe prediction

A warning is needed on the use of the `predict` method function when polynomials are used (and also splines, see Chapter 9). We illustrate this by the dataset `wtloss`, for which a more appropriate analysis is given in Chapter 8. This has a weight loss `Weight` against `Days`. Consider a quadratic polynomial regression model of `Weight` on `Days`. This may be fitted by either of

```
quad1 <- lm(Weight ~ Days + I(Days^2), wtloss)
quad2 <- lm(Weight ~ poly(Days, 2), wtloss)
```

The second uses orthogonal polynomials and is the preferred form on grounds of numerical stability.

Suppose we wished to predict future weight loss. The first step is to create a new data frame with a variable `x` containing the new values, for example,

```
new.x <- data.frame(Days = seq(250, 300, 10),
                     row.names = seq(250, 300, 10))
```

The `predict` method may now be used:

```
> predict(quad1, newdata=new.x)
    250     260     270     280     290     300
 112.51  111.47  110.58  109.83  109.21  108.74
> predict(quad2, newdata=new.x)
    250     260     270     280     290     300
 244.56  192.78  149.14  113.64  86.29  67.081
```

The first form gives correct answers but the second does not!

The reason for this is as follows. The `predict` method for `lm` objects works by attaching the estimated coefficients to a new model matrix which it constructs using the formula and the new data. In the first case the procedure will work, but in the second case the columns of the model matrix are for a *different* orthogonal polynomial basis, and so the old coefficients do not apply. The same will hold for any function used to define the model that generates mathematically different bases for old and new data, such as spline bases using `bs` or `ns`.

The remedy is to use the method function `predict.gam`:

```
> predict.gam(quad2, newdata=new.x)
    250     260     270     280     290     300
 112.51 111.47 110.58 109.83 109.21 108.74
```

This constructs a new model matrix by putting old and new data together, re-estimates the regression using the old data only and predicts using these estimates of regression coefficients. This can involve appreciable extra computation, but the results will be correct for polynomials, but not exactly so for splines since the knot positions will change. As a check, predict.gam compares the predictions with the old fitted values for the original data. If these are seriously different a warning is issued that the process has probably failed.

In our view this is a serious flaw in predict.lm. It would have been better to use the safe method as the default and provide an unsafe argument for the faster method as an option.

There is another prediction problem with factor variables. Suppose we want to predict gas consumption after insulation at three external temperatures

```
> new.white <- data.frame(Temp = 5*(0:2),
                Insul = factor(rep("After", 3)))
> predict(gasBA, new.white)
```

which gives an error. The factor variable must have the same number of levels as before. So we try again

```
> new.white <- data.frame(Temp = 5*(0:2),
                Insul = factor(rep("After", 3),
                levels=c("After", "Before")))
> predict(gasBA, new.white)
      1      2      3
 6.8538 4.8876 2.9214
```

which as Figure 6.1 shows is implausible. In fact the levels must be in the same order as for the fitted data, and no checking is done. (The levels were originally set as c("Before", "After") to give the Trellis display an appropriate sequence.) In our view this is an equally serious flaw in predict.lm.

6.5   Robust and resistant regression

There are a number of ways to perform robust regression in S-PLUS, but many have drawbacks and are not mentioned here. First consider an example. Rousseeuw & Leroy (1987) give data on annual numbers of Belgian telephone calls, given in our dataset phones.

```
phones.lm <- lm(calls ~ year, phones)
attach(phones); plot(year, calls); detach()
abline(phones.lm$coef)
abline(rlm(calls ~ year, phones, maxit=50), lty=2, col=2)
abline(lqs(calls ~ year, phones), lty=3, col=3)
legend(locator(1), legend=c("least squares", "M-estimate", "LTS"),
    lty=1:3, col=1:3)
```

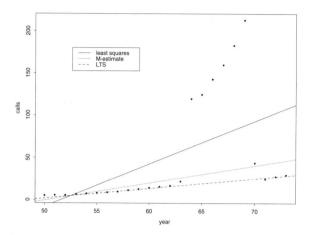

Figure 6.4: Millions of phone calls in Belgium, 1950–73, from Rousseeuw & Leroy (1987), with three fitted lines.

Figure 6.4 shows the least squares line, an M-estimated regression and the least trimmed squares regression (Section 6.5). The `lqs` line is $-56.16 + 1.16\,\text{year}$. Rousseeuw & Leroy's investigations showed that for 1964–9 the total length of calls (in minutes) had been recorded rather than the number, with each system being used during parts of 1963 and 1970.

Next some theory. In a regression problem there are two possible sources of errors, the observations y_i and the corresponding row vector of p regressors x_i. Most robust methods in regression only consider the first, and in some cases (designed experiments?) errors in the regressors can be ignored. This is the case for M-estimators, the only ones we consider in this section.

Consider a regression problem with n cases (y_i, x_i) from the model

$$y = x\beta + \epsilon$$

for a p-variate row vector x.

M-estimators

If we assume a scaled pdf $f(e/s)/s$ for ϵ and set $\rho = -\log f$, the maximum likelihood estimator minimizes

$$\sum_{i=1}^{n} \rho\left(\frac{y_i - x_i b}{s}\right) + n \log s \qquad (6.4)$$

Suppose for now that s is known. Let $\psi = \rho'$. Then the MLE b of β solves

$$\sum_{i=1}^{n} x_i \psi\left(\frac{y_i - x_i b}{s}\right) = 0 \qquad (6.5)$$

Let $r_i = y_i - x_i b$ denote the residuals.

The solution to equation (6.5) or to minimizing over (6.4) can be used to define an M-estimator of β.

A common way to solve (6.5) is by iterated re-weighted least squares, with weights

$$w_i = \psi \left(\frac{y_i - x_i b}{s} \right) \Bigg/ \left(\frac{y_i - x_i b}{s} \right) \tag{6.6}$$

The iteration is only guaranteed to converge for *convex* ρ functions, and for re-descending functions (such as those of Tukey and Hampel; page 130), equation (6.5) may have multiple roots. In such cases it is usual to choose a good starting point and iterate carefully.

Of course, in practice the scale s is not known. A simple and very resistant scale estimator is the MAD about some centre. This is applied to the residuals about zero, either to the current residuals within the loop or to the residuals from a very resistant fit (see the next subsection).

Alternatively, we can estimate s in an MLE-like way. Finding a stationary point of (6.4) with respect to s gives

$$\sum_i \psi \left(\frac{y_i - x_i b}{s} \right) \left(\frac{y_i - x_i b}{s} \right) = n$$

which is not resistant (and is biased at the normal). As in the univariate case we modify this to

$$\sum_i \chi \left(\frac{y_i - x_i b}{s} \right) = (n - p)\gamma \tag{6.7}$$

Our function `rlm`

Our main library introduces a new class `rlm` and model-fitting function `rlm`, building on `lm`. The syntax in general follows `lm`. By default Huber's M-estimator is used with tuning parameter $c = 1.345$. By default the scale s is estimated by iterated MAD, but Huber's proposal 2 can also be used.

```
> summary(lm(calls ~ year, data=phones), cor=F)
              Value Std. Error   t value  Pr(>|t|)
(Intercept) -260.059   102.607    -2.535     0.019
       year    5.041     1.658     3.041     0.006
Residual standard error: 56.2 on 22 degrees of freedom
> summary(rlm(calls ~ year, maxit=50, data=phones), cor=F)
              Value Std. Error   t value
(Intercept) -102.622    26.608    -3.857
       year    2.041     0.430     4.748
Residual standard error: 9.03 on 22 degrees of freedom
> summary(rlm(calls ~ year, scale.est="proposal 2", data=phones),
          cor=F)
Coefficients:
              Value Std. Error   t value
```

```
(Intercept) -227.925   101.874      -2.237
      year      4.453     1.646       2.705
Residual standard error: 57.3 on 22 degrees of freedom
```

As Figure 6.4 shows, in this example there is a batch of outliers from a different population in the late 1960s, and these should be rejected completely, which the Huber M-estimators do not. Let us try a re-descending estimator.

```
> summary(rlm(calls ~ year, data=phones, psi=psi.bisquare), cor=F)
Coefficients:
             Value Std. Error t value
(Intercept) -52.302    2.753   -18.999
      year    1.098    0.044    24.685
Residual standard error: 1.65 on 22 degrees of freedom
```

This happened to work well for the default least-squares start, but we might want to consider a better starting point, such as that given by `init="lts"`.

Resistant regression

M-estimators are not very resistant to outliers unless they have redescending ψ functions, in which case they need a good starting point. A succession of more resistant regression estimators was defined in the 1980s. The first to become popular was

$$\min_b \operatorname{median}_i |y_i - x_i b|^2$$

called the *least median of squares* (LMS) estimator. The square is necessary if n is even, when the central median is taken. This fit is very resistant, and needs no scale estimate. It is however very inefficient, converging at rate $1/\sqrt[3]{n}$. Furthermore, it displays marked sensitivity to central data values: see Hettmansperger & Sheather (1992) and Davies (1993, §2.3).

Rousseeuw suggested least trimmed squares (LTS) regression:

$$\min_b \sum_{i=1}^{q} |y_i - x_i b|_{(i)}^2$$

as this is more efficient, but shares the same extreme resistance. The recommended sum is over the smallest $q = \lfloor (n+p+1)/2 \rfloor$ squared residuals. (Earlier accounts differed.)

This was followed by *S-estimation*, in which the coefficients are chosen to find the solution to

$$\sum_{i=1}^{n} \chi\left(\frac{y_i - x_i b}{c_0 s}\right) = (n-p)\beta$$

with smallest scale s. Here χ is usually chosen to be the integral of Tukey's biquare function

$$\chi(u) = u^6 - 3u^4 + 3u^2, u \leqslant 1, \qquad 1, u \geqslant 1$$

$c_0 = 1.548$ and $\beta = 0.5$ is chosen for consistency at the normal distribution of errors. This gives efficiency 28.7% at the normal, which is low but better than LMS and LTS.

In only a few special cases (such as LMS for univariate regression with intercept) can these optimization problems be solved exactly, and approximate search methods are used.

S-PLUS *implementation*

Various versions of **S-PLUS** have (different) implementations of LMS and LTS regression in functions `lmsreg` and `ltsreg`[2], but as these are not fully documented and give different results in different releases, we prefer our function `lqs`. The default method is LTS.

```
> lqs(calls ~ year, data=phones)
Coefficients:
 (Intercept)    year
 -56.2         1.16
Scale estimates 1.25 1.13

> lqs(calls ~ year, data=phones, method="lms")
Coefficients:
 (Intercept)    year
 -55.9         1.15
Scale estimates 0.938 0.909

> lqs(calls ~ year, data=phones, method="S")
Coefficients:
 (Intercept)    year
 -52.5         1.1
Scale estimates 2.13
```

Two scale estimates are given for LMS and LTS: the first comes from the fit criterion, the second from the variance of the residuals of magnitude no more than 2.5 times the first scale estimate. All the scale estimates are set up to be consistent at the normal, but measure different things for highly non-normal data as here.

MM-estimation

It is possible to combine the resistance of these methods with the efficiency of M-estimation. The MM-estimator proposed by Yohai, Stahel & Zamar (1991) (see also Marazzi, 1993, §9.1.3) is an M-estimator starting at the coefficients given by the S-estimator and with fixed scale given by the S-estimator. This retains (for $c > c_0$) the high-breakdown point of the S-estimator and the high efficiency at the normal. At considerable computational expense, this gives the best of both worlds.

Our function `rlm` has an option to implement MM-estimation.

---

[2] In S-PLUS 2000 this uses 10% trimming.

```
> summary(rlm(calls ~ year, data=phones, method="MM"), cor=F)
Coefficients:
              Value Std. Error t value
(Intercept) -52.423    2.916   -17.977
      year    1.101    0.047    23.366

Residual standard error: 2.13
```

S-PLUS 4.5, 5.0 and later have a function `lmRobMM` which implements a slightly different MM-estimator with similar properties, and comes with a full set of method functions, so can be used routinely as a replacement for `lm`. Let us try it on the phones data.

```
> phones.lmr <- lmRobMM(calls ~ year, data=phones)
> summary(phones.lmr)
Final M-estimates.

Call: lmRobMM(formula = calls ~ year, data = phones)

Residuals:
    Min    1Q Median    3Q    Max
 -1.719 -0.46 0.2267 39.03 188.5

Coefficients:
              Value Std. Error  t value Pr(>|t|)
(Intercept) -52.3103   3.7851  -13.8199   0.0000
      year    1.0990   0.0636   17.2853   0.0000

Residual scale estimate: 2.027 on 22 degrees of freedom
Proportion of variation in response explained by model: 0.4898

Test for Bias
              Statistics P-value
  M-estimate       1.601   0.449
 LS-estimate       0.243   0.886
> plot(phones.lmr)
```

This works well, rejecting all the spurious observations. The 'test for bias' is of the M-estimator against the initial S-estimator; if the M-estimator appears biased the initial S-estimator is returned.

Scottish hill races revisited

We return to the data on Scottish hill races studied in the introduction and Section 6.3. There we saw one gross outlier and a number of other extreme observations.

```
> hills.lm
Coefficients:
  (Intercept)   dist    climb
```

```
      -8.992 6.218 0.011048
Residual standard error: 14.676

> hills1.lm # omitting Knock Hill
Coefficients:
 (Intercept)    dist     climb
      -13.53 6.3646 0.011855
Residual standard error: 8.8035

> rlm(time ~ dist + climb, hills)
Coefficients:
 (Intercept)    dist      climb
     -9.6067 6.5507 0.0082959
Scale estimate: 5.21

> summary(rlm(time ~ dist + climb, hills, weights=1/dist^2,
             method="MM"), cor=F)
Coefficients:
               Value Std. Error t value
(Intercept)  -1.804    1.665    -1.084
       dist   5.244    0.233    22.546
      climb   0.007    0.001     9.389
Residual standard error: 4.85 on 32 degrees of freedom

> lqs(time ~ dist + climb, data=hills, nsamp="exact")
Coefficients:
 (Intercept)      dist     climb
       -1.26      4.86   0.00851
Scale estimates 2.94 3.01
```

Notice that the intercept is no longer significant in the robust weighted fits. By default `lqs` uses a random search but here exhaustive enumeration is possible, so we use it.

If we move to the model for inverse speed:

```
> summary(hills2.lm) # omitting Knock Hill
Coefficients:
               Value Std. Error t value Pr(>|t|)
(Intercept)    4.900    0.474   10.344    0.000
       grad    0.008    0.002    5.022    0.000

Residual standard error: 1.28 on 32 degrees of freedom

> summary(rlm(ispeed ~ grad, hills), cor=F)
Coefficients:
               Value Std. Error t value
(Intercept)    5.176    0.381   13.593
       grad    0.007    0.001    5.431

Residual standard error: 0.869 on 33 degrees of freedom
```

```
# method="MM" results are very similar.
> summary(lmRobMM(ispeed ~ grad, data=hills))
Coefficients:
              Value Std. Error t value Pr(>|t|)
(Intercept) 5.0754  0.4210    12.0563  0.0000
       grad 0.0077  0.0016     4.8817  0.0000

Residual scale estimate: 0.8189 on 33 degrees of freedom

> lqs(ispeed ~ grad, data=hills)
Coefficients:
 (Intercept)     grad
 4.75         0.00805
Scale estimates 0.608 0.643
```

The results are in close agreement with the least-squares results after removing
Knock Hill.

6.6 Bootstrapping linear models

In frequentist inference we have to consider what might have happened but did
not. Linear models can arise exactly or approximately in a number of ways. The
most commonly considered form is

$$Y = X\beta + \epsilon$$

in which only ϵ is considered to be random. This supposes that in all (hypothet-
ical) repetitions the same x points would have been chosen, but the responses
would vary. This is a plausible assumption for a designed experiment such as the
N, P, K experiment on page 176 and for an observational study such as Quine's
with prespecified factors. It is less clearly suitable for the Scottish hill races, and
clearly not correct for Whiteside's gas consumption data.

Another form of regression is sometimes referred to as the *random regressor*
case in which the pairs (x_i, y_i) are thought of as a random sample from a population and we are interested in the regression function $f(x) = \mathrm{E}\left(Y \mid X = x\right)$
which is assumed to be linear. This seems appropriate for the gas consumption
data. However, it is common to perform conditional inference in this case and
condition on the observed xs, converting this to a fixed-design problem. For example, in the hill races the inferences drawn depend on whether certain races,
notably Bens of Jura, were included in the sample. As they were included, conclusions conditional on the set of races seems most pertinent. (There are other
ways that linear models can arise, including calibration problems and where both
x and y are measured with error about a true linear relationship.)

These considerations are particularly relevant when we consider bootstrap resampling. The most obvious form of bootstrapping is to randomly sample pairs

(x_i, y_i) with replacement,[3] which corresponds to randomly weighted regressions. However, this may not be appropriate in not mimicking the assumed random variation and in some examples of producing singular fits with high probability. The main alternative, *model-based resampling*, is to resample the residuals. After fitting the linear model we have

$$y_i = x_i \widehat{\beta} + e_i$$

and we create a new dataset by $y_i = x_i \widehat{\beta} + e_i^*$ where the (e_i^*) are resampled with replacement from the residuals (e_i). There are a number of possible objections to this procedure. First, the residuals need not have mean zero if there is no intercept in the model, and it is usual to subtract their mean. Second, they do not have the correct variance or even the same variance. Thus we can adjust their variance by resampling the *modified residuals* $r_i = e_1/\sqrt{1 - h_{ii}}$ which have variance σ^2 from (6.3).

We see bootstrapping as having little place in least-squares regression. If the errors are close to normal, the standard theory suffices. If not, there are better methods of fitting than least-squares, or perhaps the data should be transformed as in the quine dataset on page 182.

The distribution theory for the estimated coefficients in robust regression is based on asymptotic theory, so we could use bootstrap estimates of variability as an alternative. Resampling the residuals seems most appropriate for the phones data.

```
library(boot)
fit <- lm(calls ~ year, data=phones)
ph <- data.frame(phones, res=resid(fit), fitted=fitted(fit))
ph.fun <- function(data, i) {
  d <- data
  d$calls <- d$fitted + d$res[i]
  coef(update(fit, data=d))
}
ph.lm.boot <- boot(ph, ph.fun, R=499)
ph.lm.boot
    ....
      original     bias    std. error
t1* -260.0592   6.32092      100.3970
t2*    5.0415  -0.09289        1.6288


fit <- rlm(calls ~ year, method="MM", data=phones)
ph <- data.frame(phones, res=resid(fit), fitted=fitted(fit))
ph.rlm.boot <- boot(ph, ph.fun, R=499)
ph.rlm.boot
    ....
     original      bias   std. error
t1* -52.4231   3.232142     30.01894
t2*   1.1009  -0.013896      0.40648
Bootstrap Statistics :
```

---

[3] Davison & Hinkley (1997) call this *case-based resampling*.

Table 6.1: Layout of a classic NPK fractional factorial design.

pk	np	—	nk		n	npk	k	p
49.5	62.8	46.8	57.0		62.0	48.8	45.5	44.2
n	npk	k	p		np	—	nk	pk
59.8	58.5	55.5	56.0		52.0	51.5	49.8	48.8
p	npk	n	k		nk	np	pk	—
62.8	55.8	69.5	55.0		57.2	59.0	53.2	56.0

(The `rlm` bootstrap runs took about eight minutes on the PC, and readers might like to start with a smaller number of resamples.) These results suggest that the asymptotic theory for `rlm` is optimistic for this example, but as the residuals are clearly serially correlated the validity of the bootstrap is equally in doubt. Statistical inference really does depend on what one considers might have happened but did not.

The bootstrap results can be investigated further by using `plot`, and `boot.ci` will give confidence intervals for the coefficients. The robust results have very long tails.

6.7 Factorial designs and designed experiments

Factorial designs are powerful tools in the design of experiments. Experimenters often cannot afford to perform all the runs needed for a complete factorial experiment, or they may not all be fitted into one experimental block. To see what can be achieved, the following N, P, K (*nitrogen, phosphate, potassium*) factorial experiment on the growth of peas conducted on six blocks shown in Table 6.1. The response is yield (in lbs/(1/70)acre-plot). Half of the design (technically a fractional factorial design) is performed in each of six blocks, so each half occurs three times. (If we consider the variables to take values ± 1, the halves are defined by even or odd parity, equivalently product equal to $+1$ or -1.) Note that the NPK interaction cannot be estimated as it is confounded with block differences, specifically with $(b_2 + b_3 + b_4 - b_1 - b_5 - b_6)$. An ANOVA table may be computed by

```
> npk.aov <- aov(yield ~ block + N*P*K, npk)
> npk.aov
    ....
Terms:
                   block       N       P       K     N:P     N:K
Sum of Squares    343.29  189.28    8.40   95.20   21.28   33.14
Deg. of Freedom        5       1       1       1       1       1
                     P:K Residuals
Sum of Squares      0.48    185.29
```

```
Deg. of Freedom        1         12

Residual standard error: 3.9294
1 out of 13 effects not estimable
Estimated effects are balanced
> summary(npk.aov)
            Df Sum of Sq Mean Sq F Value    Pr(F)
block        5    343.29   68.66   4.447 0.01594
N            1    189.28  189.28  12.259 0.00437
P            1      8.40    8.40   0.544 0.47490
K            1     95.20   95.20   6.166 0.02880
N:P          1     21.28   21.28   1.378 0.26317
N:K          1     33.14   33.14   2.146 0.16865
P:K          1      0.48    0.48   0.031 0.86275
Residuals 12    185.29   15.44

> alias(npk.aov)

    ....
Complete
       (Intercept) block1 block2 block3 block4 block5 N P K
N:P:K            1   0.33   0.17   -0.3   -0.2
       N:P N:K P:K
N:P:K
> coef(npk.aov)
 (Intercept) block1 block2  block3  block4 block5
     54.875 1.7125 1.6792 -1.8229 -1.0137  0.295
       N       P       K     N:P     N:K     P:K
  2.8083 -0.59167 -1.9917 -0.94167 -1.175 0.14167
```

Note how the N:P:K interaction is silently omitted in the summary, although its absence is mentioned in printing npk.aov. The alias command shows which effect is missing (the particular combinations corresponding to the use of Helmert contrasts for the factor block).

Only the N and K main effects are significant (we ignore blocks whose terms are there precisely because we expect them to be important and so we must allow for them). For two-level factors the Helmert contrast is the same as the sum contrast (up to sign) giving -1 to the first level and $+1$ to the second level. Thus the effects of adding nitrogen and potassium are 5.62 and -3.98, respectively. This interpretation is easier to see with treatment contrasts:

```
> options(contrasts=c("contr.treatment", "contr.poly"))
> npk.aov1 <- aov(yield ~ block + N + K, npk)
> summary.lm(npk.aov1)
    ....
Coefficients:
          Value Std. Error t value Pr(>|t|)
    ....
      N   5.617    1.609     3.490    0.003
      K  -3.983    1.609    -2.475    0.025
```

```
Residual standard error: 3.94 on 16 degrees of freedom
```

Note the use of summary.lm to give the standard errors. Standard errors of contrasts can also be found from the function se.contrast. The full form is quite complex, but a simple use is:

```
> se.contrast(npk.aov1, list(N=="0", N=="1"), data=npk)
Refitting model to allow projection
[1] 1.6092
```

For highly regular designs such as this standard errors may also be found along with estimates of means, effects and other quantities using model.tables.

```
> model.tables(npk.aov1, type="means", se=T)
    ....
 N
     0     1
 52.07 57.68
    ....
Standard errors for differences of means
        block     N      K
        2.787  1.609  1.609
replic. 4.000 12.000 12.000
```

The function aov.genyates is very similar to aov but uses a generalized Yates algorithm. It can only be used with equally replicated experimental designs; that is, those in which each combination of factor levels occurs equally often, but it can be much faster than aov for very large designs. (Equal replication can be checked using the function replications.)

Generating designs

The three functions expand.grid, fac.design and oa.design can each be used to construct designs such as our example.

Of these, expand.grid is the simplest. It is used in a similar way to data.frame: the arguments may be named and the result is a data frame with those names. The columns contain all combinations of values for each argument. If the argument values are numeric the column is numeric; if they are anything else, for example, character, the column is a factor. Consider an example:

```
> mp <- c("-","+")
> NPK <- expand.grid(N=mp, P=mp, K=mp)
> NPK
   N P K
1  - - -
2  + - -
3  - + -
4  + + -
5  - - +
```

```
6 + - +
7 - + +
8 + + +
```

Note that the first column changes fastest and the last slowest. This is a single complete replicate.

Our example used three replicates, each split into two blocks so that the block comparison is confounded with the highest order interaction. We can construct such a design in stages. First we find a half-replicate to be repeated three times and form the contents of three of the blocks. The simplest way to do this is to use `fac.design`:

```
blocks13 <- fac.design(levels=c(2,2,2),
        factor=list(N=mp, P=mp, K=mp), rep=3, fraction=1/2)
```

The first two arguments give the numbers of levels and the factor names and level labels. The third argument gives the number of replications (default 1). The `fraction` argument may only be used for 2^p factorials. It may be given either as a small negative power of 2, as here, or as a *defining contrast formula*. When `fraction` is numerical the function chooses a defining contrast which becomes the `fraction` attribute of the result. For half-replicates the highest-order interaction is chosen to be aliased with the mean. To find the complementary fraction for the remaining three blocks we need to use the defining contrast formula form for `fraction`:

```
blocks46 <- fac.design(levels=c(2,2,2),
        factor=list(N=mp, P=mp, K=mp), rep=3, fraction=~ -N:P:K)
```

(This is explained in the following.) To complete our design we put the blocks together, add in the `block` factor and randomize:

```
NPK <- design(block = factor(rep(1:6, rep(4,6))),
        rbind(blocks13, blocks46))
i <- order(runif(6)[NPK$block], runif(24))
NPK <- NPK[i,]   # Randomized
```

Using `design` instead of `data.frame` creates an object of class `design` that inherits from `data.frame`. For most purposes designs and data frames are equivalent, but some generic functions such as `plot`, `formula` and `alias` have useful `design` methods.

Defining contrast formulae resemble model formulae in syntax only; the meaning is quite distinct. There is no left-hand side. The right-hand side consists of colon products of factors only, separated by + or - signs. A plus (or leading blank) specifies that the treatments with *positive* signs for that contrast are to be selected and a minus those with *negative* signs. A formula such as `~A:B:C-A:D:E` specifies a quarter-replicate consisting of the treatments that have a positive sign in the ABC interaction and a negative sign in ADE.

Box, Hunter & Hunter (1978, §12.5) consider a 2^{7-4} design used for an experiment in riding up a hill on a bicycle. The seven factors are Seat (up or down),

Dynamo (off or on), Handlebars (up or down), Gears (low or medium), Raincoat (on or off), Breakfast (yes or no) and Tyre pressure (hard or soft). A resolution III design was used, so the main effects are not aliased with each other. Such a design cannot be constructed using a numerical fraction in `fac.design` so the defining contrasts have to be known. Box *et al.* use the design relations:

$$D = AB, \quad E = AC, \quad F = BC, \quad G = ABC$$

which mean that ABD, ACE, BCF and $ABCG$ are all aliased with the mean, and form the defining contrasts of the fraction. Whether we choose the positive or negative halves is immaterial here.

```
> lev <- rep(2,7)
> factors <- list(S=mp, D=mp, H=mp, G=mp, R=mp, B=mp, P=mp)
> Bike <- fac.design(lev, factors, fraction =
    ~ S:D:G + S:H:R + D:H:B + S:D:H:P)
> Bike
  S D H G R B P
1 - - - - - - -
2 - + + + + - -
3 + - + + - + -
4 + + - - + + -
5 + + + - - - +
6 + - - + + - +
7 - + - + - + +
8 - - + - + + +

Fraction:   ~ S:D:G + S:H:R + D:H:B + S:D:H:P
```

(We chose `P` for pressure rather than `T` for tyres since `T` and `F` are reserved identifiers.)

We may check the symmetry of the design using `replications`:

```
> replications(~.^2, data = Bike)
  S D H G R B P S:D S:H S:G S:R S:B S:P D:H D:G D:R D:B D:P H:G
  4 4 4 4 4 4 4   2   2   2   2   2   2   2   2   2   2   2   2
  H:R H:B H:P G:R G:B G:P R:B R:P B:P
    2   2   2   2   2   2   2   2   2
```

Fractions may be specified either in a call to `fac.design` or subsequently using the `fractionate` function.

The third function, `oa.design`, provides some resolution III designs (also known as *main effect plans* or *orthogonal arrays*) for factors at two or three levels. Only low-order cases are provided, but these are the most useful in practice.

6.8   An unbalanced four-way layout

Aitkin (1978) discussed an observational study of S. Quine. The response is the number of days absent from school in a year by children from a large town in

rural New South Wales, Australia. The children were classified by four factors,
namely,

Age	4 levels: primary, first, second or third form
Eth	2 levels: aboriginal or non-aboriginal
Lrn	2 levels: slow or average learner
Sex	2 levels: male or female.

The dataset is included in the paper of Aitkin (1978) and is available as data frame
quine in our library MASS. This has been explored several times already, but we
now consider a more formal statistical analysis.

There were 146 children in the study. The frequencies of the combinations of
factors are

```
> attach(quine)
> table(Lrn, Age, Sex, Eth)
, , F, A                        , , F, N
    F0 F1 F2 F3                     F0 F1 F2 F3
AL   4  5  1  9                 AL   4  6  1 10
SL   1 10  8  0                 SL   1 11  9  0

, , M, A                        , , M, N
    F0 F1 F2 F3                     F0 F1 F2 F3
AL   5  2  7  7                 AL   6  2  7  7
SL   3  3  4  0                 SL   3  7  3  0
```

(The output has been slightly rearranged to save space.) The classification is
unavoidably very unbalanced. There are no slow learners in the fourth form, but
all 28 other cells are non-empty. In his paper Aitkin considers a normal analysis
on the untransformed response, but in the reply to the discussion he chooses a
transformed response, $\log(\text{Days} + 1)$.

A casual inspection of the data shows that homoscedasticity is likely to be
an unrealistic assumption on the original scale, so our first step is to plot the cell
variances and standard deviations against the cell means.

```
Means <- tapply(Days, list(Eth, Sex, Age, Lrn), mean)
Vars  <- tapply(Days, list(Eth, Sex, Age, Lrn), var)
SD <- sqrt(Vars)
par(mfrow=c(1,2))
plot(Means, Vars, xlab="Cell Means", ylab="Cell Variances")
plot(Means, SD, xlab="Cell Means", ylab="Cell Std Devn.")
```

Missing values are silently omitted from the plot. Interpretation of the result in
Figure 6.5 requires some caution because of the small and widely different degrees of freedom on which each variance is based. Nevertheless the approximate
linearity of the standard deviations against the cell means suggests a logarithmic
transformation or something similar is appropriate. (See, for example, Rao, 1973,
§6g).

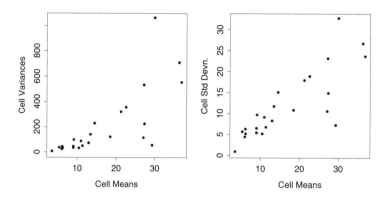

Figure 6.5: Two diagnostic plots for the Quine data.

Some further insight on the transformation needed is provided by considering a model for the transformed observations

$$y^{(\lambda)} = \begin{cases} (y^\lambda - 1)/\lambda & \lambda \neq 0 \\ \log y & \lambda = 0 \end{cases}$$

where here $y = \text{Days} + \alpha$. (The positive constant α is added to avoid problems with zero entries.) Rather than include α as a second parameter we first consider Aitkin's choice of $\alpha = 1$. Box & Cox (1964) show that the profile likelihood function for λ is

$$\widehat{L}(\lambda) = \text{const} - \tfrac{n}{2} \log \text{RSS}(z^{(\lambda)})$$

where $z^{(\lambda)} = y^{(\lambda)}/\dot{y}^{\lambda-1}$, $\dot{y}$ is the geometric mean of the observations and $\text{RSS}(z^{(\lambda)})$ is the residual sum of squares for the regression of $z^{(\lambda)}$.

Box & Cox suggest using the profile likelihood function for the largest linear model to be considered as a guide in choosing a value for λ, which will then remain fixed for any remaining analyses. Ideally other considerations from the context will provide further guidance in the choice of λ, and in any case it is desirable to choose easily interpretable values such as square-root, log or inverse.

Our MASS library function boxcox calculates and (optionally) displays the Box–Cox profile likelihood function, together with a horizontal line showing what would be an approximate 95% likelihood ratio confidence interval for λ. The function is generic and several calling protocols are allowed but a convenient one to use here is with the same arguments as lm together with an additional (named) argument, lambda, to provide the sequence at which the marginal likelihood is to be evaluated. (By default the result is extended using a spline interpolation.)

Since the dataset has four empty cells the full model Eth*Sex*Age*Lrn has a rank-deficient model matrix. Hence we must use singular.ok = T to fit the model.

```
boxcox(Days+1 ~ Eth*Sex*Age*Lrn, data = quine, singular.ok = T,
    lambda = seq(-0.05, 0.45, len = 20))
```

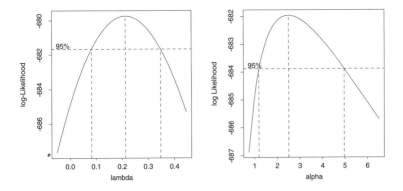

Figure 6.6: Profile likelihood for a Box–Cox transformation model with displacement $\alpha = 1$, left, and a displaced log transformation model, right.

(Alternatively the first argument may be a fitted model object that supplies all needed information apart from `lambda`.) The result is shown on the left-hand side of Figure 6.6 which strongly suggests that a log transformation is not optimal when $\alpha = 1$ is chosen. An alternative one-parameter family of transformations that could be considered in this case is

$$t(y, \alpha) = \log(y + \alpha)$$

Using the same analysis as presented in Box & Cox (1964) the profile log likelihood for α is easily seen to be

$$\widehat{L}(\alpha) = \text{const} - \tfrac{n}{2} \log \text{RSS}\{\log(y + \alpha)\} - \sum \log(y + \alpha)$$

It is interesting to see how this may be calculated directly using low-level tools, in particular the functions `qr` for the QR-decomposition and `qr.resid` for orthogonal projection onto the residual space. Readers are invited to look at our functions `logtrans.default` and `boxcox.default`.

```
logtrans(Days ~ Age*Sex*Eth*Lrn, data = quine,
    alpha = seq(0.75, 6.5, len=20), singular.ok = T)
```

The result is shown in the right-hand panel of Figure 6.6. If a displaced log transformation is chosen a value $\alpha = 2.5$ is suggested, and we adopt this in our analysis. Note that $\alpha = 1$ is outside the notional 95% confidence interval. It can also be checked that with $\alpha = 2.5$ the log transform is well within the range of reasonable Box–Cox transformations to choose.

Model selection

The complete model, `Eth*Sex*Age*Lrn`, has a different parameter for each identified group and hence contains all possible simpler models for the mean as special cases, but has little predictive or explanatory power. For a better insight into the mean structure we need to find more parsimonious models. Before considering tools to prune or extend regression models it is useful to make a few general points on the process itself.

Marginality restrictions

In regression models it is usually the case that not all terms are on an equal footing as far as inclusion or removal is concerned. For example, in a quadratic regression on a single variable x one would normally consider removing only the highest-degree term, x^2, first. Removing the first-degree term while the second-degree one is still present amounts to forcing the fitted curve to be flat at $x = 0$, and unless there were some good reason from the context to do this it would be an arbitrary imposition on the model. Another way to view this is to note that if we write a polynomial regression in terms of a new variable $x^\star = x - \alpha$ the model remains in predictive terms the same, but only the highest-order coefficient remains invariant. If, as is usually the case, we would like our model selection procedure not to depend on the arbitrary choice of origin we must work only with the highest-degree terms at each stage.

The linear term in x is said to be *marginal* to the quadratic term, and the intercept term is marginal to both. In a similar way if a second-degree term in two variables, $x_1 x_2$, is present, any linear terms in either variable or an intercept term are marginal to it.

There are circumstances where a regression through the origin does make sense, but in cases where the origin is arbitrary one would normally only consider regression models where for each term present all terms marginal to it are also present.

In the case of factors the situation is even more clear-cut. A two-factor interaction a:b is marginal to any higher-order interaction that contains a and b. Fitting a model such as a + a:b leads to a model matrix where the columns corresponding to a:b are extended to compensate for the absent marginal term, b, and the fitted values are the same as if it were present. Fitting models with marginal terms removed such as with a*b - b generates a model with no readily understood statistical meaning[4] but updating models specified in this way using update changes the model matrix so that the absent marginal term again has no effect on the fitted model. In other words removing marginal factor terms from a fitted model is either statistically meaningless or futile in the sense that the model simply changes its parametrization to something equivalent.

Variable selection for the Quine data

The anova function when given a single fitted-model object argument constructs a *sequential* analysis of variance table. That is, a sequence of models is fitted by expanding the formula, arranging the terms in increasing order of marginality

---

[4] Marginal terms are sometimes removed in this way in order to calculate what are known as 'Type III sums of squares' but we have yet to see a situation where this makes compelling statistical sense. If they are needed, they can be computed by model.tables from S-PLUS 4.5 and 5.0.

and including one additional term for each row of the table. The process is order-dependent for non-orthogonal designs and several different orders may be needed to appreciate the analysis fully if the non-orthogonality is severe. For an orthogonal design the process is not order-dependent provided marginality restrictions are obeyed.

To explore the effect of adding or dropping terms from a model our two functions `addterm` and `dropterm` are usually more convenient. These allow the effect of, respectively, adding or removing individual terms from a model to be assessed, where the model is defined by a fitted-model object given as the first argument. For `addterm` a second argument is required to specify the scope of the terms considered for inclusion. This may be a formula or an object defining a formula for a larger model. Terms are included or removed in such a way that the marginality principle for factor terms is obeyed; for purely quantitative regressors this has to be managed by the user.

Both functions are generic and compute the change in AIC (Akaike, 1974)

$$\text{AIC} = -2\,\text{maximized log-likelihood} + 2\,\#\,\text{parameters}$$

Since the log-likelihood is defined only up to a constant depending on the data, this is also true of AIC. For a regression model with n observations, p parameters and normally-distributed errors the log-likelihood is

$$L(\beta, \sigma^2; \boldsymbol{y}) = \text{const} - \tfrac{n}{2}\log \sigma^2 - \tfrac{1}{2\sigma^2}\|\boldsymbol{y} - X\boldsymbol{\beta}\|^2$$

and on maximizing over β we have

$$L(\widehat{\boldsymbol{\beta}}, \sigma^2; \boldsymbol{y}) = \text{const} - \tfrac{n}{2}\log \sigma^2 - \tfrac{1}{2\sigma^2}\text{RSS}$$

Thus if σ^2 is *known*, we can take

$$\text{AIC} = \tfrac{\text{RSS}}{\sigma^2} + 2p + \text{const}$$

but if σ^2 is *unknown*,

$$L(\widehat{\boldsymbol{\beta}}, \widehat{\sigma}^2; \boldsymbol{y}) = \text{const} - \tfrac{n}{2}\log \widehat{\sigma}^2 - \tfrac{n}{2}, \qquad \widehat{\sigma}^2 = \text{RSS}/n$$

and so

$$\text{AIC} = n\log(\text{RSS}/n) + 2p + \text{const}$$

For *known* σ^2 it is conventional to use Mallows' C_p,

$$C_p = \text{RSS}/\sigma^2 + 2p - n$$

(Mallows, 1973) and in this case `addterm` and `dropterm` label their output as C_p.

For an example consider removing the four-way interaction from the complete model and assessing which three-way terms might be dropped next.

```
> quine.hi <- aov(log(Days + 2.5) ~ .^4, quine)
> quine.nxt <- update(quine.hi, . ~ . - Eth:Sex:Age:Lrn)
> dropterm(quine.nxt, test="F")
Single term deletions
. . . .
             Df Sum of Sq    RSS     AIC F Value    Pr(F)
    <none>                64.099 -68.184
Eth:Sex:Age  3    0.9739 65.073 -71.982  0.6077 0.61125
Eth:Sex:Lrn  1    1.5788 65.678 -66.631  2.9557 0.08816
Eth:Age:Lrn  2    2.1284 66.227 -67.415  1.9923 0.14087
Sex:Age:Lrn  2    1.4662 65.565 -68.882  1.3725 0.25743
```

Clearly dropping `Eth:Sex:Age` most reduces AIC but dropping `Eth:Sex:Lrn` would increase it. Note that only non-marginal terms are included; none are significant in a conventional F-test.

Alternatively we could start from the simplest model and consider adding terms to reduce C_p; in this case the choice of scale parameter is important, since the simple-minded choice is inflated and may over-penalize complex models.

```
> quine.lo <- aov(log(Days+2.5) ~ 1, quine)
> addterm(quine.lo, quine.hi, test="F")
Single term additions
. . . .
     Df Sum of Sq    RSS     AIC F Value    Pr(F)
<none>             106.79 -43.664
Eth  1    10.682  96.11 -57.052 16.006 0.00010
Sex  1     0.597 106.19 -42.483  0.809 0.36981
Age  3     4.747 102.04 -44.303  2.202 0.09048
Lrn  1     0.004 106.78 -41.670  0.006 0.93921
```

It appears that only `Eth` and `Age` might be useful, although in fact all factors are needed since some higher-way interactions lead to large decreases in the residual sum of squares.

Automated model selection

Our function `stepAIC` may be used to automate the process of stepwise selection. It requires a fitted model to define the starting process (one somewhere near the final model is probably advantageous), a list of two formulae defining the upper (most complex) and lower (most simple) models for the process to consider and a scale estimate. If a large model is selected as the starting point, the `scope` and `scale` arguments have generally reasonable defaults, but for a small model where the process is probably to be one of adding terms, they will usually need both to be supplied. (A further argument, `direction`, may be used to specify whether the process should only add terms, only remove terms, or do either as needed.)

By default the function produces a verbose account of the steps it takes which we turn off here for reasons of space, but which the user will often care to note. The `anova` component of the result shows the sequence of steps taken and the reduction in AIC or C_p achieved.

```
> quine.stp <- stepAIC(quine.nxt,
    scope = list(upper = ~Eth*Sex*Age*Lrn, lower = ~1),
    trace = F)
> quine.stp$anova
....
            Step Df Deviance Resid. Df Resid. Dev     AIC
1                                  120    64.099 -68.184
2 - Eth:Sex:Age  3   0.9739       123    65.073 -71.982
3 - Sex:Age:Lrn  2   1.5268       125    66.600 -72.597
```

At this stage we might want to look further at the final model from a significance point of view. The result of stepAIC has the same class as its starting point argument, so in this case dropterm may be used to check each remaining non-marginal term for significance.

```
> dropterm(quine.stp, test="F")
            Df Sum of Sq    RSS     AIC F Value   Pr(F)
    <none>              66.600 -72.597
   Sex:Age  3    10.796 77.396 -56.663  6.7542 0.00029
Eth:Sex:Lrn  1     3.032 69.632 -68.096  5.6916 0.01855
Eth:Age:Lrn  2     2.096 68.696 -72.072  1.9670 0.14418
```

The term Eth:Age:Lrn is not significant at the conventional 5% significance level. This suggests, correctly, that selecting terms on the basis of AIC can be somewhat permissive in its choice of terms, being roughly equivalent to choosing an *F*-cutoff of 2. We can proceed manually

```
> quine.3 <- update(quine.stp, . ~ . - Eth:Age:Lrn)
> dropterm(quine.3, test="F")
            Df Sum of Sq    RSS     AIC F Value   Pr(F)
    <none>              68.696 -72.072
   Eth:Age  3     3.031 71.727 -71.768  1.8679 0.13833
   Sex:Age  3    11.427 80.123 -55.607  7.0419 0.00020
   Age:Lrn  2     2.815 71.511 -70.209  2.6020 0.07807
Eth:Sex:Lrn  1     4.696 73.391 -64.419  8.6809 0.00383
> quine.4 <- update(quine.3, . ~ . - Eth:Age)
> dropterm(quine.4, test="F")
            Df Sum of Sq    RSS     AIC F Value   Pr(F)
    <none>              71.727 -71.768
   Sex:Age  3    11.566 83.292 -55.942  6.987 0.000215
   Age:Lrn  2     2.912 74.639 -69.959  2.639 0.075279
Eth:Sex:Lrn  1     6.818 78.545 -60.511 12.357 0.000605
> quine.5 <- update(quine.4, . ~ . - Age:Lrn)
> dropterm(quine.5, test="F")

Model:
log(Days + 2.5) ~ Eth + Sex + Age + Lrn + Eth:Sex + Eth:Lrn
                + Sex:Age + Sex:Lrn + Eth:Sex:Lrn
            Df Sum of Sq    RSS     AIC F Value    Pr(F)
    <none>              74.639 -69.959
   Sex:Age  3     9.9002 84.539 -57.774  5.836 0.0008944
Eth:Sex:Lrn  1     6.2988 80.937 -60.130 11.140 0.0010982
```

or by setting `k = 4` in `stepAIC`. We obtain a model equivalent to `Sex/(Age + Eth*Lrn)` which is the same as that found by Aitkin (1978), apart from his choice of $\alpha = 1$ for the displacement constant. (However when we consider a negative binomial model for the same data in Section 7.4 a more extensive model seems to be needed.)

Standard diagnostic checks on the residuals from our final fitted model show no strong evidence of any failure of the assumptions, as the reader may wish to verify.

It can also be verified that had we started from a very simple model and worked forwards we would have stopped much sooner with a much simpler model, even using the same scale estimate. This is because the major reductions in the residual sum of squares only occur when the third-order interaction `Eth:Sex:Lrn` is included.

There are other tools for model selection called `stepwise` and `leaps`, but these only apply for quantitative regressors. There is also no possibility of ensuring marginality restrictions are obeyed.

6.9 Predicting computer performance

Ein-Dor & Feldmesser (1987) studied data on the performance on a benchmark of a mix of minicomputers and mainframes. The measure was normalized relative to an IBM 370/158–3. There were six machine characteristics, the cycle time (nanoseconds), the cache size (Kb), the main memory size (Kb) and number of channels. (For the latter two there are minimum and maximum possible values; what the actual machine tested had is unspecified.) The original paper gave a linear regression for the square root of performance, but log scale looks more intuitive.

We can consider the Box–Cox family of transformations.

```
boxcox(perf ~ syct+mmin+mmax+cach+chmin+chmax, data=cpus,
       lambda=seq(0, 1, 0.1))
```

which tends to suggest a power of around 0.3 (and excludes both 0 and 0.5 from its 95% confidence interval). However, this does not allow for the regressors to be transformed, and many of them would be most naturally expressed on log scale. One way to allow the variables to be transformed is to discretize them; we show a more sophisticated approach in Section 9.3.

```
cpus1 <- cpus
attach(cpus)
for(v in names(cpus)[2:6])
  cpus1[[v]] <- cut(cpus[[v]], unique(quantile(cpus[[v]])),
                    include.lowest = T)
detach()
boxcox(perf ~ syct+mmin+mmax+cach+chmin+chmax, data = cpus1,
       lambda = seq(-0.25, 1, 0.1))
```

which does give a confidence interval including zero.

The purpose of this study is to predict computer performance. We randomly select 100 examples for fitting the models and test the performance on the remaining 109 examples.

```
> set.seed(123)
> cpus0 <- cpus1[, 2:8]  # excludes names, authors' predictions
> cpus.samp <- sample(1:209, 100)
> cpus.lm <- lm(log10(perf) ~ ., data=cpus1[cpus.samp,2:8])
> test.cpus <- function(fit)
    sqrt(sum((log10(cpus0[-cpus.samp, "perf"]) -
            predict(fit, cpus0[-cpus.samp,]))^2)/109)
> test.cpus(cpus.lm)
[1] 0.20329
> cpus.lm2 <- stepAIC(cpus.lm, trace=F)
> cpus.lm2$anova
    Step Df Deviance Resid. Df Resid. Dev      AIC
1                          83     3.2300  -309.27
2 - syct  3 0.033248        86     3.2632  -314.25
test.cpus(cpus.lm2)
[1] 0.1955531
```

So selecting a smaller model does improve the prediction a little. Is the difference significant? We can test whether the prediction errors are smaller on the test set by a paired statistical test: as it is moot whether to use the absolute error or squared error, and neither is close to normally distributed, we use a rank test.

```
> res1 <- log10(cpus1[-cpus.samp, "perf"]) -
            predict(cpus.lm, cpus0[-cpus.samp,])
> res2 <- log10(cpus1[-cpus.samp, "perf"]) -
            predict(cpus.lm2, cpus0[-cpus.samp,])
> wilcox.test(res1^2, res2^2, paired=T, alternative="greater")
signed-rank normal statistic with correction Z = 3.3135,
    p-value = 5e-04
```

This is significant. We consider a variety of non-linear models for this example in later chapters.

6.10 Multiple comparisons

As we all know, the theory of p-values of hypothesis tests and of the coverage of confidence intervals applies to pre-planned analyses. However, the only circumstances in which an adjustment is routinely made for testing after looking at the data is in multiple comparisons of contrasts in designed experiments. Consider the experiment on yields of barley in our dataset `immer`.[5] This has the yields of five varieties of barley at six experimental farms in both 1931 and 1932; we average the results for the two years. An analysis of variance gives

---

[5] The Trellis dataset `barley` discussed in Cleveland (1993) is a more extensive version of the same dataset.

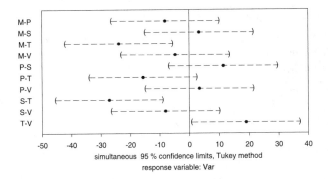

Figure 6.7: Simultaneous 95% confidence intervals for variety comparisons in the `immer` dataset.

```
> immer.aov <- aov((Y1+Y2)/2 ~ Var + Loc, data=immer)
> summary(immer.aov)
             Df  Sum of Sq  Mean Sq  F Value      Pr(F)
     Var      4       2655    663.7    5.989  0.0024526
     Loc      5      10610   2122.1   19.148  0.0000005
Residuals 20       2217    110.8
```

The interest is in the difference in yield between varieties, and there is a statistically significant difference. We can see the mean yields by a call to `model.tables`.

```
> model.tables(immer.aov, type="means", se=T, cterms="Var")
  ....
Var
     M       P       S      T       V
94.392  102.54  91.133  118.2  99.183

Standard errors for differences of means
           Var
         6.078
replic. 6.000
```

This suggests that variety `T` is different from all the others, as a pairwise significant difference at 5% would exceed $6.078 \times t_{20}(0.975) \approx 12.6$; however the comparisons to be made have been selected after looking at the fit.

Function `multicomp`[6] allows us to compute *simultaneous* confidence intervals in this problem, that is, confidence intervals such that the probability that they cover the true values for all of the comparisons considered is bounded above by 5% for 95% confidence intervals. We can also plot the confidence intervals (Figure 6.7) by

```
> multicomp(immer.aov, plot=T)
```

---

[6] Available in S-PLUS 4.0 and later.

```
95 % simultaneous confidence intervals for specified
linear combinations, by the Tukey method

critical point: 2.9925
response variable: (Y1+Y2)/2

intervals excluding 0 are flagged by '****'

      Estimate Std.Error Lower Bound Upper Bound
M-P     -8.15     6.08      -26.300      10.00
M-S      3.26     6.08      -14.900      21.40
M-T    -23.80     6.08      -42.000      -5.62 ****
M-V     -4.79     6.08      -23.000      13.40
P-S     11.40     6.08       -6.780      29.60
P-T    -15.70     6.08      -33.800       2.53
P-V      3.36     6.08      -14.800      21.50
S-T    -27.10     6.08      -45.300      -8.88 ****
S-V     -8.05     6.08      -26.200      10.10
T-V     19.00     6.08        0.828      37.20 ****
```

This does not allow us to conclude that variety T has a significantly diferent yield
than variety P.

We may want to restrict the set of comparisons, for example to comparisons
with a control treatment. The dataset `oats` is discussed on page 195; here we
ignore the split-plot structure.

```
> oats1 <- aov(Y ~ N + V + B, data=oats)
> summary(oats1)
           Df Sum of Sq Mean Sq F Value     Pr(F)
    N   3      20020  6673.5  28.460 0.000000
    V   2       1786   893.2   3.809 0.027617
    B   5      15875  3175.1  13.540 0.000000
Residuals 61   14304   234.5
> multicomp(oats1, focus="V")

95 % simultaneous confidence intervals for specified
linear combinations, by the Tukey method

critical point: 2.4022
response variable: Y

intervals excluding 0 are flagged by '****'

                        Estimate Std.Error Lower Bound
Golden.rain-Marvellous    -5.29     4.42       -15.90
   Golden.rain-Victory     6.88     4.42        -3.74
   Marvellous-Victory     12.20     4.42         1.55
                        Upper Bound
Golden.rain-Marvellous        5.33
   Golden.rain-Victory       17.50
```

```
    Marvellous-Victory        22.80 ****

> multicomp(oats1, focus="N", comparison="mcc", control=1)
  . . . .
                Estimate Std.Error Lower Bound Upper Bound
  0.2cwt-0.0cwt    19.5      5.1        7.24        31.8 ****
  0.4cwt-0.0cwt    34.8      5.1       22.60        47.1 ****
  0.6cwt-0.0cwt    44.0      5.1       31.70        56.3 ****
```

Note that we need to specify the control level: perversely by default the last level is chosen. We might also want to know if all the increases in nitrogen give significant increases in yield, which we can examine by

```
> lmat <- matrix(c(0,-1,1,rep(0, 11), 0,0,-1,1, rep(0,10),
              0,0,0,-1,1,rep(0,9)),,3, dimnames=list(NULL,
        c("0.2cwt-0.0cwt", "0.4cwt-0.2cwt", "0.6cwt-0.4cwt")))
> multicomp(oats1, lmat=lmat, bounds="lower", comparisons="none")
  . . . .
                Estimate Std.Error Lower Bound
  0.2cwt-0.0cwt    19.50     5.1        8.43 ****
  0.4cwt-0.2cwt    15.30     5.1        4.27 ****
  0.6cwt-0.4cwt     9.17     5.1       -1.90
```

There is a bewildering variety of methods for multiple comparisons reflected in the options for `multicomp`. Miller (1981a), Hsu (1996) and Yandell (1997, Chapter 6) give fuller details. Do remember that this tackles only part of the problem; the analyses here have been done after selecting a model and specific factors on which to focus: the allowance for multiple comparisons is only over contrasts of one selected factor in one selected model.

6.11   Random and mixed effects

There are several ways in S-PLUS to handle linear models with random effects, possibly as well as fixed effects.

The function `raov` may be used for balanced designs with only random effects, and gives a conventional analysis including the estimation of variance components. The function `varcomp` is more general, and may be used to estimate variance components for balanced or unbalanced mixed models.

To illustrate these functions we take a portion of the data from the data frame `coop` of our library MASS. This is a co-operative trial in analytical chemistry taken from Analytical Methods Committee (1987). Seven specimens were sent to six laboratories, each three times a month apart for duplicate analysis. The response is the concentration of (unnamed) analyte in g/kg. We use the data from Specimen 1 shown in Table 6.2.

Table 6.2: Analyte concentrations (g/kg) from six laboratories and three batches.

Batch	Laboratory					
	1	2	3	4	5	6
1	0.29	0.40	0.40	0.9	0.44	0.38
	0.33	0.40	0.35	1.3	0.44	0.39
2	0.33	0.43	0.38	0.9	0.45	0.40
	0.32	0.36	0.32	1.1	0.45	0.46
3	0.34	0.42	0.38	0.9	0.42	0.72
	0.31	0.40	0.33	0.9	0.46	0.79

The purpose of the study was to assess components of variation in co-operative trials. For this purpose laboratories and batches are regarded as random, so a model for the response for laboratory i, batch j and duplicate k is

$$y_{ijk} = \mu + \xi_i + \beta_{ij} + \epsilon_{ijk}$$

where ξ, β and ϵ are independent random variables with zero means and variances σ_L^2, σ_B^2 and σ_e^2 respectively. For l laboratories, b batches and $r = 2$ duplicates a nested analysis of variance gives:

Source of variation	Degrees of freedom	Sum of squares	Mean square	E(MS)
Between laboratories	$l - 1$	$br \sum_i (\bar{y}_i - \bar{y})^2$	MS_L	$br\sigma_L^2 + r\sigma_B^2 + \sigma_e^2$
Batches within laboratories	$l(b - 1)$	$r \sum_{ij} (\bar{y}_{ij} - \bar{y}_i)^2$	MS_B	$r\sigma_B^2 + \sigma_e^2$
Replicates within batches	$lb(r - 1)$	$\sum_{ijk} (y_{ij} - \bar{y}_{ij})^2$	MS_e	σ_e^2

So the unbiased estimators of the variance components are

$$\hat{\sigma}_L^2 = (br)^{-1}(MS_L - MS_B), \quad \hat{\sigma}_B^2 = r^{-1}(MS_B - MS_e), \quad \hat{\sigma}_e^2 = MS_e$$

The model is fitted in the same way as an analysis of variance model with `raov` replacing `aov`:

```
> summary(raov(Conc ~ Lab/Bat, data = coop, subset = Spc=="S1"))
              Df Sum of Sq Mean Sq Est. Var.
Lab            5   1.8902  0.37804  0.060168
Bat %in% Lab  12   0.2044  0.01703  0.005368
Residuals     18   0.1134  0.00630  0.006297
```

The same variance component estimates can be found using `varcomp`, but as this allows mixed models we need first to declare which factors are random effects using the `is.random` function. All factors in a data frame are declared to be random effects by

```
> coop <- coop  # make a local copy (needed in 5.x)
> is.random(coop) <- T
```

If we use `Spc` as a factor it may need to be treated as fixed. Individual factors can also have their status declared in the same way. When used as an unassigned expression `is.random` reports the status of (all the factors in) its argument:

```
> is.random(coop$Spc) <- F
> is.random(coop)
 Lab Spc Bat
   T   F   T
```

We can now estimate the variance components:

```
> varcomp(Conc ~ Lab/Bat, data = coop, subset = Spc=="S1")
Variances:
      Lab Bat %in% Lab Residuals
 0.060168    0.0053681 0.0062972
```

The fitted values for such a model are technically the grand mean, but the `fitted` method function calculates them as if it were a fixed effects model, so giving *best linear unbiased predictors* (BLUPs; see the review paper by Robinson, 1991). Residuals are also calculated as differences of observations from the BLUPs. An examination of the residuals and fitted values points up laboratory 4 batches 1 and 2 as possibly suspect, and these will inflate the estimate of σ_e^2. (However laboratory 4 overall and laboratory 6 batch 3 are outliers for the other variance components, but a simple residual analysis will not expose this. See Analytical Methods Committee, 1987.)

Variance component estimates are known to be very sensitive to aberrant observations; we can get some check on this by repeating the analysis with a robust estimating method. The result is very different:

```
> varcomp(Conc ~ Lab/Bat, data = coop, subset = Spc=="S1",
    method = c("winsor", "minque0"))
Variances:
       Lab Bat %in% Lab Residuals
 0.0040043  -0.00065681 0.0062979
```

(A related robust analysis is given in Analytical Methods Committee, 1989b.)

The possible estimation methods are `"minque0"` for minimum norm quadratic estimators (the default), `"ml"` for maximum likelihood and `"reml"` for residual (or reduced or restricted) maximum likelihood. (See, for example, Rao, 1971a,b and Rao & Kleffe, 1988.) Method `"winsor"` specifies that the data are 'cleaned' before further analysis. As in our example, a second method may be prescribed as the one to use for estimation of the variance components on the cleaned data. The method used, Winsorizing, is discussed in Section 5.5; unfortunately the tuning constant is not adjustable, and is set for rather mild data cleaning.

Multistratum models

Multistratum models occur where there is more than one source of random variation in an experiment, as in a split-plot experiment, and have been most used in agricultural field trials. The sample information and the model for the means of the observations may be partitioned into so-called *strata*. In orthogonal experiments such as split-plot designs each parameter may be estimated in one and only one stratum, but in non-orthogonal experiments such as a balanced incomplete block design with 'random blocks', treatment contrasts are estimable both from the between-block and the within-block strata. Combining the estimates from two strata is known as 'the recovery of interblock information', for which the interested reader should consult a comprehensive reference such as Scheffé (1959, pp. 170ff). In this section we consider the separate stratum analyses only.

The details of the computational scheme employed for multistratum experiments are given by Heiberger (1989).

A split-plot experiment

Our example was first used by Yates (1935) and is discussed further in Yates (1937). It has also been used by many authors since; see, for example, John (1971, §5.7). The experiment involved varieties of oats and manure (nitrogen), conducted in six blocks of three whole plots. Each whole plot was divided into four subplots. Three varieties of oats were used in the experiment with one variety being sown in each whole plot (this being a limitation of the seed drill used), while four levels of manure (0, 0.2, 0.4 and 0.6cwt per acre) were used, one level in each of the four subplots of each whole plot.

The data are shown in Table 6.3, where the blocks are labelled I–VI, V_i denotes the ith variety and N_j denotes the jth level of nitrogen. The dataset is available in our library as the data frame `oats`, with variables B, V, N and Y. Since the levels of N are ordered it is appropriate to create an ordered factor:

```
oats <- oats  # make a local copy: needed in 5.x
oats$Nf <- ordered(oats$N, levels=sort(levels(oats$N)))
```

The strata for the model we use here are:

1. a 1-dimensional stratum corresponding to the total of all observations,
2. a 5-dimensional stratum corresponding to contrasts between block totals,
3. a 12-dimensional stratum corresponding to contrasts between variety (or equivalently whole plot) totals within the same block, and
4. a 54-dimensional stratum corresponding to contrasts within whole plots.

(We use the term 'contrast' here to mean a linear function with coefficients adding to zero, and thus representing a comparison.)

Only the overall mean is estimable within stratum 1 and since no degrees of freedom are left for error, this stratum is suppressed in the printed summaries (although a component corresponding to it is present in the fitted-model object). Stratum 2 has no information on any treatment effect. Information on the V

Table 6.3: A split-plot field trial of oat varieties.

		N_1	N_2	N_3	N_4			N_1	N_2	N_3	N_4
	V_1	111	130	157	174		V_1	74	89	81	122
I	V_2	117	114	161	141	IV	V_2	64	103	132	133
	V_3	105	140	118	156		V_3	70	89	104	117
	V_1	61	91	97	100		V_1	62	90	100	116
II	V_2	70	108	126	149	V	V_2	80	82	94	126
	V_3	96	124	121	144		V_3	63	70	109	99
	V_1	68	64	112	86		V_1	53	74	118	113
III	V_2	60	102	89	96	VI	V_2	89	82	86	104
	V_3	89	129	132	124		V_3	97	99	119	121

main effect is only available from stratum 3, and the N main effect and N × V interaction are only estimable within stratum 4.

Multistratum models may be fitted using `aov` (only), and are specified by a model formula of the form

$$response ~\sim~ mean.formula ~+~ \texttt{Error}(strata.formula)$$

In our example the *strata.formula* is B/V, specifying strata 2 and 3; the fourth stratum is included automatically as the "within" stratum, the residual stratum from the strata formula.

The fitted-model object is of class `aovlist`, which is a list of fitted-model objects corresponding to the individual strata. Each component is an object of class `aov` (although in some respects incomplete). There is no `anova` method function for class `aovlist`. The appropriate display function is `summary` which displays separately the ANOVA tables for each stratum (except the first).

```
> oats.aov <- aov(Y ~ Nf*V + Error(B/V), data = oats, qr = T)
> summary(oats.aov)

Error: B
          Df Sum of Sq Mean Sq F Value Pr(F)
Residuals  5     15875  3175.1

Error: V %in% B
          Df Sum of Sq Mean Sq F Value    Pr(F)
V          2    1786.4  893.18  1.4853 0.27239
Residuals 10    6013.3  601.33

Error: Within
          Df Sum of Sq Mean Sq F Value  Pr(F)
```

```
Nf          3      20020   6673.5  37.686 0.0000
Nf:V        6        322     53.6   0.303 0.9322
Residuals 45       7969    177.1
```

There is a clear nitrogen effect, but no evidence of variety differences nor interactions. Since the levels of Nf are quantitative, it is natural to consider partitioning the sums of squares for nitrogen and its interactions into polynomial components. Here the details on factor coding become important.

Since the levels of nitrogen are equally spaced, and we chose an ordered factor, the default contrast matrix is appropriate. We can obtain an analysis of variance table with the degrees of freedom for nitrogen partitioned into components by giving an additional argument to the summary function:

```
> summary(oats.aov, split=list(Nf=list(L=1, Dev=2:3)))
    ....
Error: Within
           Df Sum of Sq Mean Sq F Value    Pr(F)
Nf          3     20020    6673   37.69 0.00000
    Nf: L   1     19536   19536  110.32 0.00000
    Nf: Dev 2       484     242    1.37 0.26528
Nf:V        6       322      54    0.30 0.93220
    Nf:V: L 2       168      84    0.48 0.62476
  Nf:V: Dev 4       153      38    0.22 0.92786
Residuals  45      7969     177
```

The split argument is a list with the name of each component that of some factor appearing in the model. Each component is also a list with components of the form name=int.seq where name is the name for the table entry and int.seq is an integer sequence giving the contrasts to be grouped under that name in the ANOVA table.

Residuals in multistratum analyses: Projections

Residuals and fitted values from the individual strata are available in the usual way by accessing each component as a fitted-model object. Thus fitted(oats.aov[[4]]) and resid(oats.aov[[4]]) are vectors of length 54 representing fitted values and residuals from the last stratum, based on 54 orthonormal linear functions of the original data vector. It is not possible to associate them uniquely with the plots of the original experiment, which for field trials would sometimes be very useful.

The function proj takes a fitted model object and finds the projections of the original data vector onto the subspaces defined by each line in the analysis of variance tables (including, for multistratum objects, the suppressed table with the grand mean only). The result is a list of matrices, one for each stratum, where the column names for each are the component names from the analysis of variance tables. As the argument qr=T has been set when the model was initially fitted this calculation is considerably faster. Two diagnostic plots are shown in Figure 6.8, computed by

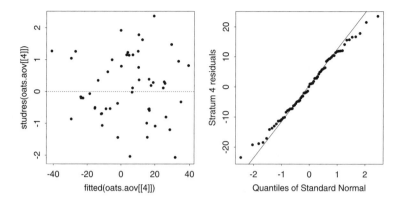

Figure 6.8: Two diagnostic plots for the oats data. (The exact plot obtained varies among releases of S-PLUS.)

```
plot(fitted(oats.aov[[4]]), studres(oats.aov[[4]]))
abline(h=0, lty=2)
oats.pr <- proj(oats.aov)
qqnorm(oats.pr[[4]][,"Residuals"], ylab="Stratum 4 residuals")
qqline(oats.pr[[4]][,"Residuals"])
```

Tables of means and components of variance

A better appreciation of the results of the experiment comes from looking at the estimated marginal means. The function `model.tables` calculates tables of the effects, means or residuals and optionally their standard errors; tables of means are really only well defined and useful for balanced designs[7], and the standard errors calculated are for differences of means. We now refit the model omitting the `V:N` interaction, but retaining the `V` main effect (since that might be considered a blocking factor). Since the `model.tables` calculation also requires the projections we should fit the model with either `qr=T` or `proj=T` set to avoid later refitting.

```
> oats.aov <- aov(Y ~ N + V + Error(B/V), data = oats, qr = T)
> model.tables(oats.aov, type = "means", se = T)
Tables of means
   . . . .
N
0.0cwt 0.2cwt 0.4cwt 0.6cwt
 79.39  98.89  114.2  123.4
V
Golden.rain Marvellous Victory
      104.5      109.8   97.63
```

---

[7] In S-PLUS 4.5 and later one definition of marginal means in unbalanced designs is provided called 'adjusted means', but we feel this is of limited utility.

```
Standard errors for differences of means
                N       V
              4.25    7.079
replic.      18.00   24.000
```

When interactions are present the table is a little more complicated.

Finally we use `varcomp` to estimate the variance components associated with the lowest three strata. To do this we need to declare B to be a random factor:

```
> is.random(oats$B) <- T
> varcomp(Y ~ N + V + B/V, data = oats)
Variances:
      B V %in% B Residuals
 214.48    109.69    162.56
    . . . .
```

In this simple balanced design the estimates are the same as those obtained by equating the residual mean squares to their expectations. The result suggests that the main blocking scheme, at least, has been reasonably effective.

Linear mixed effects models

S-PLUS 3.4 and later contain the `nlme` software of Pinheiro, Bates and Lindstrom for fitting linear and nonlinear mixed effects models, but a later version (3.0) that will handle multiple error strata is included in S-PLUS 2000, and that version is described here (see page 471 for how to obtain it).

The simplest case where a mixed effects model might be considered is when there are two stages of sampling. At the first stage units are selected at random from a population and at the second stage a number of measurements is made on each unit sampled in the first stage. For example, in the simple split-plot analysis of variance model the blocks are regarded as being chosen at random from a population of blocks and the plots within the blocks give the repeated measurements within the primary sampled unit. Another common situation giving rise to a mixed effects model is when measurements are made sequentially in time or space on the primary sampling units. Common names for such datasets analysed by mixed effects models include *repeated measures* data and *longitudinal* data. A good reference for both linear and nonlinear mixed effects models is Davidian & Giltinan (1995); Pinheiro, Bates and Lindstrom give Laird & Ware (1982) as their primary reference.

The *gasoline* data (Prater, 1956) were originally used to build an estimation equation for yield of the refining process. The possible predictor variables were the specific gravity, vapour pressure and ASTM 10% point made on the crude oil itself and the volatility of the desired product measured as the ASTM end point. The dataset is included in our library as `petrol`. Several authors (including Daniel & Wood, 1980 and Hand *et al.*, 1993) have noted that the three measurements made on the crude oil occur in only 10 discrete sets and have inferred that only 10 crude oil samples were involved and several yields with varying ASTM end points were measured on each crude oil sample. We adopt this view here.

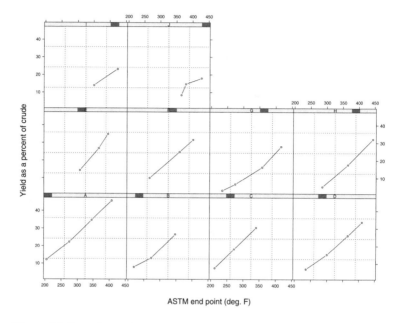

Figure 6.9: Prater's gasoline data: yield versus ASTM end point within samples.

A natural way to inspect the data is to plot the yield against the ASTM end point for each sample, for example, using Trellis graphics by

```
xyplot(Y ~ EP | No, data = petrol,
    xlab = "ASTM end point (deg. F)",
    ylab = "Yield as a percent of crude",
    panel = function(x, y) {
        m <- sort.list(x)
        panel.grid()
        panel.xyplot(x[m], y[m], type = "b", cex = 0.5)
    })
```

The result shown in Figure 6.9 shows a fairly consistent and generally linear rise in yield with the ASTM end point but with some variation in intercept that may correspond to differences in the covariates made on the crude oil itself. In fact the intercepts appear to be steadily decreasing and since the samples are arranged so that the value of V10 is increasing this suggests a significant regression on V10 with negative coefficient.

If we regard the 10 crude oil samples as fixed the most general model we might consider would have separate simple linear regressions of Y on EP for each crude oil sample. First we centre the determining variables so that the origin is in the centre of the design space, thus making the intercepts simpler to interpret.

```
> Petrol <- petrol
> names(Petrol)
[1] "No"  "SG"  "VP"  "V10" "EP"  "Y"
> Petrol[, 2:5] <- scale(as.matrix(Petrol[, 2:5]), scale = F)
> pet1.lm <- lm(Y ~ No/EP - 1, Petrol)
> matrix(round(coef(pet1.lm),2), 2, 10, byrow = T, dimnames =
    list(c("b0","b1"),levels(Petrol$No)))

       A     B     C     D     E     F     G     H     I    J
b0 32.75 23.62 28.99 21.13 19.82 20.42 14.78 12.68 11.62 6.18
b1  0.17  0.15  0.18  0.15  0.23  0.16  0.14  0.17  0.13 0.13
```

There is a large variation in intercepts, as expected, and some variation in slopes. We test if the slopes can be considered constant; as they can we adopt the simpler parallel-line regression model.

```
> pet2.lm <- lm(Y ~ No - 1 + EP, Petrol)
> anova(pet2.lm, pet1.lm)
       Terms RDf    RSS Df Sum of Sq F Value    Pr(F)
No - 1 + EP   21 74.132
  No/EP - 1   12 30.329  9    43.803  1.9257   0.1439
```

We still need to explain the variation in intercepts; we try a regression on SG, VP and V10.

```
> pet3.lm <- lm(Y ~ SG + VP + V10 + EP, Petrol)
> anova(pet3.lm, pet2.lm)
               Terms RDf    RSS Df Sum of Sq F Value     Pr(F)
SG + VP + V10 + EP    27 134.80
       No - 1 + EP    21  74.13  6    60.672  2.8645  0.033681
```

(Notice that SG, VP and V10 are constant within the levels of No so these two models are genuinely nested.) The result suggests that differences between intercepts are *not* adequately explained by such a regression.

A promising way of generalizing the model is to assume that the 10 crude oil samples form a random sample from a population where the intercepts after regression on the determining variables depend on the sample. The model we investigate has the form

$$y_{ij} = \mu + e_i + \beta_1 \, \text{SG}_i + \beta_2 \, \text{VP}_i + \beta_3 \, \text{V10}_i + \beta_4 \, \text{EP}_{ij} + \epsilon_{ij}$$

where i denotes the sample and j the observation on that sample, and $e_i \sim N(0, \sigma_1^2)$ and $\epsilon_{ij} \sim N(0, \sigma^2)$, independently. This allows for the possibility that predictions from the fitted equation will need to cope with two sources of error, one associated with the sampling process for the crude oil sample and the other with the measurement process within the sample itself.

The fitting function for linear mixed effects models is `lme`. Its arguments are numerous but the main ones are the first three that specify the fixed effects model, the random effects including the factor(s) defining the 'clusters' over which the random effects vary, both as formulae, and the usual data frame. For our example we can use

```
library(nlme3, first=T)          # must have first=T
pet3.lme <- lme(Y ~ SG + VP + V10 + EP,
                random = ~ 1 | No, data = Petrol)
```

(Fits by `lme` are slower than those by `lm`.)

```
> summary(pet3.lme)
Linear mixed-effects model fit by REML
 Data: Petrol
      AIC    BIC  logLik
   166.38 175.45 -76.191

Random effects:
 Formula: ~1 | No
         (Intercept) Residual
 StdDev:      1.4447   1.8722

Fixed effects: Y ~ SG + VP + V10 + EP
               Value Std.Error DF t-value p-value
(Intercept)  19.707   0.56827 21  34.679  <.0001
         SG   0.219   0.14694  6   1.493  0.1860
         VP   0.546   0.52052  6   1.049  0.3347
        V10  -0.154   0.03996  6  -3.860  0.0084
         EP   0.157   0.00559 21  28.128  <.0001
 . . . .
```

The estimate of the residual variance within clusters is $\hat{\sigma}^2 = 1.8722^2 = 3.5052$ and the variance of the random effects between clusters is $\hat{\sigma}_1^2 = 1.444^2 = 2.0873$.

The estimation method used is REML. The parameters in the variance structure (such as the variance components) are estimated by maximizing the marginal likelihood of the residuals from a least-squares fit of the linear model, and then the fixed effects are estimated by maximum likelihood assuming that the variance structure is known, which amounts to fitting by generalized least squares. (This is a reasonable procedure since taking the residuals from *any* generalized least-squares fit will lead to the same marginal likelihood: McCullagh & Nelder (1989, pp. 247, 282–3).) With such a small sample and so few clusters the difference between REML and maximum likelihood estimation is appreciable, especially in the variance estimates and hence the standard error estimates.

```
> pet3.lme <- update(pet3.lme, method="ML")
> summary(pet3.lme)
Linear mixed-effects model fit by maximum likelihood
 Data: Petrol
      AIC    BIC  logLik
```

```
    149.38 159.64 -67.692

Random effects:
 Formula:   ~ 1 | No
         (Intercept) Residual
 StdDev:     0.92889   1.8273

Fixed effects: Y ~ SG + VP + V10 + EP
              Value Std.Error DF t-value p-value
(Intercept) 19.694     0.478 21  41.188   0.000
        SG   0.221     0.123  6   1.802   0.122
        VP   0.549     0.441  6   1.246   0.259
       V10  -0.153     0.034  6  -4.469   0.004
        EP   0.156     0.006 21  26.620   0.000
   . . . .
```

Both tables of fixed effects suggest that only V10 and EP may be useful predictors (although the `z ratio` columns assume that the asymptotic theory is appropriate, in particular that the estimates of variance components are accurate). We can check this by fitting it as a submodel and performing a likelihood ratio test. However for testing hypotheses on the fixed effects the estimation method *must* be set to ML to ensure the test compares likelihoods based on the same data.

```
> pet4.lme <- update(pet3.lme, fixed = Y ~ V10 + EP)
> anova(pet4.lme, pet3.lme)
          Model df     AIC     BIC  logLik    Test Lik.Ratio
pet4.lme      1  5 149.61 156.94 -69.806
pet3.lme      2  7 149.38 159.64 -67.692 1 vs. 2    4.2285
          p-value
pet4.lme
pet3.lme  0.1207
> coef(pet4.lme)
    (Intercept)     V10      EP
A        21.054 -0.21081 0.15759
   . . . .
```

Note how the `coef` method combines the slopes (the fixed effects) and the (random effect) intercept for each sample. The coefficients of the fixed effects may be extracted by the function `fixed.effects` (short form `fixef`).

The AIC column gives (as in Section 6.8) minus twice the log-likelihood plus twice the number of parameters, and BIC refers to the criterion of Schwarz (1978) with penalty $\log n$ times the number of parameters.

Finally we check if we need both random regression intercepts and slopes on EP, so we fit the model

$$y_{ij} = \mu + e_i + \beta_3 \text{V10}_i + (\beta_4 + \eta_i) \text{EP}_{ij} + \epsilon_{ij}$$

where (e_i, η_i) and ϵ_{ij} are independent, but e_i and η_i can be correlated.

```
> pet5.lme <- update(pet4.lme, random = ~ 1 + EP | No)
> anova(pet4.lme, pet5.lme)
           Model df     AIC     BIC  logLik      Test Lik.Ratio
pet4.lme       1  5 149.61 156.94 -69.806
pet5.lme       2  7 153.61 163.87 -69.805 1 vs. 2 0.0025194
           p-value
pet4.lme
pet5.lme  0.9987
```

The test is non-significant, leading us to retain the simpler model. Notice that the degrees of freedom for testing this hypothesis are $7 - 5 = 2$. The additional two parameters are the variance of the random slopes and the covariance between the random slopes and random intercepts.

Prediction and fitted values

A little thought must be given to fitted and predicted values for a mixed effects model. Do we want the prediction for a new data point in an existing cluster, or from a new cluster; in other words, which of the random effects are the same? This is handled by the argument `level`. For our final model `pet4.lme` the fitted values might be

$$\hat{\mu} + \hat{\beta}_3 \, \mathsf{V10}_i + \hat{\beta}_4 \, \mathsf{EP}_{ij} \qquad \text{and} \qquad \hat{\mu} + \hat{e}_i + \hat{\beta}_3 \, \mathsf{V10}_i + \hat{\beta}_4 \, \mathsf{EP}_{ij}$$

given by `level=0` and `level=1` (the default). Random effects are set either to zero or to their BLUP values.

Relation to `raov` *and multistratum models*

The `raov` and multistratum models are also linear mixed effects models, and showing how they might be fitted with `lme` allows us to demonstrate some further features of that function. In both cases we need more than one level of random effects, which we specify by a nested model in the `random` formula.

For the cooperative trial `coop` we can use

```
> lme(Conc ~ 1, random = ~1 | Lab/Bat, data = coop,
      subset = Spc=="S1")
Linear mixed-effects model fit by REML
  Data: coop
  Log-restricted-likelihood: 21.022
  Fixed: Conc ~ 1
  (Intercept)
      0.50806

Random effects:
 Formula:  ~ 1 | Lab
         (Intercept)
StdDev:     0.24529

 Formula:  ~ 1 | Bat %in% Lab
```

```
          (Intercept) Residual
StdDev:    0.073267 0.079355

Number of Observations: 36
Number of Groups:
 Lab Bat %in% Lab
    6          18
```

Notice that these are REML estimates which agree with the minque default for varcomp as the design is balanced.

The final model we used for the oats dataset corresponds to the linear model

$$y_{bnv} = \mu + \alpha_n + \beta_v + \eta_b + \zeta_{bv} + \epsilon_{bnv}$$

with three random terms corresponding to plots, subplots and observations. We can code this in lme by

```
options(contrasts = c("contr.treatment", "contr.poly"))
oats.lme <- lme(Y ~ N + V, random = ~1 | B/V, data = oats)
summary(oats.lme)
 Data: oats
      AIC    BIC  logLik
   586.07 605.78 -284.03

Random effects:
 Formula:  ~ 1 | B
         (Intercept)
StdDev:      14.645

 Formula:  ~ 1 | V %in% B
         (Intercept) Residual
StdDev:      10.473    12.75

Fixed effects: Y ~ N + V
                Value Std.Error DF t-value p-value
(Intercept)    79.917    8.2203 51   9.722  <.0001
    NO.2cwt    19.500    4.2500 51   4.588  <.0001
    NO.4cwt    34.833    4.2500 51   8.196  <.0001
    NO.6cwt    44.000    4.2500 51  10.353  <.0001
VMarvellous     5.292    7.0789 10   0.748  0.4720
   VVictory    -6.875    7.0789 10  -0.971  0.3544
    . . . .
```

The variance components are estimated as $14.645^2 = 214.48$, $10.4735^2 = 109.69$ and $12.75^2 = 162.56$. The standard errors for treatment differences also agree.

It is time to specify precisely what models lme considers. For a single level of random effects, let i index the clusters and j the measurements on each cluster. Then the linear mixed-effects model is

$$y_{ij} = \mu_{ij} + z_{ij}\eta_i + \epsilon_{ij}, \qquad \mu_{ij} = x_{ij}\beta \qquad (6.8)$$

Here x and z are row vectors of explanatory variables associated with each measurement. The ϵ_{ij} are independent for each cluster but possibly correlated within clusters, so

$$\text{var}(\epsilon_{ij}) = \sigma^2 g(\mu_{ij}, z_{ij}, \theta), \qquad \text{corr}(\epsilon_{ij}) = \Gamma(\alpha) \qquad (6.9)$$

Thus the variances are allowed to depend on the means and other covariates. The random effects η_i are independent of the ϵ_{ij} with a variance matrix

$$\text{var}(\eta_i) = D(\alpha_\eta) \qquad (6.10)$$

Almost all examples will be much simpler than this, and by default $g \equiv 1$, $\Gamma = I$ and D is unrestricted; in the previous subsection we imposed restrictions on D.

Full details of how to specify the components of a linear mixed-effects model can be obtained by `help(lme)`. It is possible for users to supply their own functions to compute many of these components. The free parameters $(\sigma^2, \theta, \alpha, \alpha_\eta)$ in the specification of the error distribution are estimated, by REML by default.

Diggle, Liang & Zeger (1994) give an example of repeated measurements on the log-size[8] of 79 Sitka spruce trees, 54 of which were grown in ozone-enriched chambers and 25 of which were controls. The size was measured five times in 1988, at roughly monthly intervals. (The time is given in days since 1 January 1988. There were a further eight measurements in 1989 given in data frame `Sitka89`.) We consider a random effect of the tree and serial correlation of the residual terms for each tree.

We first consider a general curve on the five times, for each of the two groups, then a linear difference between the two groups. Taking `ordered(Time)` parametrizes the curve by polynomial components.

```
> sitka.lme <- lme(size ~ treat*ordered(Time),
                   random = ~1 | tree, data = Sitka)
> summary(sitka.lme)
Linear mixed-effects model fit by REML
 Data: Sitka
      AIC    BIC  logLik
   79.901 127.34 -27.95

Random effects:
 Formula:  ~ 1 | tree
         (Intercept) Residual
 StdDev:     0.61011  0.16105

Fixed effects: size ~ treat * ordered(Time)
                        Value Std.Error  DF t-value p-value
       (Intercept)     4.9851   0.12287 308  40.572  <.0001
             treat   -0.2112   0.14861  77  -1.421  0.1594
 ordered(Time).L      1.1971   0.03221 308  37.166  <.0001
 ordered(Time).Q     -0.1341   0.03221 308  -4.162  <.0001
```

---

[8] By convention this is log height plus twice log diameter.

```
         ordered(Time).C -0.0409   0.03221 308  -1.268  0.2056
         ordered(Time) ^ 4 -0.0273   0.03221 308  -0.848  0.3974
    treatordered(Time).L -0.1786   0.03896 308  -4.583  <.0001
    treatordered(Time).Q -0.0264   0.03896 308  -0.679  0.4977
    treatordered(Time).C -0.0142   0.03896 308  -0.366  0.7148
treatordered(Time) ^ 4  0.0124   0.03896 308   0.318  0.7504

    ....
> Sitka <- Sitka
> attach(Sitka)
> Sitka$treatslope <- Time * (treat=="ozone")
> detach()
> sitka.lme2 <- update(sitka.lme,
      fixed = size ~ ordered(Time) + treat + treatslope)
> summary(sitka.lme2)
Linear mixed-effects model fit by REML
 Data: Sitka
      AIC    BIC   logLik
   69.269 104.92 -25.635

Random effects:
 Formula:  ~ 1 | tree
         (Intercept) Residual
StdDev:      0.61015   0.1604

Fixed effects: size ~ ordered(Time) + treat + treatslope
                  Value Std.Error  DF t-value p-value
      (Intercept) 4.9851   0.12287 311  40.572  <.0001
  ordered(Time).L 1.1976   0.03204 311  37.372  <.0001
  ordered(Time).Q -0.1455   0.01810 311  -8.037  <.0001
  ordered(Time).C -0.0506   0.01805 311  -2.804  0.0054
ordered(Time) ^ 4 -0.0167   0.01805 311  -0.926  0.3549
            treat 0.2217   0.17561  77   1.262  0.2107
       treatslope -0.0021   0.00046 311  -4.626  <.0001

    ....
# fitted curve for ozone-enriched
fitted(sitka.lme2, level=0)[1:5]
 4.0606 4.4709 4.8427 5.1789 5.3167
fitted(sitka.lme2, level=0)[301:305]
 4.164 4.6213 5.0509 5.4427 5.6467
```

This shows a significant difference between the two groups, which varies linearly with time. (This can be confirmed by maximum-likelihood-based tests.)

We could try to allow for the effect of serial correlation by

```
sitka.lme <- lme(size ~ treat*ordered(Time), random = ~1 | tree,
              data = Sitka, corr = corCAR1(, ~Time | tree))
summary(sitka.lme)
```

but this fails to optimize over the correlation. Setting an initial value by corCAR1(0.9, Time | tree) produces a fit in which the correlation is essentially one (0.9989) but the standard errors are hardly changed.

6.12   Exercises

6.1.   Apply regression diagnostics to the fits to the `whiteside` energy consumption data. Note (Figure 6.1) that the evidence for a quadratic fit to the 'after' data stems from points with high leverage, so also try resistant fits.

6.2.   Dataset `cabbages` gives the results of a field trial on the growth of cabbages (Rawlings, 1988, p. 219). Analyse this trial.

6.3.   The data frame `rubber` in the library `MASS` gives 30 measurements of rubber loss under accelerated testing together with the hardness and tensile strength of the rubber itself. Explore the data in `brush`, then fit linear and quadratic regressions of `loss` on `hard` and `tens`. Select a suitable submodel of the quadratic model, and inspect the fitted surface by a perspective plot.

6.4.   Criminologists are interested in the effect of punishment regimes on crime rates. This has been studied using aggregate data on 47 states of the USA for 1960, available in data frame `UScrime` (Ehrlich, 1973; Vandaele, 1978; Raftery, 1995). The response variable is the rate of crimes in a particular category per head of population. There are 15 explanatory variables; most of these and the response variable have been rescaled to convenient numbers.

(a) Analyse these data. In your report pay particular attention to how your model was selected.

(b) Comment on the effect of the last two explanatory variables in relation to the criminologists' interest in the effect of punishment.

(c) Comment critically on the assumptions needed to draw conclusions from aggregate studies such as this.

6.5.   Susan Prosser collected data on the concentration of a chemical GAG in the urine of 314 children aged from 0 to 17 years. The data are in data frame `GAGurine`. Analyse these data, and produce a chart to help a paediatrican to assess if a child's GAG concentration is 'normal'.

6.6.   The Janka hardness data in data frame `janka` gives the density (`Dens`) and hardness (`Hard`) of a sample of Australian Eucalypt hardwoods. The problem is to build a prediction equation for hardness in terms of density.

6.7.   The `Cars93` data frame (Lock, 1993) gives data on 93 new car models on sale in the USA in 1993. Use this dataset to predict fuel consumption from the remaining variables. [Hint: The fuel consumption is in miles per US gallon. In metric units fuel consumption is expressed in litres/100km, a reciprocal scale.]

6.8.   The data in Table 6.4 (from Scheffé, 1959, and in data frame `genotype`) refer to rat litters that were separated from their natural mothers at birth and given to foster mothers to rear. The rats were classified into one of four genotypes, A, B, I and J. The response is the litter average weight gain, in grams, over the time of the study. The aim is to test whether the litters' and mothers' genotypes act

Table **6.4**: The rat genotype data.

Litter	A	B	I	J
		Foster mother		
A	61.5	55.0	52.5	42.0
	68.2	42.0	61.8	54.0
	64.0	60.2	49.5	61.0
	65.0		52.7	48.2
	59.7			39.6
B	60.3	50.8	56.5	51.3
	51.7	64.7	59.0	40.5
	49.3	61.7	47.2	
	48.0	64.0	53.0	
		62.0		
I	37.0	56.3	39.7	50.0
	36.3	69.8	46.0	43.8
	68.0	67.0	61.3	54.5
			55.3	
			55.7	
J	59.0	59.5	45.2	44.8
	57.4	52.8	57.0	51.5
	54.0	56.0	61.4	53.0
	47.0			42.0
				54.0

additively and if this may be retained to test for differences in litter and mother genotype effects.

6.9. Extend the analysis of the coop dataset to all the specimens.

6.10. Add the fitted lines for the final model for the petrol data to Figure 6.9.

6.11. Find a way to plot the Sitka data that facilitates comparison of the growth curves for the two treatment groups.

Add to your plot the fitted mean growth curve and some 95% confidence intervals.

6.12. Consider how to explore the assumptions made for the lme model for the Sitka data. Are the plot methods for lme objects helpful in this?

6.13. The object Sitka89 contains the 1989 data on the same 79 Sitka trees measured on eight days in 1989. Analyse the 1989 data separately, and then in conjunction with the 1988 data.

Chapter 7

Generalized Linear Models

Generalized linear models (GLMs) extend linear models to accommodate both non-normal response distributions and transformations to linearity. (We assume that Chapter 6 has been read before this chapter.) The essay by Firth (1991) gives a good introduction to GLMs; the comprehensive reference is McCullagh & Nelder (1989).

A generalized linear model may be described by the following assumptions.

- There is a response y observed independently at fixed values of stimulus variables $x_1, \ldots, x_p$.
- The stimulus variables may only influence the distribution of y through a single linear function called the *linear predictor* $\eta = \beta_1 x_1 + \cdots + \beta_p x_p$.
- The distribution of y has density of the form

$$f(y_i; \theta_i, \varphi) = \exp\left[A_i \left\{ y_i \theta_i - \gamma(\theta_i) \right\} / \varphi + \tau(y_i, \varphi/A_i) \right] \qquad (7.1)$$

where φ is a *scale parameter* (possibly known), A_i is a *known* prior weight and parameter θ_i depends upon the linear predictor.

- The mean μ is a smooth invertible function of the linear predictor:

$$\mu = m(\eta), \qquad \eta = m^{-1}(\mu) = \ell(\mu)$$

The inverse function $\ell(\cdot)$ is called the *link function*.

Note that θ is also an invertible function of μ, in fact $\theta = (\gamma')^{-1}(\mu)$ as we show in the following. If φ were known the distribution of y would be a one-parameter canonical exponential family. An unknown φ is handled as a nuisance parameter by moment methods.

GLMs allow a unified treatment of statistical methodology for several important classes of models. We consider a few examples.

Gaussian For a normal distribution $\varphi = \sigma^2$ and we can write

$$\log f(y) = \frac{1}{\varphi} \left\{ y\mu - \tfrac{1}{2}\mu^2 - \tfrac{1}{2}y^2 \right\} - \tfrac{1}{2}\log(2\pi\varphi)$$

so $\theta = \mu$ and $\gamma(\theta) = \theta^2/2$.

Table 7.1: Families and link functions. The default link is denoted by D.

Link	binomial	Gamma	gaussian	inverse.-gaussian	poisson
logit	D				
probit	●				
cloglog	●				
identity		●	D		●
inverse		D			
log		●			D
1/mu^2				D	
sqrt					●

Poisson For a Poisson distribution with mean μ we have

$$\log f(y) = y \log \mu - \mu - \log(y!)$$

so $\theta = \log \mu$, $\varphi = 1$ and $\gamma(\theta) = \mu = e^\theta$.

Binomial For a binomial distribution with fixed number of trials a and parameter p we take the response to be $y = s/a$ where s is the number of 'successes'. The density is

$$\log f(y) = a \left[y \log \frac{p}{1-p} + \log(1-p) \right] + \log \binom{a}{ay}$$

so we take $A_i = a_i$, $\varphi = 1$, θ to be the logit transform of p and $\gamma(\theta) = -\log(1-p) = \log(1 + e^\theta)$.

The functions supplied with S for handling generalized linear modelling distributions include gaussian, binomial, poisson, inverse.gaussian and Gamma.

Each response distribution allows a variety of link functions to connect the mean with the linear predictor. Those automatically available are given in Table 7.1. The combination of response distribution and link function is called the *family* of the generalized linear model.

For n observations from a GLM the log-likelihood is

$$l(\theta, \varphi; Y) = \sum_i [A_i \{y_i \theta_i - \gamma(\theta_i)\} / \varphi + \tau(y_i, \varphi/A_i)] \qquad (7.2)$$

and this has score function for θ of

$$U(\theta) = A_i \{y_i - \gamma'(\theta_i)\} / \varphi \qquad (7.3)$$

From this it is easy to show that

$$\mathrm{E}(y_i) = \mu_i = \gamma'(\theta_i) \qquad \mathrm{var}(y_i) = \frac{\varphi}{A_i} \gamma''(\theta_i)$$

Table 7.2: Canonical (default) links and variance functions.

Family	Canonical link	Name	Variance	Name
`binomial`	$\log(\mu/(1-\mu))$	`logit`	$\mu(1-\mu)$	`mu(1-mu)`
`Gamma`	$-1/\mu$	`inverse`	μ^2	`mu^2`
`gaussian`	μ	`identity`	1	`constant`
`inverse.gaussian`	$-2/\mu^2$	`1/mu^2`	μ^3	`mu^3`
`poisson`	$\log\mu$	`log`	μ	`mu`

(See, for example, McCullagh & Nelder, 1989, §2.2.) It follows immediately that

$$\mathrm{E}\left(\frac{\partial^2 l(\theta,\varphi;y)}{\partial\theta\,\partial\varphi}\right)=0$$

Hence θ and φ, or more generally β and φ, are *orthogonal parameters*.

The function defined by $V(\mu)=\gamma''(\theta(\mu))$ is called the *variance function*.

For each response distribution the link function $\ell=(\gamma')^{-1}$ for which $\theta\equiv\eta$ is called the *canonical link*. If X is the model matrix, so $\eta=X\beta$, it is easy to see that with φ known, $A=\mathrm{diag}A_i$ and the canonical link that $X^T Ay$ is a minimal sufficient statistic for β. Also using (7.3) the score equations for the regression parameters β reduce to

$$X^T Ay = X^T A\widehat{\mu} \qquad (7.4)$$

This relation is sometimes described by saying that the "observed and fitted values have the same (weighted) marginal totals." Equation (7.4) is the basis for certain simple fitting procedures, for example, Stevens' algorithm for fitting an additive model to a non-orthogonal two-way layout by alternate row and column sweeps (Stevens, 1948), and the Deming & Stephan (1940) or *iterative proportional scaling* algorithm for fitting log-linear models to frequency tables (see Darroch & Ratcliff, 1972).

Table 7.2 shows the canonical links and variance functions. The canonical link function is the default link for the families catered for by the S software (except for the inverse Gaussian, where the factor of two is dropped).

Iterative estimation procedures

Since explicit expressions for the maximum likelihood estimators are not usually available estimates must be calculated iteratively. It is convenient to give an outline of the iterative procedure here; for a more complete description the reader is referred to McCullagh & Nelder (1989, §2.5, pp. 40ff) or Firth (1991, §3.4). The scheme is sometimes known by the acronym IWLS, for *iterative weighted least squares*.

An initial estimate to the linear predictor is found by some guarded version of $\widehat{\eta}_0 = \ell(y)$. (Guarding is necessary to prevent problems such as taking logarithms of zero.) Define *working weights* W and *working values* z by

$$W = \frac{A}{V}\left(\frac{d\mu}{d\eta}\right)^2, \qquad z = \eta + \frac{y - \mu}{d\mu/d\eta}$$

Initial values for W_0 and z_0 can be calculated from the initial linear predictor.

At iteration k a new approximation to the estimate of β is found by a weighted regression of the working values z_k on X with weights W_k. This provides a new linear predictor and hence new working weights and values which allow the iterative process to continue. The difference between this process and a Newton–Raphson scheme is that the Hessian matrix is replaced by its expectation. This statistical simplification was apparently first used in Fisher (1925), and is often called *Fisher scoring*. The estimate of the large sample variance of $\widehat{\beta}$ is $\widehat{\varphi}(X^T\widehat{W}X)^{-1}$, which is available as a by-product of the iterative process. It is easy to see that the iteration scheme for $\widehat{\beta}$ does not depend on the scale parameter φ.

This iterative scheme depends on the response distribution only through its mean and variance functions. This has led to ideas of *quasi-likelihood*, implemented in the family `quasi`.

The analysis of deviance

A *saturated model* is one in which the parameters θ_i, or almost equivalently the linear predictors η_i, are free parameters. It is then clear from (7.3) that the maximum likelihood estimator of θ_i is obtained by $y_i = \gamma'(\widehat{\theta}_i) = \widehat{\mu}_i$ which is also a special case of (7.4). Denote the saturated model by S.

Assume temporarily that the scale parameter φ is known and has value 1. Let M be a model involving $p < n$ regression parameters, and let $\hat{\theta}_i$ be the maximum likelihood estimate of θ_i under M. Twice the log-likelihood ratio statistic for testing M within S is given by

$$D_M = 2\sum_{i=1}^{n} A_i \left[\left\{y_i\theta(y_i) - \gamma\left(\theta(y_i)\right)\right\} - \left\{y_i\widehat{\theta}_i - \gamma(\widehat{\theta}_i)\right\}\right] \qquad (7.5)$$

The quantity D_M is called the *deviance* of model M, even when the scale parameter is unknown or is known to have a value other than one. In the latter case D_M/φ, the difference in twice the log-likelihood, is known as the *scaled deviance*. (Confusingly, sometimes either is called the *residual deviance*, for example, McCullagh & Nelder, 1989, p. 119.)

For a Gaussian family with identity link the scale parameter φ is the variance and D_M is the residual sum of squares. Hence in this case the scaled deviance has distribution

$$D_M/\varphi \sim \chi^2_{n-p} \qquad (7.6)$$

leading to the customary unbiased estimator

$$\widehat{\varphi} = \frac{D_M}{n - p} \tag{7.7}$$

An alternative estimator is the sum of squares of the standardized residuals divided by the residual degrees of freedom

$$\widetilde{\varphi} = \frac{1}{n - p} \sum_i \frac{(y_i - \hat{\mu}_i)^2}{V(\hat{\mu}_i)/A_i} \tag{7.8}$$

where $V(\mu)$ is the variance function. Note that $\widetilde{\varphi} = \widehat{\varphi}$ for the Gaussian family, but in general differs.

In other cases the distribution (7.6) for the deviance under M may be approximately correct suggesting $\widehat{\varphi}$ as an approximately unbiased estimator of φ. It should be noted that sufficient (if not always necessary) conditions under which (7.6) becomes approximately true are that the individual distributions for the components y_i should become closer to normal form and the link effectively closer to an identity link. The approximation will often *not* improve as the sample size n increases since the number of parameters under S also increases and the usual likelihood ratio approximation argument does not apply. Nevertheless, (7.6) may sometimes be a good approximation, for example, in a binomial GLM with large values of a_i. Firth (1991, p. 69) discusses this approximation, including the extreme case of a binomial GLM with only one trial per case, that is, with $a_i = 1$.

Let $M_0 \subset M$ be a submodel with $q < p$ regression parameters and consider testing M_0 within M. If φ is known, by the usual likelihood ratio argument under M_0 we have a test given by

$$\frac{D_{M_0} - D_M}{\varphi} \overset{.}{\sim} \chi^2_{p-q} \tag{7.9}$$

where $\overset{.}{\sim}$ denotes "is approximately distributed as." The distribution is exact only in the Gaussian family with identity link. If φ is not known, by analogy with the Gaussian case it is customary to use the approximate result

$$\frac{(D_{M_0} - D_M)}{\widehat{\varphi}(p - q)} \overset{.}{\sim} F_{p-q, n-p} \tag{7.10}$$

although this must be used with some caution in non-Gaussian cases.

7.1 Functions for generalized linear modelling

The linear predictor part of a generalized linear model may be specified by a model formula using the same notation and conventions as linear models. Generalized linear models also require the family to be specified, that is, the response distribution, the link function and perhaps the variance function for `quasi` models.

The fitting function is `glm` for which main arguments are

```
glm(formula, family, data, weights, control)
```

The `family` argument is usually given as the name of one of the standard family functions listed under "Family Name" in Table 7.1. Where there is a choice of links, the name of the link may also be supplied in parentheses as a parameter, for example `binomial(link=probit)`. (The variance function for the `quasi` family may also be specified in this way.) For user-defined families (such as our `neg.bin` discussed in Section 7.4) other arguments to the family function may be allowed or even required.

Prior weights A_i may be specified using the `weights` argument. For binomial models these are implied and should not be specified separately.

The iterative process can be controlled by many parameters. The only ones that are at all likely to need altering are `maxit`, which controls the maximum number of iterations and whose default value of 10 is occasionally too small, `trace` which will often be set as `trace=T` to trace the iterative process, and `epsilon` which controls the stopping rule for convergence. The convergence criterion is to stop if

$$|\text{deviance}^{(i)} - \text{deviance}^{(i-1)}| < \epsilon(\text{deviance}^{(i-1)} + \epsilon)$$

(This comes from reading the S code and is not as described in Chambers & Hastie, 1992, p. 243.) It is quite often necessary to reduce the tolerance `epsilon` whose default value is 10^{-4}. Under some circumstances the convergence of the IWLS algorithm can be extremely slow, when the change in deviance at each step can be small enough for premature stopping to occur with the default ϵ.

Generic functions with methods for `glm` objects include `coef`, `resid`, `print`, `summary` and `deviance`. It is useful to have a `glm` method function to extract the variance-covariance matrix of the estimates. This can be done using part of the result of `summary`:

```
vcov.glm <- function(obj) {
    so <- summary(obj, corr = F)
    so$dispersion * so$cov.unscaled
}
```

Our library `MASS` contains the generic function, `vcov`, and methods for objects of classes `lm` and `nls` as well as `glm`.

For `glm` fitted-model objects the `anova` function allows an additional argument `test` to specify which test is to be used. Two possible choices are `test="Chisq"` for chi-squared tests using (7.9) and `test="F"` for F-tests using (7.10). The default is `test="Chisq"` for the binomial and Poisson families, otherwise `test="F"`.

The scale parameter ϕ is only used within the `summary` method. It can be supplied via the `dispersion` argument, and defaults to 1 for binomial and Poisson fits, and to (7.8) otherwise. To allow ϕ to be estimated for binomial and Poisson fits set `dispersion=0`.

Prediction and residuals

The `predict` method function for `glm` has a `type` argument to specify what is to be predicted. The default is `type="link"` which produces predictions of the linear predictor η. Predictions on the scale of the mean μ (for example, the fitted values $\widehat{\mu}_i$) are specified by `type="response"`. There is a `se.fit` argument which if true asks for standard errors to be returned. The system version of `predict.glm` will always estimate the scale parameter by $\widetilde{\phi}$; this is corrected in the version in our library `MASS` to use the same defaults as the `summary` method.

For `glm` models there are four types of residual that may be requested, known as *deviance, working, Pearson* and *response* residuals. The response residuals are simply $y_i - \widehat{\mu}_i$. The Pearson residuals are a standardized version of the response residuals, $(y_i - \widehat{\mu}_i)/\sqrt{\widehat{V}_i}$. The working residuals come from the last stage of the iterative process, $(y_i - \widehat{\mu}_i) \ / \ \mathrm{d}\mu_i/\mathrm{d}\eta_i$. The deviance residuals d_i are defined as the signed square roots of the summands of the deviance (7.5) taking the same sign as $y_i - \widehat{\mu}_i$.

For Gaussian families all four types of residual are identical. For binomial and Poisson GLMs the sum of the squared Pearson residuals is the Pearson chi-squared statistic, which often approximates the deviance, and the deviance and Pearson residuals are usually then very similar.

Method functions for the `resid` function have an argument `type` that defaults to `type="deviance"` for objects of class `glm`. Other values are `"response"`, `"pearson"` or `"working"`; these may be abbreviated to the initial letter. Deviance residuals are the most useful for diagnostic purposes.

Concepts of leverage and its effect on the fit are as important for GLMs as they are in linear regression, and are discussed in some detail by Davison & Snell (1991) and extensively for binomial GLMs by Collett (1991). On the other hand, they seem less often used, as GLMs are most often used either for simple regressions or for contingency tables where, as in designed experiments, high leverage can not occur.

Model formulae

The model formula language for GLMs is slightly more general than that described in Section 6.2 for linear models in that the function `offset` may be used. Its effect is to evaluate its argument and to add it to the linear predictor, equivalent to enclosing the argument in `I( )` and forcing the coefficient to be one.

Note that `offset` can be used with the formulae of `lm` or `aov`, but it is completely ignored, without any warning. Of course, with ordinary linear models the same effect can be achieved by subtracting the offset from the dependent variable, but if the more intuitive formula with offset is desired, it can be used with `glm` and the `gaussian` family. For example, our library contains a data set `anorexia` that contains the pre- and post-treatment weights for a clinical trial of three treatment methods for patients with anorexia. A natural model to consider would be

```
ax.1 <- glm(Postwt ~ Prewt + Treat + offset(Prewt),
            family = gaussian, data = anorexia)
```

Sometimes as here a variable is included both as a free regressor and as an offset to allow a test of the hypothesis that the regression coefficient is one or, by extension, any specific value. (In this example the hypothesis is rejected; see Exercise 7.1.)

If we had fitted this model omitting the offset but using `Postwt - Prewt` on the left side of the formula, predictions from the model would have been of weight gains, not the actual post-treatment weights. Hence another reason to use an offset rather than an adjusted dependent variable is to allow direct predictions from the fitted model object.

The default Gaussian family

A call to `glm` with the default `gaussian` family achieves the same purpose as a call to `lm` but less efficiently. The `gaussian` family is not provided with a choice of links, so no argument is allowed. If a problem requires a Gaussian family with a non-identity link, this can usually be handled using the `quasi` family. (Indeed yet another, even more inefficient, way to emulate `lm` with `glm` is to use the family `quasi(link = identity, variance = constant)`.) Although the `gaussian` family is the default, it is virtually never used in practice other than when an offset is needed.

7.2   Binomial data

Consider first a small example. Collett (1991, p. 75) reports an experiment on the toxicity to the tobacco budworm *Heliothis virescens* of doses of the pyrethroid *trans*-cypermethrin to which the moths were beginning to show resistance. Batches of 20 moths of each sex were exposed for three days to the pyrethroid and the number in each batch that were dead or knocked down was recorded. The results were

				Dose		
Sex	1	2	4	8	16	32
Male	1	4	9	13	18	20
Female	0	2	6	10	12	16

The doses were in μg. We fit a logistic regression model using $\log_2$ (dose) since the doses are powers of two. To do so we must specify the numbers of trials of a_i. This is done using `glm` with the `binomial` family in one of three ways.

1. If the response is a numeric vector it is assumed to hold the data in ratio form, $y_i = s_i/a_i$, in which case the a_is must be given as a vector of weights using the `weights` argument. (If the a_i are all one the default `weights` suffices.)

2. If the response is a logical vector or a two-level factor it is treated as a 0/1 numeric vector and handled as previously.

   If the response is a multi-level factor, the first level is treated as 0 (failure) and all others as 1 (success).

3. If the response is a two-column matrix it is assumed that the first column holds the number of successes, s_i, and the second holds the number of failures, $a_i - s_i$, for each trial. In this case no `weights` argument is required.

The less-intuitive third form allows the fitting function to select a better starting value, so we tend to favour it.

In all cases the response is the relative frequency $y_i = s_i/a_i$, so the means μ_i are the probabilities p_i. Hence `fitted` yields probabilities, not binomial means.

Since we have binomial data we use the second possibility:

```
> options(contrasts=c("contr.treatment", "contr.poly"))
> ldose <- rep(0:5, 2)
> numdead <- c(1, 4, 9, 13, 18, 20, 0, 2, 6, 10, 12, 16)
> sex <- factor(rep(c("M", "F"), c(6, 6)))
> SF <- cbind(numdead, numalive=20-numdead)
> budworm.lg <- glm(SF ~ sex*ldose, family=binomial)
> summary(budworm.lg, cor=F)
    ....
Coefficients:
                Value Std. Error   t value
(Intercept) -2.99354    0.55253  -5.41789
        sex  0.17499    0.77816   0.22487
      ldose  0.90604    0.16706   5.42349
  sex:ldose  0.35291    0.26994   1.30735
    ....
     Null Deviance: 124.88 on 11 degrees of freedom
Residual Deviance: 4.9937 on 8 degrees of freedom
    ....
```

This shows slight evidence of a difference in slope between the sexes. Note that we use treatment contrasts to make interpretation easier. Since female is the first level of `sex` (they are in alphabetical order) the parameter for `sex:ldose` represents the increase in slope for males just as the parameter for `sex` measures the increase in intercept. We can plot the data and the fitted curves by

```
plot(c(1,32), c(0,1), type="n", xlab="dose",
   ylab="prob", log="x")
text(2^ldose, numdead/20, labels=as.character(sex))
ld <- seq(0, 5, 0.1)
lines(2^ld, predict(budworm.lg, data.frame(ldose=ld,
   sex=factor(rep("M", length(ld)), levels=levels(sex))),
   type="response"), col=3)
lines(2^ld, predict(budworm.lg, data.frame(ldose=ld,
   sex=factor(rep("F", length(ld)), levels=levels(sex))),
   type="response"), lty=2, col=2)
```

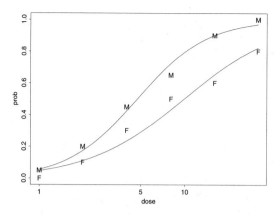

Figure 7.1: Tobacco budworm destruction versus dosage of *trans*-cypermethrin by sex.

see Figure 7.1. Note that when we set up a factor for the new data we must specify all the levels or both lines would refer to level one of `sex`. (Of course, here we could have predicted for both sexes and plotted separately, but it helps to understand the general mechanism needed.)

The apparently non-significant sex effect in this analysis has to be interpreted carefully. Since we are fitting separate lines for each sex, it tests the (uninteresting) hypothesis that the lines do not differ at zero log-dose. If we re-parametrize to locate the intercepts at dose 8 we find

```
> budworm.lgA <- update(budworm.lg, . ~ sex*I(ldose-3))
> summary(budworm.lgA, cor=F)$coefficients
                   Value Std. Error t value
   (Intercept) -0.27543    0.23049 -1.1950
           sex  1.23373    0.37694  3.2730
   I(ldose - 3)  0.90604    0.16706  5.4235
sex:I(ldose - 3)  0.35291    0.26994  1.3074
```

which shows a significant difference between the sexes at dose 8. The model fits very well as judged by the residual deviance (4.9937 is a small value for a χ^2_8 variate, and the estimated probabilities are based on a reasonable number of trials), so there is no suspicion of curvature. We can confirm this by an analysis of deviance:

```
> anova(update(budworm.lg, . ~ . + sex*I(ldose^2)), test="Chisq")
....
Terms added sequentially (first to last)
              Df Deviance Resid. Df Resid. Dev Pr(Chi)
        NULL                     11     124.88
         sex  1     6.08         10     118.80 0.01370
       ldose  1   112.04          9       6.76 0.00000
  I(ldose^2)  1     0.91          8       5.85 0.34104
   sex:ldose  1     1.24          7       4.61 0.26552
sex:I(ldose^2)  1    1.44          6       3.17 0.23028
```

This isolates a further two degrees of freedom which if curvature were appreciable would most likely be significant, but are not. Note how `anova` when given a single fitted-model object produces a sequential analysis of deviance, which will nearly always be order-dependent for `glm` objects. The additional argument `test="Chisq"` to the `anova` method may be used to specify tests using equation (7.9). The default corresponds to `test="none"`. The other possible choices are `"F"` and `"Cp"`, neither of which is appropriate here.

Our analysis so far suggests a model with parallel lines (on logit scale) for each sex. We estimate doses corresponding to a given probability of death. The first step is to re-parametrize the model to give separate intercepts for each sex and a common slope.

```
> budworm.lg0 <- glm(SF ~ sex + ldose - 1, family=binomial)
> summary(budworm.lg0, cor=F)$coefficients
          Value Std. Error t value
sexF  -3.4732    0.46829 -7.4166
sexM  -2.3724    0.38539 -6.1559
ldose  1.0642    0.13101  8.1230
```

Let ξ_p be the log-dose for which the probability of response is p. (Historically $2^{\xi_{0.5}}$ was called the "50% lethal dose" or LD50.) Clearly

$$\xi_p = \frac{\ell(p) - \beta_0}{\beta_1}, \qquad \frac{\partial \xi_p}{\partial \beta_0} = -\frac{1}{\beta_1}, \qquad \frac{\partial \xi_p}{\partial \beta_1} = -\frac{\ell(p) - \beta_0}{\beta_1^2} = -\frac{\xi_p}{\beta_1}$$

where β_0 and β_1 are the slope and intercept. Our library MASS contains functions to calculate and print $\widehat{\xi}_p$ and its asymptotic standard error, namely,

```
dose.p <- function(obj, cf = 1:2, p = 0.5) {
  eta <- family(obj)$link(p)
  b <- coef(obj)[cf]
  x.p <- (eta - b[1])/b[2]
  names(x.p) <- paste("p = ", format(p), ":", sep = "")
  pd <-  - cbind(1, x.p)/b[2]
  SE <- sqrt(((pd %*% vcov(obj)[cf, cf]) * pd) %*% c(1, 1))
  structure(x.p, SE = SE, p = p, class = "glm.dose")
}
print.glm.dose <- function(x, ...) {
  M <- cbind(x, attr(x, "SE"))
  dimnames(M) <- list(names(x), c("Dose", "SE"))
  x <- M
  NextMethod("print")
}
```

Notice how the `family` function can used to extract the link function, which can then be used to obtain values of the linear predictor. For females the values of ξ_p at the quartiles may be calculated as

```
> dose.p(budworm.lg0, cf = c(1,3), p = 1:3/4)
            Dose      SE
p = 0.25: 2.2313 0.24983
p = 0.50: 3.2636 0.22971
p = 0.75: 4.2959 0.27462
```

For males the corresponding log-doses are lower.

In biological assays the *probit* link used to be more conventional and the technique was called *probit analysis*. Unless the estimated probabilities are concentrated in the tails, probit and logit links tend to give similar results with the values of ξ_p near the centre almost the same. We can demonstrate this for the budworm assay by fitting a probit model and comparing the estimates of ξ_p for females.

```
> dose.p(update(budworm.lg0, family=binomial(link=probit)),
          cf = c(1,3), p = 1:3/4)
            Dose      SE
p = 0.25: 2.1912 0.23841
p = 0.50: 3.2577 0.22405
p = 0.75: 4.3242 0.26685
```

The differences are insignificant. This occurs because the logistic and standard normal distributions can approximate each other very well, at least between the 10th and 90th percentiles, by a simple scale change in the abscissa. (See, for example, Cox & Snell, 1989, p. 21.) The `menarche` data frame in our `MASS` library has data with a substantial proportion of the responses at very high probabilities and provides an example where probit and logit models appear noticeably different.

A binary data example: Low birth weight in infants

Hosmer & Lemeshow (1989) give a dataset on 189 births at a US hospital, with the main interest being in low birth weight. The following variables are available in our data frame `birthwt`:

`low`	birth weight less than 2.5 kg (0/1),
`age`	age of mother in years,
`lwt`	weight of mother (lbs) at last menstrual period,
`race`	white / black / other,
`smoke`	smoking status during pregnancy (0/1),
`ptl`	number of previous premature labours,
`ht`	history of hypertension (0/1),
`ui`	has uterine irritability (0/1),
`ftv`	number of physician visits in the first trimester,
`bwt`	actual birth weight (grams).

Although the actual birth weights are available, we concentrate here on predicting if the birth weight is low from the remaining variables. The dataset contains a small number of pairs of rows that are identical apart from the ID; it is possible that these refer to twins but identical birth weights seem unlikely.

We use a logistic regression with a binomial (in fact 0/1) response. It is worth considering carefully how to use the variables. It is unreasonable to expect a linear response with `ptl`. Since the numbers with values greater than one are so small we reduce it to a indicator of past history. Similarly, `ftv` can be reduced to three levels. With non-Gaussian GLMs it is usual to use treatment contrasts.

```
> options(contrasts=c("contr.treatment", "contr.poly"))
> attach(birthwt)
> race <- factor(race, labels=c("white", "black", "other"))
> table(ptl)
  0  1 2 3
159 24 5 1
> ptd <- factor(ptl > 0)
> table(ftv)
  0  1  2 3 4 6
100 47 30 7 4 1
> ftv <- factor(ftv)
> levels(ftv)[-(1:2)] <- "2+"
> table(ftv)  # as a check
  0  1 2+
100 47 42
> bwt <- data.frame(low=factor(low), age, lwt, race,
    smoke=(smoke>0), ptd, ht=(ht>0), ui=(ui>0), ftv)
> detach(); rm(race, ptd, ftv)
```

We can then fit a full logistic regression, and omit the rather large correlation matrix from the summary.

```
> birthwt.glm <- glm(low ~ ., family=binomial, data=bwt)
> summary(birthwt.glm, cor=F)
    ....
Coefficients:
                Value Std. Error   t value
(Intercept)  0.823013  1.2440732   0.66155
       age -0.037234  0.0386777  -0.96267
       lwt -0.015653  0.0070759  -2.21214
 raceblack  1.192409  0.5357458   2.22570
 raceother  0.740681  0.4614609   1.60508
     smoke  0.755525  0.4247645   1.77869
       ptd  1.343761  0.4804445   2.79691
        ht  1.913162  0.7204344   2.65557
        ui  0.680195  0.4642156   1.46526
      ftv1 -0.436379  0.4791611  -0.91071
     ftv2+  0.179007  0.4562090   0.39238

    Null Deviance: 234.67 on 188 degrees of freedom
Residual Deviance: 195.48 on 178 degrees of freedom
```

Since the responses are binary, even if the model is correct there is no guarantee that the deviance will have even an approximately chi-squared distribution, but since the value is about in line with its degrees of freedom there seems no serious reason to question the fit. Rather than select a series of sub-models by hand, we make use of the `stepAIC` function. By default argument `trace` is true and produces voluminous output.

```
> birthwt.step <- stepAIC(birthwt.glm, trace=F)
> birthwt.step$anova
Initial Model:
low ~ age + lwt + race + smoke + ptd + ht + ui + ftv
Final Model:
low ~ lwt + race + smoke + ptd + ht + ui

   Step Df Deviance Resid. Df Resid. Dev    AIC
1                         178     195.48 217.48
2 -ftv  2   1.3582        180     196.83 214.83
3 -age  1   1.0179        181     197.85 213.85
> birthwt.step2 <- stepAIC(birthwt.glm, ~ .^2 + I(scale(age)^2)
     + I(scale(lwt)^2), trace = F)
> birthwt.step2$anova
Initial Model:
low ~ age + lwt + race + smoke + ptd + ht + ui + ftv
Final Model:
low ~ age + lwt + smoke + ptd + ht + ui + ftv + age:ftv
     + smoke:ui

        Step Df Deviance Resid. Df Resid. Dev    AIC
1                            178     195.48 217.48
2   +age:ftv  2   12.475     176     183.00 209.00
3  +smoke:ui  1    3.057     175     179.94 207.94
4      -race  2    3.130     177     183.07 207.07

> summary(birthwt.step2, cor=F)$coef
                Value Std. Error    t value
(Intercept) -0.582520   1.418729   -0.41059
        age  0.075535   0.053880    1.40190
        lwt -0.020370   0.007465   -2.72878
      smoke  0.780057   0.419249    1.86061
        ptd  1.560205   0.495741    3.14722
         ht  2.065549   0.747204    2.76437
         ui  1.818252   0.664906    2.73460
       ftv1  2.920800   2.278843    1.28170
      ftv2+  9.241693   2.631548    3.51188
     ageftv1 -0.161809   0.096472   -1.67726
    ageftv2+ -0.410873   0.117548   -3.49537
    smoke:ui -1.916401   0.970786   -1.97407
> table(bwt$low, predict(birthwt.step2) > 0)
   FALSE TRUE
0    116   14
1     28   31
```

Note that although both `age` and `ftv` were previously dropped, their interaction is now included, the slopes on `age` differing considerably within the three `ftv` groups. The AIC criterion penalizes terms less severely than a likelihood ratio or Wald's test would, and so although adding the term `smoke:ui` reduces the AIC its t-statistic is only just significant at the 5% level. We also considered three-way interactions, but none were chosen.

Residuals are not always very informative with binary responses but at least none are particularly large here.

We can examine the linearity in age and mother's weight more flexibly using generalized additive models with smooth terms. This is considered in Section 9.1 after generalized additive models have been discussed, but no serious evidence of non-linearity on the logistic scale arises, suggesting the model chosen here is acceptable.

An alternative approach is to predict the actual live birth weight and later threshold at 2.5 kilograms. This is left as an exercise for the reader; surprisingly it produces somewhat worse predictions with around 52 errors.

Problems with binomial GLMs

There is a little-known phenomenon for binomial GLMs that was pointed out by Hauck & Donner (1977). The standard errors and t values derive from the Wald approximation to the log-likelihood, obtained by expanding the log-likelihood in a second-order Taylor expansion at the maximum likelihood estimates. If there are some $\widehat{\beta}_i$ which are large, the curvature of the log-likelihood at $\widehat{\beta}$ can be much less than near $\beta_i = 0$, and so the Wald approximation underestimates the change in log-likelihood on setting $\beta_i = 0$. This happens in such a way that as $|\widehat{\beta}_i| \to \infty$, the t statistic tends to zero. Thus highly significant coefficients according to the likelihood ratio test may have non-significant t ratios. (Curvature of the log-likelihood surface may to some extent be explored *post hoc* using the `glm` method for the `profile` generic function and the associated `plot` and `pairs` methods supplied with our `MASS` library. The `profile` generic function is discussed in Chapter 8 on page 251.)

There is one fairly common circumstance in which both convergence problems and the Hauck–Donner phenomenon can occur. This is when the fitted probabilities are extremely close to zero or one. Consider a medical diagnosis problem with thousands of cases and around 50 binary explanatory variables (which may arise from coding fewer categorical factors); one of these indicators is rarely true but always indicates that the disease is present. Then the fitted probabilities of cases with that indicator should be one, which can only be achieved by taking $\widehat{\beta}_i = \infty$. The result from `glm` will be warnings and an estimated coefficient of around ± 10. (Such cases are not hard to spot, but might occur during a series of stepwise fits.) There has been fairly extensive discussion of this in the statistical literature, usually claiming the non-existence of maximum likelihood estimates; see Santer & Duffy (1989, p. 234). However, the phenomenon was discussed much earlier (as a desirable outcome) in the pattern recognition literature (Duda & Hart, 1973; Ripley, 1996).

It is straightforward to fit binomial GLMs by direct maximization (see page 268) and this provides a route to study several extensions of logistic regression.

7.3 Poisson and multinomial models

The canonical link for the Poisson family is `log`, and the major use of this family is to fit surrogate Poisson log-linear models to what are actually multinomial frequency data. Such log-linear models have a large literature. (For example, Plackett, 1974; Bishop, Fienberg & Holland, 1975; Haberman, 1978, 1979; Goodman, 1978; Whittaker, 1990.) Poisson log-linear models are often applied directly to rates; see Exercises 7.7 and 7.8.

It is convenient to divide the factors classifying a multi-way frequency table into *response* and *stimulus* factors. Stimulus factors have their marginal totals fixed in advance (or for the purposes of inference). The main interest lies in the conditional probabilities of the response factor given the stimulus factors.

It is well known that the conditional distribution of a set of independent Poisson random variables given their sum is multinomial with probabilities given by the ratios of the Poisson means to their total. This result is applied to the counts for the multi-way response within each combination of stimulus factor levels. This allows models for multinomial data with a multiplicative probability specification to be fitted and tested using Poisson log-linear models.

Identifying the multinomial model corresponding to any surrogate Poisson model is straightforward. Suppose A, B, ... are factors classifying a frequency table. The *minimum model* is the interaction of all stimulus factors, and must be included for the analysis to respect the fixed totals over response factors. This model is usually of no interest, corresponding to a uniform distribution over response factors independent of the stimulus factors. Interactions between response and stimulus factors indicate interesting structure. For large models it is often helpful to use a graphical structure to represent the dependency relations (Whittaker, 1990; Lauritzen, 1996).

A four-way frequency table example

As an example of a surrogate Poisson model analysis consider the data in Table 7.3. This shows a four-way classification of 1 681 householders in Copenhagen who were surveyed on the *type* of rental accommodation they occupied, the degree of *contact* they had with other residents, their feeling of *influence* on apartment management and their level of *satisfaction* with their housing conditions. The data were originally given by Madsen (1976) and later analysed by Cox & Snell (1984). In our MASS library the data frame `housing` gives the same data set with four factors and a numeric column of frequencies.

Table 7.3: A four-way frequency table of 1 681 householders from a study of satisfaction with housing conditions in Copenhagen.

Contact		Low			High		
Satisfaction		Low	Med.	High	Low	Med.	High
Housing	**Influence**						
Tower blocks	Low	21	21	28	14	19	37
	Medium	34	22	36	17	23	40
	High	10	11	36	3	5	23
Apartments	Low	61	23	17	78	46	43
	Medium	43	35	40	48	45	86
	High	26	18	54	15	25	62
Atrium houses	Low	13	9	10	20	23	20
	Medium	8	8	12	10	22	24
	High	6	7	9	7	10	21
Terraced houses	Low	18	6	7	57	23	13
	Medium	15	13	13	31	21	13
	High	7	5	11	5	6	13

Cox & Snell (1984, pp. 155ff) present an analysis with type of housing as the only explanatory factor and contact, influence and satisfaction treated symmetrically as response factors and mention an alternative analysis with satisfaction as the only response variable, and hence influence and contact as (conditional) explanatory variables. We consider such an alternative analysis.

Our initial model may be described as having the conditional probabilities for each of the three satisfaction classes the same for all type $\times$ contact $\times$ influence groups. In other words satisfaction is independent of the other explanatory factors.

```
> names(housing)
[1] "Sat"   "Infl" "Type" "Cont" "Freq"
> house.glm0 <- glm(Freq ~ Infl*Type*Cont + Sat, family=poisson,
                    data=housing)
> summary(house.glm0, cor=F)
    ....
    Null Deviance: 833.66 on 71 degrees of freedom
Residual Deviance: 217.46 on 46 degrees of freedom
```

The high residual deviance clearly indicates that this simple model is inadequate, so the probabilities do appear to vary with the explanatory factors. We now consider adding the simplest terms to the model that allow for some variation of this kind.

```
> addterm(house.glm0, ~. + Sat:(Infl+Type+Cont), test="Chisq")
```

```
        . . . .
        Df Deviance    AIC    LRT Pr(Chi)
<none>        217.46 269.46
Sat:Infl  4   111.08 171.08 106.37 0.00000
Sat:Type  6   156.79 220.79  60.67 0.00000
Sat:Cont  2   212.33 268.33   5.13 0.07708
```

The 'influence' factor achieves the largest single term reduction in the AIC and our next step would be to incorporate the term `Sat:Infl` in our model and re-assess. It turns out that all three terms are necessary, so we now update our initial model by including all three at once.

```
> house.glm1 <- update(house.glm0, . ~ . + Sat:(Infl+Type+Cont))
> summary(house.glm1, cor=F)
    . . . .
    Null Deviance: 833.66 on 71 degrees of freedom
Residual Deviance: 38.662 on 34 degrees of freedom
> 1 - pchisq(deviance(house.glm1), house.glm1$df.resid)
[1] 0.26714
```

The deviance indicates a satisfactorily fitting model, but we might look to see if some adjustments to the model might be warranted.

```
> dropterm(house.glm1, test="Chisq")
    . . . .
                Df Deviance    AIC    LRT Pr(Chi)
<none>              38.66 114.66
Sat:Infl       4   147.78 215.78 109.12 0.00000
Sat:Type       6   100.89 164.89  62.23 0.00000
Sat:Cont       2    54.72 126.72  16.06 0.00033
Infl:Type:Cont 6    43.95 107.95   5.29 0.50725
```

Note that the final term here is part of the minimum model and hence may *not* be removed. Only terms that contain the response factor, `Sat`, are of any interest to us for this analysis. Now consider adding possible interaction terms.

```
> addterm(house.glm1, ~. + Sat:(Infl+Type+Cont)^2, test = "Chisq")
    . . . .
              Df Deviance    AIC    LRT Pr(Chi)
<none>            38.662 114.66
Infl:Sat:Type 12  16.107 116.11 22.555 0.03175
Infl:Sat:Cont  4  37.472 121.47  1.190 0.87973
Type:Sat:Cont  6  28.256 116.26 10.406 0.10855
```

The first term, a type × influence interaction, appears to be mildly significant, but as it increases the AIC we choose not include it on the grounds of simplicity, although in some circumstances we might view this decision differently.

We have now shown (subject to checking assumptions) that the three explanatory factors, type, influence and contact do affect the probabilities of each of the satisfaction classes in a simple 'main effect' only way. Our next step is to look at these estimated probabilities under the model and assess what effect these factors are having. The picture becomes clear if we normalize the means to probabilities.

Table 7.4: Estimated probabilities from a main effects model for the Copenhagen housing conditions study.

Contact		Low			High		
Satisfaction		Low	Med.	High	Low	Med.	High
Housing	**Influence**						
Tower blocks	Low	0.40	0.26	0.34	0.30	0.28	0.42
	Medium	0.26	0.27	0.47	0.18	0.27	0.54
	High	0.15	0.19	0.66	0.10	0.19	0.71
Apartments	Low	0.54	0.23	0.23	0.44	0.27	0.30
	Medium	0.39	0.26	0.34	0.30	0.28	0.42
	High	0.26	0.21	0.53	0.18	0.21	0.61
Atrium houses	Low	0.43	0.32	0.25	0.33	0.36	0.31
	Medium	0.30	0.35	0.36	0.22	0.36	0.42
	High	0.19	0.27	0.54	0.13	0.27	0.60
Terraced houses	Low	0.65	0.22	0.14	0.55	0.27	0.19
	Medium	0.51	0.27	0.22	0.40	0.31	0.29
	High	0.37	0.24	0.39	0.27	0.26	0.47

```
hnames <- lapply(housing[, -5], levels) # omit Freq
house.pm <- predict(house.glm1, expand.grid(hnames),
                    type = "response")  # poisson means
house.pm <- matrix(house.pm, ncol=3, byrow=T,
                   dimnames=list(NULL, hnames[[1]]))
house.pr <- house.pm/drop(house.pm %*% rep(1, 3))
cbind(expand.grid(hnames[-1]), round(house.pr, 2))
```

The result is shown in conventional typeset form in Table 7.4. The message of the fitted model is now clear. The factor having most effect on the probabilities is influence, with an increase in influence reducing the probability of low satisfaction and increasing that of high. The next most important factor is the type of housing itself. Finally, as contact with other residents rises, the probability of low satisfaction tends to fall and that of high to rise, but the effect is relatively small.

The reader should compare the model-based probability estimates with the relative frequencies from the original data. In a few cases the smoothing effect of the model is perhaps a little larger than might have been anticipated, but there are no very surprising differences. A conventional residual analysis also shows up no serious problem with the assumptions underlying the model, and the details are left as an informal exercise.

Fitting by iterative proportional scaling

The function `loglin` fits log-linear models by iterative proportional scaling. This starts with an array of fitted values that has the correct multiplicative structure, (for example with all values equal to 1) and makes multiplicative adjustments so that the observed and fitted values come to have the same marginal totals, in accordance with equation (7.4). (See Darroch & Ratcliff, 1972.) This is usually very much faster than GLM fitting but is less flexible.

The function provided in **S-PLUS**, `loglin`, is rather awkward to use as it requires the frequencies to be supplied as a multi-way array. Our front-end to `loglin` called `loglm` (in library MASS) accepts the frequencies in a variety of forms and can be used with essentially the same convenience as any model fitting function. Methods for standard generic functions to deal with fitted-model objects are also provided.

```
> loglm(Freq ~ Infl*Type*Cont + Sat*(Infl+Type+Cont),
        data=housing)
Statistics:
                      X^2 df P(> X^2)
Likelihood Ratio 38.662 34  0.26714
        Pearson 38.908 34  0.25823
```

The output shows that the deviance is the same as for the Poisson log-linear model and the (Pearson) chi-squared statistic is very close, as would be expected.

Fitting as a multinomial model

We can fit a multinomial model directly rather than use a surrogate Poisson model by using our function `multinom` from library nnet. No interactions are needed:

```
> library(nnet)
> house.mult <- multinom(Sat ~ Infl + Type + Cont, weights=Freq,
                         data=housing)
> house.mult
Coefficients:
       (Intercept)  Infl.L   Infl.Q TypeAtrium TypeApartment
Medium   -0.048783 0.47018 -0.093025    0.13137      -0.43568
  High    0.643750 1.14030  0.058341   -0.40797      -0.73563
       TypeTerrace    Cont
Medium    -0.66657 0.36085
  High    -1.41232 0.48182

Residual Deviance: 3470.1
AIC: 3498.1
```

Here the deviance is comparing with the model that correctly predicts each person, not the multinomial response for each cell of the minimum model: we can compare with the usual saturated model by

```
> house.mult2 <- multinom(Sat ~ Infl*Type*Cont, weights=Freq,
                          data=housing)
> anova(house.mult, house.mult2, test="none")
  . . . .
              Model Resid. df Resid. Dev    Test    Df LR stat.
1 Infl + Type + Cont    130      3470.1
2 Infl * Type * Cont     96      3431.4 1 vs 2    34   38.662
```

A table of fitted probabilities can be found by

```
house.pm <- predict(house.mult, expand.grid(hnames[-1]),
                    type = "probs")
cbind(expand.grid(hnames[-1]), round(house.pm, 2))
```

A proportional-odds model

Since the response in this example is ordinal a natural model to consider would be a proportional-odds model. Under such a model the odds ratio for cumulative probabilities for low and medium satisfaction does not depend on the cell to which the three probabilities belong. For a definition and good discussion see the seminal paper of McCullagh (1980) and McCullagh & Nelder (1989, §5.2.2). A proportional-odds logistic regression for a response factor with K levels has

$$\text{logit } P(Y \leqslant k \mid \boldsymbol{x}) = \zeta_k + \eta$$

for $\zeta_0 = -\infty < \zeta_1 < \cdots < \zeta_K < \zeta_{K+1} = \infty$ and η the usual linear predictor.

One check to see if the proportional-odds assumption is reasonable is to look at the differences of logits of the cumulative probabilities either using the simple relative frequencies or the model-based estimates. Since the multinomial model appears to fit well, we use the latter here.

```
> house.cpr <- apply(house.pr, 1, cumsum)
> logit <- function(x) log(x/(1-x))
> house.ld <- logit(house.cpr[2, ]) - logit(house.cpr[1, ])
> sort(drop(house.ld))
 [1] 0.93573 0.98544 1.05732 1.06805 1.07726 1.08036 1.08249
 [8] 1.09988 1.12000 1.15542 1.17681 1.18664 1.20915 1.24350
[15] 1.27241 1.27502 1.28499 1.30626 1.31240 1.39047 1.45401
[22] 1.49478 1.49676 1.60688
> mean(.Last.value)
[1] 1.2238
```

The average log-odds ratio is about 1.2 and variations from it are not great, so such a model may prove a useful simplification.

There are no standard S-PLUS fitting functions for proportional-odds models, but our MASS library contains a function, polr, for proportional-odds logistic regression that behaves in a very similar way to the standard fitting functions.

```
> house.plr <- polr(Sat ~ Infl + Type + Cont,
                    data = housing, weights = Freq)
> house.plr
    ....
Coefficients:
 InflMedium InflHigh TypeApartment TypeAtrium TypeTerrace
   0.56639   1.2888      -0.57235   -0.36619      -1.091
     Cont
  0.36029

Intercepts:
 Low|Medium Medium|High
   -0.49613      0.6907


Residual Deviance: 3479.1
AIC: 3495.1
```

The residual deviance is comparable with that of the multinomial model fitted before, showing that the increase is only 9.0 for a reduction from 14 to 8 parameters. The AIC criterion is consequently much reduced. Note that the difference in intercepts, 1.19, agrees fairly well with the average logit difference of 1.22 found previously.

We may calculate a matrix of estimated probabilities comparable with the one we obtained from the surrogate Poisson model by

```
house.pr1 <- predict(house.plr, expand.grid(hnames[-1]),
                     type = "probs")
cbind(expand.grid(hnames[-1]), round(house.pr1, 2))
```

These fitted probabilities are close to those of the multinomial model given in Table 7.4.

Finally we can calculate directly the likelihood ratio statistic for testing the proportional-odds model within the multinomial:

```
> Fr <- matrix(housing$Freq, ncol = 3, byrow=T)
> 2*sum(Fr*log(house.pr/house.pr1))
[1] 9.0654
```

which agrees with the difference in deviance noted previously. For six degrees of freedom this is clearly not significant.

The advantage of the proportional-odds model is not just that it is so much more parsimonious than the multinomial, but with the smaller number of parameters the action of the covariates on the probabilities is much easier to interpret and to describe. However, as it is more parsimonious, stepAIC will select a more complex model linear predictor:

```
> house.plr2 <- stepAIC(house.plr, ~.^2)
> house.plr2$anova
    ....
        Step Df Deviance Resid. Df Resid. Dev    AIC
```

1				1673	3479.1	3495.1
2 + Infl:Type	6	22.509		1667	3456.6	3484.6
3 + Type:Cont	3	7.945		1664	3448.7	3482.7

7.4 A negative binomial family

Once the link function and its derivative, the variance function, the deviance function and some method for obtaining starting values are known the fitting procedure is the same for all generalized linear models. This particular information is all taken from the *family object*, which in turn makes it fairly easy to handle a new GLM family by writing a *family object generator* function. We illustrate the procedure with a family for negative binomial models with known shape parameter, a restriction that is relaxed later.

Using the negative binomial distribution in modelling is important in its own right and has a long history. An excellent modern reference is Lawless (1987). The variance is greater than the mean, suggesting that it might be useful for describing frequency data where this is a prominent feature, often loosely called 'overdispersed Poisson data'.

The negative binomial can arise from a two-stage model for the distribution of a discrete variable Y. We suppose there is an unobserved random variable E having a gamma distribution gamma$(\theta)/\theta$, that is with mean 1 and variance $1/\theta$. Then the model postulates that conditionally on E, Y is Poisson with mean μE. Thus:

$$Y \mid E \sim \text{Poisson}(\mu E), \qquad \theta E \sim \text{gamma}(\theta)$$

The marginal distribution of Y is then negative binomial with probability function, mean and variance given by

$$\text{E}\left(Y\right) = \mu, \quad \text{var}\left(Y\right) = \mu + \mu^2/\theta, \quad f_Y(y; \theta, \mu) = \frac{\Gamma(\theta + y)}{\Gamma(\theta)\, y!} \frac{\mu^y\, \theta^\theta}{(\mu + \theta)^{\theta+y}}$$

If θ is known, as for the present we assume it is, this distribution has the general form (7.1).

A function `make.family` is available to piece together the information required for the `family` argument of the `glm` fitting function. It requires three main arguments;

`name`   a character string giving the name of the family,

`link`   a list supplying information about the link function, its inverse and derivative and an initialization expression,

`variance`   a list specifying the variance and deviance functions.

The two datasets `glm.links` and `glm.variances` are matrices of lists with examples that can be used as templates. We have in mind a negative binomial model for the Quine data introduced in Section 6.8, so we consider only a `log` link. The following function provides the family, passing the parameter θ as a required argument.

```
neg.bin <- function(theta = stop("theta must be given"))
{
   nb.lnk <- list(names = "Log: log(mu)",
      link = function(mu) log(mu),
      inverse = function(eta) exp(eta),
      deriv = function(mu) 1/mu,
      initialize = expression(mu <- y + (y == 0)/6) )
   nb.var <- list(
      names = "mu + mu^2/theta",
      variance = substitute(function(mu, th = .Theta)
         mu * (1 + mu/th), list(.Theta = theta)),
      deviance = substitute(
         function(mu, y, A, residuals = F, th = .Theta)
         {
            devi <- 2 * A * (y * log(pmax(1, y)/mu) -
               (y + th) * log((y + th)/(mu + th)))
            if(residuals) sign(y - mu) * sqrt(abs(devi))
            else sum(devi)
         }, list(.Theta = theta) )
   )
   make.family("Negative Binomial", link = nb.lnk,
      variance = nb.var)
}
```

The `names` components are cosmetic and only used by the `print` method.

A negative binomial model for the Quine data

A Poisson model for the Quine data has an excessively large deviance:

```
> glm(Days ~ .^4, family=poisson, data=quine)
   ....
Degrees of Freedom: 146 Total; 118 Residual
Residual Deviance: 1173.9
```

Inspection of the mean–variance relationship in Figure 6.5 suggests a negative binomial model with $\theta \approx 2$ might be appropriate. We assume at first that $\theta = 2$ is known. A negative binomial model may be fitted by

```
quine.nb <- glm(Days ~ .^4, family=neg.bin(2), data=quine)
```

The standard generic functions may now be used to fit sub-models, produce analysis of variance tables, and so on. For example, let us check the final model found in Chapter 6. (The output has been edited.)

```
> quine.nb0 <- update(quine.nb, . ~ Sex/(Age + Eth*Lrn))
> anova(quine.nb0, quine.nb, test="Chi")
  Resid. Df Resid. Dev   Test  Df Deviance  Pr(Chi)
1       132     198.51
2       118     171.98 1 vs. 2  14   26.527 0.022166
```

which suggests that model to be an over-simplification in the fixed-θ negative binomial setting.

Consider now what happens when θ is estimated rather than held fixed. The function `negative.binomial` supplied with our library MASS is similar to `neg.bin` defined before, but allows more links. We have also included a function `glm.nb`, a modification of `glm` that incorporates maximum likelihood estimation of θ. This has summary and anova methods; the latter produces likelihood ratio tests for the sequence of fitted models. (For deviance tests to be applicable the θ parameter has to be held constant for all fitted models.)

The following models summarize the results of a manual selection (`step` does not work with `glm.nb`, although `stepAIC` does), and indicate model `quine.nb2`.

```
> quine.nb <- glm.nb(Days ~ .^4, data=quine)
> quine.nb2 <- stepAIC(quine.nb)
    . . . .
                df      AIC
      <none> 23 -10227.57
-Sex:Age:Lrn 21 -10225.86
-Eth:Age:Lrn 21 -10225.80
-Eth:Sex:Lrn 22 -10224.53
> quine.nb2$anova
Initial Model:
Days ~ (Eth + Sex + Age + Lrn)^4

Final Model:
Days ~ Eth + Sex + Age + Lrn + Eth:Sex + Eth:Age + Eth:Lrn +
        Sex:Age + Sex:Lrn + Age:Lrn + Eth:Sex:Lrn +
        Eth:Age:Lrn + Sex:Age:Lrn


                 Step Df Deviance Resid. Df Resid. Dev    AIC
1                                       118     -10278 -10222
2 -Eth:Sex:Age:Lrn  2   1.4038       120     -10276 -10224
3     -Eth:Sex:Age  3   2.6809       123     -10274 -10228
```

This model is `Lrn/(Age + Eth + Sex)^2`, but AIC tends to overfit, so we test the terms `Sex:Age:Lrn` and `Eth:Sex:Lrn`.

```
> dropterm(quine.nb2, test="Chisq")
            Df    AIC    LRT  Pr(Chi)
    <none>      -10228
Eth:Sex:Lrn  1 -10225 5.0384 0.024791
Eth:Age:Lrn  2 -10226 5.7685 0.055896
Sex:Age:Lrn  2 -10226 5.7021 0.057784
```

This indicates that we might try removing both terms,

```
> quine.nb3 <-
    update(quine.nb2, . ~ . - Eth:Age:Lrn - Sex:Age:Lrn)
> anova(quine.nb2, quine.nb3)
   theta Resid. df  2 x log-lik   Test  df LR stat.  Pr(Chi)
1 1.7250       127       10264
2 1.8654       123       10274  1 vs 2   4   10.022 0.040058
```

which suggests removing either term but not both! Clearly there is a close-run choice among models with one, two or three third-order interactions.

The estimate of θ and its standard error are available from the summary or as components of the fitted model object:

```
> c(theta=quine.nb2$theta, SE=quine.nb2$SE)
  theta       SE
 1.8654  0.25801
```

We can perform some diagnostic checks by examining the deviance residuals:

```
rs <- resid(quine.nb2, type="deviance")
plot(predict(quine.nb2), rs, xlab="Linear predictors",
    ylab="Deviance residuals")
abline(h=0, lty=2)
qqnorm(rs, ylab="Deviance residuals")
qqline(rs)
```

Such plots show nothing untoward for any of the candidate models.

7.5   Exercises

7.1.   Explore the `anorexia` data example introduced in the discussion of offsets on page 217 and report your final linear model.

Begin with a Trellis display of the data showing post-treatment weight against pre-treatment weight for the three treatment groups. In each panel include the individual regression line, the parallel regression line and the parallel regression line with slope 1 as well as the points.

7.2.   Analyse the `menarche` dataset on the proportions of female children in Warsaw at various ages during adolescence who have reached menarche (Milicer & Szczotka, 1966) using both logit and probit links.

7.3.   Knight & Skagen (1988) collected the data shown in the table (and in data frame `eagles`) during a field study on the foraging behaviour of wintering Bald Eagles in Washington State, USA. The data concern 160 attempts by one (pirating) Bald Eagle to steal a chum salmon from another (feeding) Bald Eagle. The abbreviations used are

L = large   S = small;   A = adult   I = immature

Number of successful attempts	Total number of attempts	Size of pirating eagle	Age of pirating eagle	Size of feeding eagle
17	24	L	A	L
29	29	L	A	S
17	27	L	I	L
20	20	L	I	S
1	12	S	A	L
15	16	S	A	S
0	28	S	I	L
1	4	S	I	S

Report on factors that explain the success of the pirating attempt, and give a prediction formula for the probability of success.

7.4.   The following data are part of a survey by Dr Mutch of low-weight births in Scotland between 1981 and 1988. The table refers to 661 children with birth weights between 650g and 1749g all of whom survived for at least one year. The variables of interest are:

Cardiac: mild heart problems of the mother during pregnancy;

Comps: gynaecological problems during pregnancy;

Smoking: mother smoked at least one cigarette per day during the first 6 months of pregnancy;

BW: was the birth weight less than 1250g?

Cardiac		Yes				No			
Comps		Yes		No		Yes		No	
Smoking		Yes	No	Yes	No	Yes	No	Yes	No
BW	Yes	10	25	12	15	18	12	42	45
	No	7	5	22	19	10	12	202	205

Analyse this table.

7.5.   A survey was made of bicycle and other traffic in the neighbourhood of the Berkeley campus of the University of California in 1993 (Gelman *et al.*, 1995, p. 91). Sixty city streets were selected at random, with a stratification into three levels of activity and whether the street had a marked bicycle lane. The counts observed in one hour are shown in the table: for two of the streets the data were lost.

Type of street	Bike lane?		Counts									
Residential	yes	bikes	16	9	10	13	19	20	18	17	35	55
		other	58	90	48	57	103	57	86	112	273	64
Residential	no	bikes	12	1	2	4	9	7	9	8		
		other	113	18	14	44	208	67	29	154		
Side	yes	bikes	8	35	31	19	38	47	44	44	29	18
		other	29	415	425	42	180	675	620	437	47	462
Side	no	bikes	10	43	5	14	58	15	0	47	51	32
		other	557	1258	499	601	1163	700	90	1093	1459	1086
Main	yes	bikes	60	51	58	59	53	68	68	60	71	63
		other	1545	1499	1598	503	407	1494	1558	1706	476	752
Main	no	bikes	8	9	6	9	19	61	31	75	14	25
		other	1248	1246	1596	1765	1290	2498	2346	3101	1918	2318

Report on these data, paying particular attention to the effects of bicycle lanes.

7.6. To study the relative survival capacities of two species of native and exotic snails, here labelled A and B, groups of 20 animals were held in controlled laboratory conditions for periods of 1, 2, 3 or 4 weeks. At the end of the period the animals were checked for whether they had survived, but as the check itself is a destructive process a longitudinal study with the same animals was not possible. The groups were held in chambers where the temperature and relative humidity were held fixed at three and four levels, respectively. There were thus $2 \times 4 \times 3 \times 4 = 96$ groups laid out in a complete factorial design.

The data are shown in Table 7.5, where each entry is the number who did not survive out of the 20 test animals. The data set is also available as the data frame `snails` in library MASS. Variable `Species` is a two-level factor but treat the other stimulus variables as quantitative.

(a) Fit separate logistic regression models on exposure, relative humidity and temperature for each species, that is, a logistic regression of the form
    `Species/(Exposure + Rel.Hum + Temp)`.

(b) Fit parallel logistic regressions for the two species on the three stimulus variables and show that it may be retained when tested within the separate regressions model.

(c) There are no deaths for either species for the 1 week exposure time. This suggests a quadratic term in `Exposure` might be warranted. Repeat the analysis including such a quadratic term.

Table 7.5: The snail mortality data.

Rel. Hum.	Temp. (° C)	Species A Exposure				Species B Exposure			
		1	2	3	4	1	2	3	4
60.0%	10	0	0	1	7	0	0	7	12
	15	0	1	4	7	0	3	11	14
	20	0	1	5	7	0	2	11	16
65.8%	10	0	0	0	4	0	0	4	10
	15	0	1	2	4	0	2	5	12
	20	0	0	4	7	0	1	9	12
70.5%	10	0	0	0	3	0	0	2	5
	15	0	0	2	3	0	0	4	7
	20	0	0	3	5	0	1	6	9
75.8%	10	0	0	0	2	0	1	2	4
	15	0	0	1	3	0	0	3	5
	20	0	0	2	3	0	1	5	7

(d) Because deaths are so sparse a residual analysis is fairly meaningless. Nevertheless look at the residuals to see how they appear for this kind of data set.

(e) Is there a significant difference between the survival rates of the two species? Describe qualitatively how the probability of death depends upon the stimulus variables. Summarize your conclusions.

7.7. An experiment was performed in Sweden in 1961–2 to assess the effect of speed limits on the motorway accident rate (Svensson, 1981). The experiment was conducted on 92 days in each year, matched so that day j in 1962 was comparable to day j in 1961. On some days the speed limit was in effect and enforced, whereas on other days there was no speed limit and cars tended to be driven faster. The speed limit days tended to be in contiguous blocks.

The data set is given in the data frame `Traffic` with factors `year`, `day` and `limit` and the response is the daily traffic accident count `y`.

Fit Poisson log-linear models and summarize what you discover.

You might assume `day` occurs as a main effect only (fitting models with interaction terms involving factors of 92 levels may take some time and memory!), but assess if an interaction between `limit` and `year` is needed.

Check if the deviance residuals provide any hint of irregular behaviour.

7.8. The data given in data frame `Insurance` consist of the numbers of policy-holders n of an insurance company who were exposed to risk, and the numbers of car insurance claims made by those policyholders in the third quarter of 1973 (Baxter, Coutts & Ross, 1980, Aitkin *et al.*, 1989. The data are cross-classified by `District` (four levels), `Group` of car (four levels), and `Age` of driver (four ordered levels). The other variables in the data frame are the numbers of `Holders` and `Claims`.

The relevant model is taken to be a Poisson log-linear model with `offset` $\log n$.

(a) Fit an initial model with all terms present up to the three-way interaction; that is,

```
Claims ~ District*Group*Age - District:Group:Age
           + offset(log(Holders))
```

(b) Using `stepAIC`, or otherwise, prune the model of unjustified terms and report your findings.

Present your results as a table of estimated claim rates per policy holder for each category of holder.

(c) It is not strictly valid to regard such data as having the obvious binomial distribution, since some policyholders may make multiple claims. Nevertheless it should be a reasonable approximation. Repeat the analysis with a binomial model and compare the outcomes on estimated claim rates (or in this case, estimated probabilities of making a claim).

Chapter 8

Non-linear Models

In linear regression the mean surface is a plane in sample space; in non-linear regression it may be an arbitrary curved surface but in all other respects the models are the same. Fortunately the mean surface in most non-linear regression models met in practice will be approximately planar in the region of highest likelihood, allowing some good approximations based on linear regression to be used, but non-linear regression models can still present tricky computational and inferential problems.

A thorough treatment of non-linear regression is given in Bates & Watts (1988). Another encyclopaedic reference is Seber & Wild (1989), and the books of Gallant (1987) and Ross (1990) also offer some practical statistical advice. The S software is described by Bates & Chambers (1992) who state that its methods are based on those described in Bates & Watts (1988). An alternative approach based on their nls2 library is described by Huet *et al.* (1996).

In this chapter we consider first non-linear regression and then more general minimization and maximum likelihood estimation. Non-linear mixed models are considered in Section 8.8.

8.1 An introductory example

Obese patients on a weight reduction programme tend to lose adipose tissue at a diminishing rate. Our dataset wtloss has been supplied by Dr T Davies (personal communication). The two variables are Days, the time (in days) since start of the programme, and Weight, the patient's weight in kilograms measured under standard conditions. The dataset pertains to a male patient, aged 48, height 193 cm ($6'4''$) with a large body frame. The results are illustrated in Figure 8.1, produced by

```
attach(wtloss)
# alter margin 4; others are default
oldpar <- par(mar=c(5.1, 4.1, 4.1, 4.1))
plot(Days, Weight, type="p", ylab="Weight (kg)")
Wt.lbs <- pretty(range(Weight*2.205))
axis(side=4, at=Wt.lbs/2.205, lab=Wt.lbs, srt=90)
```

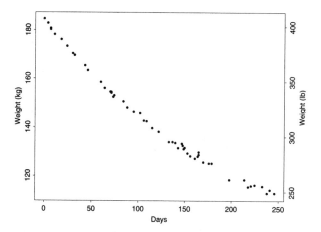

Figure 8.1: Weight loss from an obese patient.

```
mtext("Weight (lb)", side=4, line=3)
par(oldpar) # restore settings
```

Although polynomial regression models may describe such data very well within the observed range, they can fail spectacularly outside this range (see Exercise 8.1 on page 277). A more useful model with some theoretical and empirical support is non-linear in the parameters, of the form

$$y = \beta_0 + \beta_1 2^{-t/\theta} + \epsilon \qquad (8.1)$$

Notice that all three parameters have a ready interpretation, namely

β_0 is the ultimate lean weight, or asymptote,
β_1 is the total amount to be lost and
θ is the time taken to lose half the amount remaining to be lost,

which allows us to find rough initial estimates directly from the plot of the data.

The parameters β_0 and β_1 are called *linear parameters* since the second partial derivative of the model function with respect to them is identically zero. The parameter, θ, for which this is not the case, is called a *non-linear parameter*.

8.2   Fitting non-linear regression models

The general form of a non-linear regression model is

$$y = \eta(\boldsymbol{x}, \boldsymbol{\beta}) + \epsilon \qquad (8.2)$$

where $\boldsymbol{x}$ is a vector of covariates, $\boldsymbol{\beta}$ is a p-component vector of unknown parameters and ϵ is a $\mathrm{N}(0, \sigma^2)$ error term. In the weight loss example the parameter vector is $\beta = (\beta_0, \beta_1, \theta)^T$. (As x plays little part in the discussion that follows we often omit it from the notation.)

Suppose y is a sample vector of size n and $\eta(\beta)$ is its mean vector. It is easy to show that the maximum likelihood estimate of β is a least-squares estimate, that is, a minimizer of $\|y - \eta(\beta)\|^2$. The variance parameter σ^2 is then estimated by the residual mean square as in linear regression.

For varying β the vector $\eta(\beta)$ traces out a p-dimensional surface in $\mathbb{R}^n$ that we refer to as the *solution locus*. The parameters β define a coordinate system within the solution locus. From this point of view a linear regression model is one for which the solution locus is a plane through the origin and the coordinate system within it defined by the parameters is affine; that is, it has no curvature. The computational problem in both cases is then to find the coordinates of the point on the solution locus closest to the sample vector y in the sense of Euclidean distance.

The process of fitting non-linear regression models in S is similar to that for fitting linear models, with two important differences:

1. there is no explicit formula for the estimates, so iterative procedures are required, for which initial values must be supplied:

2. linear model formulae that only define the model matrix are not adequate to specify non-linear regression models. A more flexible protocol is needed.

The primary S function for fitting a non-linear regression model is `nls`. We can fit the weight loss model by

```
> wtloss.st <- c(b0=90, b1=95, th=120)
> wtloss.fm <- nls(Weight ~ b0 + b1*2^(-Days/th),
     data = wtloss, start = wtloss.st, trace = T)
67.5435 : 90 95 120
40.1808 : 82.7263 101.305 138.714
39.2449 : 81.3987 102.658 141.859
39.2447 : 81.3737 102.684 141.911
> wtloss.fm
Residual sum of squares : 39.245
parameters:
     b0      b1      th
 81.374 102.68 141.91
formula: Weight ~ b0 + b1 * 2^( - Days/th)
52 observations
```

The arguments to `nls` are the following.

`formula` A non-linear model formula. The form is `response ~ mean`, where the right-hand side can have either of two forms. The standard form as used previously is an ordinary algebraic expression containing both parameters and determining variables. Note that the operators now have their usual arithmetical meaning. (The second form is used with the `plinear` fitting algorithm, discussed in Section 8.3 on page 248.)

`data` An optional data frame for the variables (and sometimes parameters).

`start` A list or numeric vector specifying the starting values for the parameters in the model.

The `names` of the components of `start` are also used to specify which of the variables occurring on the right-hand side of the model formula are parameters. All other variables are then assumed to be to be determining variables.[1]

`control` An optional argument allowing some features of the default iterative procedure to be changed.

`algorithm` An optional character string argument allowing a particular fitting algorithm to be specified. The default procedure is simply `"default"`.

`trace` An argument allowing tracing information from the iterative procedure to be printed. By default none is printed.

In our example the names of the parameters were specified as `b0`, `b1` and `th`. The initial values of 90, 95 and 120 were found by inspection of Figure 8.1. From the trace output the procedure is seen to converge in three iterations.

Weighted data

The `nls` function has no `weights` argument, but non-linear regressions with *known* weights may be handled by writing the formula as ~ `sqrt(W)*(y - M)` rather than `y ~ M`. (The algorithm minimizes the sum of squared differences between left- and right-hand sides and an empty left-hand side counts as zero.) If `W` contains unknown parameters to be estimated the log-likelihood function has an extra term and the problem must be handled by the more general optimization methods such as those discussed in Section 8.7 on pages 261ff. See Problem 8.5.

Using function derivative information

Most non-linear regression fitting algorithms operate in outline as follows. The first-order Taylor-series approximation to η_k at an initial value $\beta^{(0)}$ is

$$\eta_k(\beta) \approx \eta_k(\beta^{(0)}) + \sum_{j=1}^{p} (\beta_j - \beta_j^{(0)}) \left. \frac{\partial \eta_k}{\partial \beta_j} \right|_{\beta=\beta^{(0)}}$$

In vector terms these may be written

$$\eta(\beta) \approx \omega^{(0)} + Z^{(0)}\beta \qquad (8.3)$$

where

$$Z_{kj}^{(0)} = \left. \frac{\partial \eta_k}{\partial \beta_j} \right|_{\beta=\beta^{(0)}} \qquad \text{and} \qquad \omega_k^{(0)} = \eta_k(\beta^{(0)}) - \sum_{j=1}^{p} \beta_j^{(0)} Z_{kj}^{(0)}$$

---

[1] In S-PLUS 3.4 and earlier (and possibly later!) there is a bug that may be avoided if the order in that the parameters appear in the `start` vector is the same as the order in which they first appear in the model. It is as if the order in the `names` attribute were ignored.

Equation (8.3) defines the tangent plane to the surface at the coordinate point $\beta = \beta^{(0)}$. The process consists of regressing the observation vector y onto the tangent plane defined by $Z^{(0)}$ with *offset* vector $\omega^{(0)}$ to give a new approximation, $\beta = \beta^{(1)}$, and iterating to convergence. For a linear regression the offset vector is 0 and the matrix $Z^{(0)}$ is the model matrix X, a constant matrix, so the process converges in one step. In the non-linear case the next approximation is

$$\beta^{(1)} = \left(Z^{(0)\,T} Z^{(0)}\right)^{-1} Z^{(0)\,T} \left(y - \omega^{(0)}\right)$$

With the `default` algorithm the Z matrix is computed approximately by numerical methods unless formulae for the first derivatives are supplied. Providing derivatives often (but not always) improves convergence.

Derivatives can be provided as an attribute of the model. The standard way to do this is to write an S function to calculate the mean vector η and the Z matrix. The result of the function is η with the Z matrix included as a `gradient` attribute.

For our simple example the three derivatives are

$$\frac{\partial \eta}{\partial \beta_0} = 1, \qquad \frac{\partial \eta}{\partial \beta_1} = 2^{-x/\theta}, \qquad \frac{\partial \eta}{\partial \theta} = \frac{\log(2)\,\beta_1 x 2^{-x/\theta}}{\theta^2}$$

so an S function to specify the model including derivatives is

```
expn <- function(b0, b1, th, x) {
      temp <- 2^(-x/th)
      model.func <- b0 + b1 * temp
      Z <- cbind(1, temp, (b1 * x * temp * log(2))/th^2)
      dimnames(Z) <- list(NULL, c("b0","b1","th"))
      attr(model.func, "gradient") <- Z
      model.func
}
```

Note that the gradient matrix must have column names matching those of the corresponding parameters.

We can fit our model again using first derivative information:

```
> wtloss.gr <- nls(Weight ~ expn(b0, b1, th, Days),
      data = wtloss, start = wtloss.st, trace = T)
67.5435 : 90 95 120
40.1808 : 82.7263 101.305 138.714
39.2449 : 81.3987 102.658 141.859
39.2447 : 81.3738 102.684 141.911
```

This appears to make no difference to the speed of convergence, but tracing the function `expn` shows that only 6 evaluations are required when derivatives are supplied compared with 21 if they are not supplied.

Functions such as `expn` can often be generated automatically using the symbolic differentiation function `deriv`. It is called with three arguments:

(a) the model formula, with the left-hand side optionally left blank,

(b) a character vector giving the names of the parameters and

(c) an empty function with an argument specification as required for the result.

An example makes the process clearer. For the weight loss data with the exponential model, we can use:

```
expn1 <- deriv(y ~ b0 + b1 * 2^(-x/th), c("b0", "b1", "th"),
               function(b0, b1, th, x) {})
```

The result is the function

```
expn1 <- function(b0, b1, th, x)
{
    .expr3 <- 2^(( - x)/th)
    .value <- b0 + (b1 * .expr3)
    .grad <- array(0, c(length(.value), 3),
                   list(NULL, c("b0", "b1", "th")))
    .grad[, "b0"] <- 1
    .grad[, "b1"] <- .expr3
    .grad[, "th"] <- b1 *
        (.expr3 * (0.693147180559945 * (x/(th^2))))
    attr(.value, "gradient") <- .grad
    .value
}
```

Self-starting non-linear regressions

Very often reasonable starting values for a non-linear regression can be calculated by some simple and fairly automatic procedure. Setting up such a *self-starting* non-linear model is somewhat technical and requires some attention to detail.

(a) The response function for the non-linear model must be specified as an explicit S function.

(b) This model *function* is given (old-style) class "selfStart", and an initial attribute which is another S function defining the initial value procedure. The initial value function must have an identical argument sequence to the model function, and return a named list of initial values with the same names as the parameters used on the call.

(c) The initial value procedure will usually require the response variable as well. This may be obtained as the object .nls.initial.response, guaranteed to exist in frame 1 when needed.

The function selfStart can be used like deriv to generate a self-starting model function, including first derivative information.

Consider again a negative exponential decay model such as that used in the weight loss example but this time written in the more usual exponential form:

$$y = \beta_0 + \beta_1 \exp(-x/\theta) + \epsilon$$

One effective initial value procedure follows.

(i) Fit an initial quadratic regression in x.

(ii) Find the fitted values, say, y_0, y_1 and y_2 at three equally spaced points x_0, $x_1 = x_0 + \delta$ and $x_2 = x_0 + 2\delta$.

(iii) Equate the three fitted values to their expectation under the non-linear model to give an initial value for θ as

$$\theta_0 = \delta \big/ \log\left(\frac{y_0 - y_1}{y_1 - y_2}\right)$$

(iv) Initial values for β_0 and β_1 can then be obtained by linear regression of y on $\exp(-x/\theta_0)$.

An S function to implement this procedure (with a few extra checks) called[2] negexp.ival is supplied in the MASS library. We do not reproduce the function here, but interested readers should study it carefully.

We can make a self-starting model with both first derivative information and this initial value routine by

```
negexp <- selfStart(model = ~ b0 + b1*exp(-x/th),
    initial = negexp.ival, parameters = c("b0", "b1", "th"),
    template = function(x, b0, b1, th) {})
```

where the first, third and fourth arguments are the same as for deriv. We may now fit the model without explicit initial values.

```
> wtloss.ss <- nls(Weight ~ negexp(Days, B0, B1, theta),
                    data = wtloss, trace = T)
      B0      B1    theta
  82.713  101.49  200.16
39.5453 : 82.7131 101.495 200.160
39.2450 : 81.3982 102.659 204.652
39.2447 : 81.3737 102.684 204.734
```

(The first two lines of output come from the initial value procedure and the last three from the nls trace.)

The standard collection of software contains four examples of self-starting models, a two-term exponential model, a first-order compartmental model and three- and four-parameter logistic models.

8.3 Non-linear fitted model objects and method functions

The result of a call to nls is an object of class nls. The standard method functions are available, the most important of which are

print to print an abbreviated synopsis (usually called implicitly),

---

[2] Use negexp.SSival if version 3 of the nlme library is in use.

summary to print a summary of the fitting process results,

coef to extract the estimated mean parameters,

fitted to extract the fitted value,

predict to estimate values for the mean at new values of the covariates, optionally with standard errors,

anova to compare non-linear regression models, usually with an approximate
F-test.

residuals to extract residuals,

profile to explore the least-squares surface in the vicinity of the estimate.

For the preceding example the summary function gives:

```
> summary(wtloss.gr)
Formula: Weight ~ expn1(b0, b1, th, Days)
Parameters:
      Value Std. Error t value
b0 81.374     2.2690  35.863
b1 102.684    2.0828  49.302
th 141.911    5.2945  26.803
Residual standard error: 0.894937 on 49 degrees of freedom
Correlation of Parameter Estimates:
       b0      b1
b1 -0.989
th -0.986  0.956
```

Surprisingly, no working deviance method function exists but such a method
function is easy to write and is included in the MASS library. It merely requires

```
> deviance.nls <- function(object) sum(object$residuals^2)
> deviance(wtloss.gr)
[1] 39.245
```

Library MASS also has a generic function vcov that will extract the estimated
variance matrix of the mean parameters:

```
> vcov(wtloss.gr)
         b0       b1      th
b0   5.1484  -4.6745 -11.841
b1  -4.6745   4.3379  10.543
th -11.8414  10.5432  28.032
```

Taking advantage of linear parameters

If all non-linear parameters were known the model would be linear and standard linear regression methods could be used. This simple idea lies behind the
"plinear" algorithm. It requires a different form of model specification that
combines aspects of linear and non-linear model formula protocols. In this case
the right-hand side expression specifies a *matrix* whose columns are functions of

the non-linear parameters. The linear parameters are then implied as the regression coefficients for the columns of the matrix. Initial values are only needed for the non-linear parameters. Unlike the linear model case there is no implicit intercept term.

There are several advantages in using the partially linear algorithm. It can be much more stable than methods that do not take advantage of linear parameters, it requires fewer initial values and it can often converge from poor starting positions where other procedures fail.

Asymptotic regressions with different asymptotes

As an example of a case where the partially linear algorithm is very convenient, we consider a data set first discussed in Linder, Chakravarti & Vuagnat (1964). The object of the experiment was to assess the influence of calcium in solution on the contraction of heart muscle in rats. The left auricle of 21 rat hearts was isolated and on several occasions electrically stimulated and dipped into various concentrations of calcium chloride solution, after which the shortening was measured. The data frame `muscle` in the `MASS` library contains the data as variables `Strip`, `Conc` and `Length`.

The particular model posed by the authors is of the form

$$\log y_{ij} = \alpha_j + \beta \rho^{x_{ij}} + \varepsilon_{ij} \tag{8.4}$$

where i refers to the concentration and j to the muscle strip. This model has 1 non-linear and 22 linear parameters. We take the initial estimate for ρ to be 0.1. Our first step is to construct a matrix to select the appropriate α.

```
> A <- model.matrix(~ Strip - 1, data=muscle)
> rats.nls1 <- nls(log(Length) ~ cbind(A, rho^Conc),
    data = muscle, start = c(rho=0.1), algorithm="plinear")
> B <- coef(rats.nls1); B
      rho
 0.077778 3.0831 3.3014 3.4457 2.8047 2.6084 3.0336 3.523
 3.3871 3.4671 3.8144 3.7388 3.5133 3.3974 3.4709 3.729 3.3186
 3.3794 2.9645 3.5847 3.3963 3.37 -2.9601
```

The linear parameters are unnamed components at the end of the coefficient vector pertaining in order to the columns of the matrix specified in the model. We can now use this coefficient vector as a starting value for a fit using the conventional algorithm.

```
> st <- list(alpha = B[2:22], beta = B[23], rho = B[1])
> rats.nls2 <- nls(log(Length) ~ alpha[Strip] + beta*rho^Conc,
                  data = muscle, start = st)
```

Notice that if a parameter in the non-linear regression is indexed, such as `alpha[Strip]` here, the starting values must be supplied as a list with named separate components.

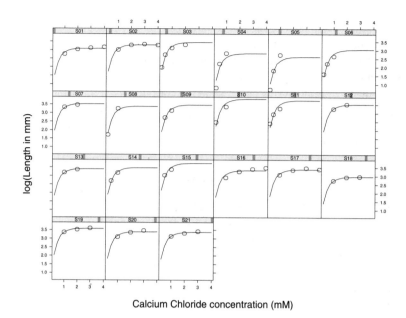

Figure 8.2: The heart contraction data of Linder *et al.* (1964): points and fitted model.

A trace will show that this converges in one step with unchanged parameter estimates, as expected.

Having the fitted model object in standard rather than `"plinear"` form allows us to predict from it using the standard generic function `predict`. We now put data and predicted values in a Trellis display.

```
attach(muscle)
Muscle <- expand.grid(Conc = sort(unique(Conc)),
                      Strip = levels(Strip))
Muscle$Yhat <- predict(rats.nls2, Muscle)
Muscle$logLength <- rep(NA, nrow(Muscle))
ind <- match(paste(Strip, Conc),
             paste(Muscle$Strip, Muscle$Conc))
Muscle$logLength[ind] <- log(Length)
detach()

xyplot(Yhat ~ Conc | Strip, Muscle, as.table = T,
   ylim = range(c(Muscle$Yhat, Muscle$logLength), na.rm=T),
   subscripts = T, xlab = "Calcium Chloride concentration (mM)",
   ylab = "log(Length in mm)", panel =
   function(x, y, subscripts, ...) {
      lines(spline(x, y))
      panel.xyplot(x, Muscle$logLength[subscripts], ...)
   })
```

The result is shown in Figure 8.2. The model appears to describe the situation fairly well, but for some purposes a non-linear mixed effects model might be more suitable (Exercise 8.8).

8.4 Confidence intervals for parameters

For some non-linear regression problems the standard error may be an inadequate summary of the uncertainty of a parameter estimate and an asymmetric confidence interval may be more appropriate. In non-linear regression it is customary to invert the "extra sum of squares" to provide such an interval, although the result is usually almost identical to the likelihood ratio interval.

Suppose the parameter vector is $\boldsymbol{\theta} = \widehat{\boldsymbol{\theta}}(\theta_1, \boldsymbol{\theta}_2)^T$ and without loss of generality we wish to test an hypothesis $\mathrm{H}_0 : \theta_1 = \theta_{10}$. Let $\widehat{\boldsymbol{\theta}}$ be the overall least-squares estimate, $\widehat{\boldsymbol{\theta}}_{2|1}$ be the conditional least-squares estimate of $\boldsymbol{\theta}_2$ with $\theta_1 = \theta_{10}$ and put $\widehat{\boldsymbol{\theta}}(\theta_{10}) = (\theta_{10}, \widehat{\boldsymbol{\theta}}_{2|1})^T$. The extra sum of squares principle then leads to the test statistic:

$$F(\theta_{10}) = \frac{\mathrm{RSS}\{\widehat{\boldsymbol{\theta}}(\theta_{10})\} - \mathrm{RSS}(\widehat{\boldsymbol{\theta}})}{s^2}$$

which under the null hypothesis has an approximately $F_{1, n-p}$-distribution. (Here RSS denotes the residual sum of squares.) Similarly the *signed square root*

$$\tau(\theta_{10}) = \mathrm{sign}(\theta_{10} - \widehat{\theta}_1)\sqrt{F(\theta_{10})} \qquad (8.5)$$

has an approximate t_{n-p}-distribution under H_0. The confidence interval for θ_1 is then the set $\{\theta_1 \mid -t < \tau(\theta_1) < t\}$, where t is the appropriate percentage point from the t_{n-p}-distribution.

The `profile` function is generic. The method for `nls` objects explores the residual sum of squares surface by taking each parameter θ_i in turn and fixing it at a series of values above and below $\widehat{\theta}_i$ so that (if possible) $\tau(\theta_i)$ varies from 0 by at least some pre-set value t in both directions. (If the sum of squares function "flattens out" in either direction, or if the fitting process fails in some way, this may not be achievable, in which case the profile is incomplete.) The value returned is a list of data frames, one for each parameter, and named by the parameter names. Each data frame component consists of

(a) a vector, `tau`, of values of $\tau(\theta_i)$ and

(b) a matrix, `par.vals`, giving the values of the corresponding parameter estimates $\widehat{\boldsymbol{\theta}}(\theta_i)$, sometimes called the 'parameter traces'.

The argument `which` of `profile` may be used to specify only a subset of the parameters to be so profiled. (Note that profiling is not available for fitted model objects in which the `plinear` algorithm is used.)

Finding a confidence interval then requires the `tau` values for that parameter to be interpolated and the set of θ_i values to be found (here always of the form

$\underline{\theta}_i < \theta_i < \overline{\theta}_i$). The function `confint` in the MASS library is an old-style generic function to perform this interpolation. It works for `glm` or `nls` fitted models, although the latter must be fitted with the default fitting method. The critical steps in the method function are

```
confint.profile.nls <-
  function(object, parm = seq(along = pnames),
           level = 0.95) {
    ....
  for(pm in parm) {
    pro <- object[[pm]]
    sp <- spline(x = pro[, "par.vals"][, pm], y = pro$tau)
    ci[pnames[pm], ] <- approx(sp$y, sp$x, xout = cutoff)$y
  }
  drop(ci)
}
```

The three steps inside the loop are, respectively, extract the appropriate component of the *profile* object, do a spline interpolation in the t-statistic and linearly interpolate in the digital spline to find the parameter values at the two cutoffs. If the function is used on the fitted object itself the first step is to make a profile object from it and this will take most of the time. If a profiled object is already available it should be used instead.

The weight loss data, continued

For the person on the weight loss programme an important question is how long it might be before he achieves some goal weight. If δ_0 is the time to achieve a predicted weight of w_0, then solving the equation $w_0 = \beta_0 + \beta_1 2^{-\delta_0/\theta}$ gives

$$\delta_0 = -\theta \log_2\{(w_0 - \beta_0)/\beta_1\}$$

We now find a confidence interval for δ_0 for various values of w_0 by the outlined method.

The first step is to re-write the model function using δ_0 as one of the parameters; the most convenient parameter for it to replace is θ and the new expression of the model becomes

$$y = \beta_0 + \beta_1 \left(\frac{w_0 - \beta_0}{\beta_1}\right)^{x/\delta_0} + \epsilon \qquad (8.6)$$

In order to find a confidence interval for δ_0 we need to fit the model with this parametrization. First we build a model function:

```
expn2 <- deriv(~b0 + b1*((w0 - b0)/b1)^(x/d0),
               c("b0","b1","d0"), function(b0, b1, d0, x, w0) {})
```

It is also convenient to have a function that builds an initial value vector from the estimates of the present fitted model:

```
wtloss.init <- function(obj, w0) {
  p <- coef(obj)
  d0 <-  - log((w0 - p["b0"])/p["b1"])/log(2) * p["th"]
  c(p[c("b0", "b1")], d0 = as.vector(d0))
}
```

(Note the use of `as.vector` to remove unwanted names.) The main calculations are done in a short loop:

```
> out <- NULL
> w0s <- c(110, 100, 90)
> for(w0 in w0s) {
    fm <- nls(Weight ~ expn2(b0, b1, d0, Days, w0),
            wtloss, start = wtloss.init(wtloss.gr, w0))
    out <- rbind(out, c(coef(fm)["d0"], confint(fm, "d0")))
  }

Waiting for profiling to be done...
  ....
> dimnames(out)[[1]] <- paste(w0s,"kg:")
> out
          d0   2.5%  97.5%
110 kg: 261.51 256.23 267.50
100 kg: 349.50 334.74 368.02
 90 kg: 507.09 457.56 594.97
```

As the weight decreases the confidence intervals become much wider and more asymmetric. For weights closer to the estimated asymptotic weight the profiling procedure can fail.

A bivariate region: The Stormer viscometer data

The following example comes from Williams (1959). The Stormer viscometer measures the viscosity of a fluid by measuring the time taken for an inner cylinder in the mechanism to perform a fixed number of revolutions in response to an actuating weight. The viscometer is calibrated by measuring the time taken with varying weights while the mechanism is suspended in fluids of accurately known viscosity. The dataset comes from such a calibration, and theoretical considerations suggest a non-linear relationship between time T, weight w and viscosity V of the form

$$T = \frac{\beta_1 v}{w - \beta_2} + \epsilon$$

where β_1 and β_2 are unknown parameters to be estimated. Note that β_1 is a linear parameter and β_2 is non-linear. The dataset is given in Table 8.1.

Williams suggested that a suitable initial value may be obtained by writing the regression model in the form

$$wT = \beta_1 v + \beta_2 T + (w - \beta_2)\epsilon$$

Table 8.1: The Stormer viscometer calibration data. The body of the table shows the times in seconds.

Weight (grams)	Viscosity (poise)								
	14.7	27.5	42.0	75.7	89.7	146.6	158.3	161.1	298.3
20	35.6	54.3	75.6	121.2	150.8	229.0	270.0		
50	17.6	24.3	31.4	47.2	58.3	85.6	101.1	92.2	187.2
100				24.6	30.0	41.7	50.3	45.1	89.0,86.5

and regressing wT on v and T using ordinary linear regression. With the data available in a data frame `stormer` with variables `Viscosity`, `Wt` and `Time`, we may proceed as follows.

```
> fm0 <- lm(Wt*Time ~ Viscosity + Time - 1,  data=stormer)
> b0 <- coef(fm0);  names(b0) <- c("b1", "b2")
> b0
      b1      b2
 28.876 2.8437
> storm.fm <- nls(Time ~ b1*Viscosity/(Wt-b2), data=stormer,
            start=b0, trace=T)
885.365 : 28.8755 2.84373
825.110 : 29.3935 2.23328
825.051 : 29.4013 2.21823
```

Since there are only two parameters we can display a confidence region for the regression parameters as a contour map. To this end put:

```
bc <- coef(storm.fm)
se <- sqrt(diag(vcov(storm.fm)))
dv <- deviance(storm.fm)
```

Define $d(\beta_1, \beta_2)$ as the sum of squares function:

$$d(\beta_1, \beta_2) = \sum_{i=1}^{23} \left(T_i - \frac{\beta_1 v_i}{w_i - \beta_2} \right)^2$$

Then `dv` contains the minimum value, $d_0 = d(\widehat{\beta}_1, \widehat{\beta}_2)$, the residual sum of squares or model deviance.

If β_1 and β_2 are the true parameter values the "extra sum of squares" statistic

$$F(\beta_1, \beta_2) = \frac{(d(\beta_1, \beta_2) - d_0)/2}{d_0/21}$$

is approximately distributed as $F_{2,21}$. An approximate confidence set contains those values in parameter space for which $F(\beta_1, \beta_2)$ is less than the 95% point

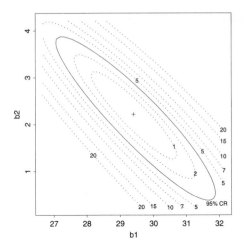

Figure 8.3: The Stormer data. The F-statistic surface and a confidence region for the regression parameters.

of the $F_{2,21}$ distribution. We construct a contour plot of the $F(\beta_1, \beta_2)$ function and mark off the confidence region. The result is shown in Figure 8.3.

A suitable region for the contour plot is three standard errors either side of the least-squares estimates in each parameter. Since these ranges are equal in their respective standard error units it is useful to make the plotting region square.

```
par(pty = "s")
b1 <- bc[1] + seq(-3*se[1], 3*se[1], length = 51)
b2 <- bc[2] + seq(-3*se[2], 3*se[2], length = 51)
bv <- expand.grid(b1, b2)
```

The simplest way to calculate the sum of squares function is to use the apply function:

```
ssq <- function(b)
        sum((Time - b[1] * Viscosity/(Wt-b[2])))^2)
dbetas <- apply(bv, 1, ssq)
```

However using a function such as outer makes better use of the vectorizing facilities of S, and in this case a direct calculation is the most efficient:

```
attach(stormer)
cc <- matrix(Time - rep(bv[,1],rep(23, 2601)) *
        Viscosity/(Wt - rep(bv[,2], rep(23, 2601))), 23)
dbetas <- matrix(drop(rep(1, 23) %*% cc^2), 51)
```

The F-statistic array is then:

```
fstat <- matrix( ((dbetas - dv)/2) / (dv/21), 51, 51)
```

We can now produce a contour map of the F-statistic, taking care that the contours occur at relatively interesting levels of the surface. Note that the confidence region contour is at about 3.5:

```
> qf(0.95, 2, 21)
[1] 3.4668
```

Our intention is to produce a slightly non-standard contour plot and for this the old style plotting functions are more flexible than Trellis graphics functions. Rather than use `contour` to set up the plot directly, we begin with a call to `plot`:

```
plot(b1, b2, type="n")
lev <- c(1, 2, 5, 7, 10, 15, 20)
contour(b1, b2, fstat, levels=lev, labex=0.75, lty=2, add=T)
contour(b1, b2, fstat, levels=qf(0.95,2,21), add=T, labex=0)
text(31.6, 0.3, labels="95% CR", adj=0, cex=0.75)
points(bc[1], bc[2], pch=3, mkh=0.1)
par(pty = "m")
```

Since the likelihood function has the same contours as the F-statistic, the near elliptical shape of the contours is an indication that the approximate theory based on normal linear regression is probably accurate, although more than this is needed to be confident. (See the next section.) Given the way the axis scales have been chosen, the elongated shape of the contours shows that the estimates $\widehat{\beta}_1$ and $\widehat{\beta}_2$ are highly (negatively) correlated.

Note that finding a bivariate confidence region for two regression parameters where there are several others present can be a difficult calculation, at least in principle, since each point of the F-statistic surface being contoured must be calculated by optimizing with respect to the other parameters.

Bootstrapping

An alternative way to explore the distribution of the parameter estimates is to use the bootstrap. This was a designed experiment, so we illustrate model-based bootstrapping using the functions from library `boot`. As this is a non-linear model the residuals may have a non-zero mean, which we remove.

```
> library(boot)
> storm.fm <- nls(Time ~ b*Viscosity/(Wt - c), stormer,
                  start = c(b=29.401, c=2.2183))
> summary(storm.fm)$parameters
    Value Std. Error t value
b 29.4010    0.91553 32.1135
c  2.2183    0.66553  3.3332
> st <- cbind(stormer, fit=fitted(storm.fm))
> storm.bf <- function(rs, i) {
      st <- st # for 5.x
      st$Time <-  st$fit + rs[i]
      nls(Time ~ b * Viscosity/(Wt - c), st,
```

```
              start = coef(storm.fm))$parameters
   }
> rs <- scale(resid(storm.fm), scale = F) # remove the mean
> storm.boot <- boot(rs, storm.bf, R = 4999)
> storm.boot
   ....
Bootstrap Statistics :
     original     bias    std. error
t1*  28.7156   0.72479     0.85811
t2*   2.4799  -0.28268     0.62431
> boot.ci(storm.boot, index=1,
           type=c("norm", "basic", "perc", "bca"))
Intervals :
Level        Normal                 Basic
95%   (26.37, 29.68 )    (26.42, 29.77 )

Level        Percentile             BCa
95%   (27.66, 31.01 )    (26.23, 29.64 )
Calculations and Intervals on Original Scale
Warning : BCa Intervals used Extreme Quantiles
> boot.ci(storm.boot, index=2,
           type=c("norm", "basic", "perc", "bca"))
Intervals :
Level        Normal                 Basic
95%   ( 1.541,  3.962 )    ( 1.557,  4.006 )

Level        Percentile             BCa
95%   ( 0.953,  3.403 )    ( 1.585,  4.096 )
Calculations and Intervals on Original Scale
Some BCa intervals may be unstable
```

The 'original' here is not the original fit as the residuals were adjusted. Figure 8.3 suggests that a likelihood-based analysis would support the percentile intervals. Note that despite the very large number of bootstrap samples, this is still not enough for BC_a intervals.

8.5 Profiles

One way of assessing departures from the planar assumption (both of the surface itself and of the coordinate system within the surface) is to look at the low-dimensional profiles of the sum of squares function, as discussed in Section 8.4. Indeed this is the original intended use for the `profile` function.

For coordinate directions along which the approximate linear methods are accurate, a plot of the non-linear t-statistics, $\tau(\beta_i)$ against β_i over several standard deviations on either side of the maximum likelihood estimate should be straight. (The $\tau(\beta_i)$ were defined in equation (8.5) on page 251.) Any deviations from straightness serve as a warning that the linear approximation may be misleading in that direction.

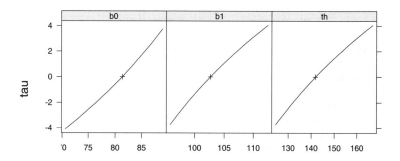

Figure 8.4: Profile plots for the negative exponential weight loss model.

Appropriate plot methods for objects produced by the `profile` function are available (and our library `MASS` contains an enhanced version using Trellis). For the weight loss data we can use (Figure 8.4)

```
plot(profile(wtloss.gr))
```

A limitation of these plots is that they only examine each coordinate direction separately. Profiles can also be defined for two or more parameters simultaneously. These present much greater difficulties of computation and to some extent of visual assessment. An example is given in the previous section with the two parameters of the `stormer` data model. In Figure 8.3 an assessment of the linear approximation requires checking both that the contours are approximately elliptical and that each one-dimensional cross-section through the minimum is approximately quadratic. In one dimension it is much easier visually to check the linearity of the signed square root than the quadratic nature of the sum of squares itself.

The profiling process itself actually accumulates much more information than that shown in the plot. As well as recording the value of the optimized likelihood for a fixed value of a parameter β_j, it also finds and records the values of all other parameters for which the conditional optimum is achieved. The `pairs` method function for `profile` objects supplied in our `MASS` library can display this so-called *profile trace* information in graphical form, which can shed considerably more light on the local features of the log-likelihood function near the optimum. This is discussed in more detail in the on-line complements.

8.6 Constrained non-linear regression

All the functions for regression we have seen so far assume that the parameters can take any real value. Constraints on the parameters can be often incorporated by reparametrization (such as $\beta_i = e^\theta$, possibly thereby making a linear model non-linear), but this can be undesirable if, for example, the constraint is $\beta_i \geqslant 0$ and 0 is expected to occur. Two functions for constrained regression are provided in S-PLUS.

The function `nnls.fit` performs linear least squares subject to the constraint that all the coefficients are non-negative. If some of the coefficients are unconstrained, we can use the usual 'trick' of including both x and -x. For the whiteside data of Section 6.1 we can use

```
> attach(whiteside)
> Gas <- Gas[Insul=="Before"]; Temp <- -Temp[Insul=="Before"]
> nnls.fit(cbind(1, -1, Temp), Gas)
$coefficients:
                  Temp
 6.8538 0 0.39324
    . . . .
```

This constrains the slope (of -Temp) to be non-negative, but does not constrain the intercept. The fitted object includes residuals and some diagnostics of the algorithm.

The function `nlregb` performs non-linear regression with range bounds on the parameters, so the parameters are constrained to lie within a hypercube. The user must specify a function giving the *residuals* whose sum of squares is to be minimized. First derivatives of the residuals (called the *Jacobian* in the function documentation) may be supplied, otherwise the derivatives are approximated by finite differences. We can use `nlregb` for the weight loss example by

```
attach(wtloss)
nlregb(nrow(wtloss), c(90,95,120), function(x)
    Weight-x[1]-x[2]*2^(-Days/x[3])), lower=rep(0,3))$parameters
[1]  81.374 102.684 141.910
detach()
```

Of course, in this example the positivity constraints on the parameters are unlikely to be needed. It is also possible to pass the data via extra arguments, which we illustrate while supplying derivatives.

```
wtloss.r <- function(x, Weight, Days)
    Weight - x[1] - x[2] * 2^(-Days/x[3])
wtloss.rg <- function(x, Weight, Days) {
    temp <- 2^(-Days/x[3])
    -cbind(1, temp, x[2]*Days*temp*log(2)/x[3]^2)
}
wtloss.nl <- nlregb(nrow(wtloss), c(90, 95, 120),
    wtloss.r,  wtloss.rg, lower = rep(0,3),
    Weight = wtloss$Weight, Days = wtloss$Days)
```

No facility is provided for finding the estimated variance matrix of the parameters in a `nlregb` fit. The non-constant part of the negative log-likelihood is $E = \sum_i r_i(\theta)^2/2\sigma^2$ for residuals $r_i(\theta)$, when the error variance σ^2 is assumed known. Then

$$\frac{\partial^2 E}{\partial\theta_s\partial\theta_t} = \frac{1}{\sigma^2}\left[\sum_i \frac{\partial r_i}{\partial\theta_s}\frac{\partial r_i}{\partial\theta_t} + \sum_i r_i\frac{\partial^2 r_i}{\partial\theta_s\partial\theta_t}\right]$$

The first term is the Gauss–Newton approximation, and under the usual modelling assumptions is the Fisher information (as the residuals have mean zero). We can base an estimated variance matrix on this term (as used by `summary.nls` and hence `vcov.nls`) by

```
vcov1 <- function(object) {
   gr <- object$jacobian
   df <- length(object$resid) - length(object$param)
   sum(object$resid^2)/df * solve(t(gr) %*% gr)
}
```

As usual, we substitute $\hat{\theta}$ for θ and s^2 for σ^2.

We can use a finite-difference approximation to the second derivative to find the observed information more precisely, using the function `vcov.nlreg` in our library. (Note: there is no class `nlreg`, so this must be called directly.)

```
> sqrt(diag(vcov1(wtloss.nl)))
[1] 2.2690 2.0828 5.2945
> sqrt(diag(vcov.nlregb(wtloss.nl, method="Fisher")))
[1] 2.2690 2.0828 5.2945
> sqrt(diag(vcov.nlregb(wtloss.nl, method="observed")))
[1] 2.2653 2.0795 5.2856
```

In our example (and many others) the correction is negligible. Where there is a difference, the approach via the observed information is thought to be more accurate (see Efron & Hinkley, 1978).

The rationale for using either the observed or Fisher information is based on asymptotic theory which assumes that the data were in fact generated by the non-linear regression for a particular set of parameters. If we drop this (dubious) assumption we can regard the estimate of the 'least false' parameter (that which fits best the true data-generating mechanism), and use the appropriate asymptotic theory developed by Huber (1967) and White (1982). (See also Ripley, 1996, pp. 31–36.) We can use

```
> sqrt(diag(vcov.nlregb(wtloss.nl, method="Huber")))
[1] 2.2616 2.0763 5.2766
```

Note that the theory used here is unreliable if the solution $\hat{\theta}$ is close to or on the boundary; the asymptotic theory is delicate in that case and `vcov.nlreg` issues a warning.

The function `vcov.nlreg` is quite long and employs some intricate S programming ideas, so may be of interest to S programmers. If the `nlregb` fit did not supply the Jacobian, a local linear approximation is used, and only the Fisher information method is available.

```
> wtloss.nl0 <-  nlregb(nrow(wtloss), c(90,95,120),
      wtloss.r, lower = rep(0,3),
      Weight = wtloss$Weight, Days = wtloss$Days)
> sqrt(diag(vcov.nlregb(wtloss.nl0)))
[1] 2.2690 2.0827 5.2945
```

8.7   General optimization and maximum likelihood estimation

Each of `nls` and `nlregb` has an analogue for general minimization, and there
are also functions for univariate searches. Most are documented in the section
`Non-linear Regression` of the help system, rather than `Optimization`.

Univariate functions

The functions `optimize` and `uniroot` work with continuous functions of a single variable: `optimize` can find a (local) minimum (the default) or a (local)
maximum whereas `uniroot` finds a zero. Both search within an interval specified either by the argument `integer` or arguments `lower` and `upper`. Both are
based on safeguarded polynomial interpolation (Brent, 1973; Nash, 1990).

Using ms

The `ms` function is a general function for minimizing quantities that can be written as sums of one or more terms. The call to `ms` is very similar to that for
`nls`, but the interpretation of the formula is slightly different. It has no response,
and the quantity on the right-hand side specifies the entire sum to be minimized.
Hence the parameter estimates from the call

```
obj.nls <- nls(y ~ fn(a,b,c,x), start=p0, data=dat)
```

should in principle be the same as those from

```
obj.ms <- ms( ~ (y - fn(a,b,c,x))^2, start=p0, data=dat)
```

The resulting object is of class `ms`, for which few method functions are available.
The four main ones are

`print` for basic printing (mostly used implicitly),

`coef` for the parameter estimates,

`summary` for a succinct report of the fitted results,

`profile` for an exploration of the region near the minimum.

The fitted object has additional components, two of independent interest being

`obj.ms$value` the minimum value of the sum, and

`obj.ms$pieces` the vector of summands at the minimum.

   Note that `fitted` and `resid` are not available since in this general context
they have no unique meaning.

   The most common statistical use of `ms` is to minimize negative log-likelihood
functions, thus finding maximum likelihood estimates. The fitting algorithm
can make use of first- and second-derivative information, the first- and second-
derivative arrays of the summands rather than of a model function. In the case
of a negative log-likelihood function, these are arrays whose "row sums" are the

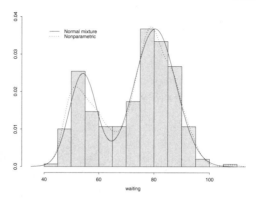

Figure 8.5: Histogram for the waiting times between successive eruptions for the "Old Faithful" geyser, with non-parametric and parametric estimated densities superimposed.

negative score vector and observed information matrix, respectively. The first-derivative information is supplied as a `gradient` attribute of the formula, as described for `nls` in Section 8.2. Second-derivative information may be supplied via a `hessian` attribute. The function `deriv3` (written by David Smith and included with permission in the `MASS` library) extends `deriv` by generating a model function with both `gradient` and `hessian` attributes to the value.

The convergence criterion for `ms` is that *one of* the relative changes in the parameter estimates is small, the relative change in the achieved value is small *or* that the optimized function value is close to zero. The latter is only appropriate for deviance-like quantities, and can be eliminated by setting the argument

```
control = ms.control(f.tolerance=-1)
```

Fitting a mixture model

The waiting times between eruptions in the data `geyser` are strongly bimodal. Figure 8.5 shows a histogram with a density estimate superimposed (with bandwidth chosen by the Sheather–Jones method described in Chapter 5).

```
attach(geyser)
truehist(waiting, xlim=c(35,110), ymax=0.04, h=5)
width.SJ(waiting)
[1] 10.239
wait.dns <- density(waiting, 200, width=10.24)
lines(wait.dns, lty=2)
```

From inspection of this figure a mixture of two normal distributions would seem to be a reasonable descriptive model for the marginal distribution of waiting times. We now consider how to fit this by maximum likelihood. The observations are not independent since successive waiting times are strongly negatively correlated. In this section we propose to ignore this, both for simplicity and because ignoring this aspect is not likely to misrepresent seriously the information in the sample on the marginal distribution in question.

Useful references for mixture models include Everitt & Hand (1981), Titterington, Smith & Makov (1985) and McLachlan & Basford (1988). Everitt & Hand describe the EM algorithm for fitting a mixture model, which is simple (and provides an interesting S programming exercise), but we consider here a more direct function minimization method which can be faster and more reliable (Redner & Walker, 1984; Ingrassia, 1992).

If y_i, $i = 1, 2, \ldots, n$ is a sample waiting time the log-likelihood function for a mixture of two normal components is

$$L(\pi, \mu_1, \sigma_1, \mu_2, \sigma_2) = \sum_{i=1}^{n} \log \left[\frac{\pi}{\sigma_1} \phi \left(\frac{y_i - \mu_1}{\sigma_1} \right) + \frac{1 - \pi}{\sigma_2} \phi \left(\frac{y_i - \mu_2}{\sigma_2} \right) \right]$$

We estimate the parameters by minimizing $-L$.

It is helpful in this example to use both first- and second-derivative information and we use `deriv3` to produce a model function. Both `deriv` and `deriv3` use the system function `D` to generate symbolic derivatives but as it stands it does not differentiate functions involving the normal distribution or density functions. To overcome this we need to make minor changes to the function `D`, which does the actual differentiation, and the ancillary function, `make.call`. Extended versions are supplied in our library `MASS`, which must be attached with `first=T` to make them effective. Comparing these with the system versions will show how to make similar extensions if needed in the future. At most two or three statements are involved. Note that these extensions only allow calls to `pnorm` and `dnorm` with one argument. Thus calls such as `pnorm(x,u,s)` must be written as `pnorm((x-u)/s)`.

We can now generate a function that calculates the summands of $-L$ and also returns first- and second-derivative information for each summand by:

```
lmix2 <- deriv3(
        ~ -log(p*dnorm((x-u1)/s1)/s1 + (1-p)*dnorm((x-u2)/s2)/s2),
        c("p", "u1", "s1", "u2", "s2"),
        function(x, p, u1, s1, u2, s2) NULL)
```

which took 13 seconds on the PC.

Initial values for the parameters could be obtained by the method of moments described in Everitt & Hand (1981, pp. 31ff) but for well-separated components as we have here, simply choosing initial values by reference to the plot suffices. For the initial value of π we take the proportion of the sample below the density low point at 70.

```
> p0 <- c(p=mean(waiting < 70), u1=50, s1=5, u2=80, s2=5)
> p0
          p u1 s1 u2 s2
   0.36120 50  5 80  5
```

We can trace the iterative process, using a trace function that produces just one line of output per iteration:

```
tr.ms <- function(info, theta, grad, scale, flags, fit.pars) {
    cat(round(info[3], 3), ":", signif(theta), "\n")
    invisible()
}
```

Note that the trace function must have the same arguments as the standard trace function, `trace.ms`.

We can now fit the mixture model:

```
> wait.mix2 <- ms(~ lmix2(waiting, p, u1, s1, u2, s2),
    start=p0, data = geyser, trace = tr.ms)
1233.991 : 0.361204 50 5 80 5
1176.955 : 0.358276 55.0387 4.87174 80.9926 5.35572
1165.579 : 0.350125 54.921 5.06088 80.9345 5.95334
1158.498 : 0.318598 54.5326 5.14273 80.6561 6.86238
1157.574 : 0.309889 54.2604 4.99206 80.424 7.38264
1157.542 : 0.307688 54.2049 4.95342 80.3632 7.50246
1157.542 : 0.307594 54.2027 4.952 80.3603 7.50763
```

For future reference, this took 0.8 seconds. Without any derivative information ms converges from this starting point in 32 iterations and 1.6 seconds whereas with first derivatives only it requires 36 iterations and 1.4 seconds.

We can now add the parametric density estimate to our original histogram plot.

```
dmix2 <- function(x, p, u1, s1, u2, s2)
            p * dnorm(x, u1, s1) + (1-p) * dnorm(x, u2, s2)
cf <- coef(wait.mix2)
attach(structure(as.list(cf), names = names(cf)))
wait.fdns <- list(x = wait.dns$x,
            y = dmix2(wait.dns$x, p, u1, s1, u2, s2))
lines(wait.fdns)
par(usr = c(0,1,0,1))
legend(0.1, 0.9, c("Normal mixture", "Nonparametric"),
    lty = c(1,2), bty = "n")
```

The computations for Figure 8.5 are now complete.

The parametric and nonparametric density estimates are in fair agreement on the right component, but there is a suggestion of disparity on the left. We can informally check the adequacy of the parametric model by a Q-Q plot. First we solve for the quantiles using a reduced-step Newton method:

```
pmix2 <- deriv(~ p*pnorm((x-u1)/s1) + (1-p)*pnorm((x-u2)/s2),
            "x", function(x, p, u1, s1, u2, s2) {})
pr0 <- (seq(along = waiting) - 0.5)/length(waiting)
x0 <- x1 <- as.vector(sort(waiting)) ; del <- 1; i <- 0
while((i <- 1 + i) < 10 && abs(del) > 0.0005) {
  pr <- pmix2(x0, p, u1, s1, u2, s2)
  del <- (pr - pr0)/attr(pr, "gradient")
  x0 <- x0 - 0.5*del
```

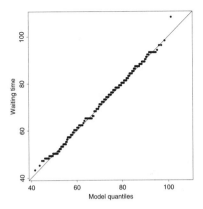

Figure 8.6: Sorted waiting times against normal mixture model quantiles for the 'Old Faithful' eruptions data.

```
    cat(format(del <- max(abs(del))), "\n")
}
detach()
par(pty = "s")
plot(x0, x1, xlim = range(x0, x1), ylim = range(x0, x1),
    xlab = "Model quantiles", ylab = "Waiting time")
abline(0,1)
par(pty = "m")
```

The plot is shown in Figure 8.6 and confirms the adequacy of the mixture model except perhaps for one outlier in the right tail.

Another advantage of using second-derivative information is that it provides an estimate of the (large sample) variance matrix of the parameters. The fitted model object has a component, hessian, giving the sum of the second derivative array over the observations and since the function we minimized is the negative of the log-likelihood, this component is the observed information matrix. Its inverse is calculated (but incorrectly labelled[3]) by summary.ms

```
vmat <- summary(wait.mix2)$Information
cbind(coef(wait.mix2), sqrt(diag(vmat)))
         [,1]     [,2]
 p   0.30759 0.030438
 u1 54.20265 0.683066
 s1  4.95200 0.518231
 u2 80.36031 0.633387
 s2  7.50764 0.507095
```

We saw in Figure 5.11 (on page 139) that the distribution of waiting time depends on the previous duration, and we can easily enhance our model to allow for this with the proportions depending logistically on the previous duration. We start at the fitted values of the simpler model.

---

[3] And differently from the help page which has information.

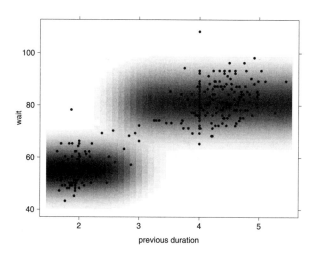

Figure 8.7: Waiting times for the next eruption against the previous duration, with the density of the conditional distribution predicted by a mixture model.

```
lmix2r <- deriv3(
      ~ -log((exp(a+b*y)*dnorm((x-u1)/s1)/s1 +
              dnorm((x-u2)/s2)/s2) / (1+exp(a+b*y)) ),
      c("a", "b", "u1", "s1", "u2", "s2"),
      function(x, y, a, b, u1, s1, u2, s2) NULL)
p1 <- wait.mix2$par; tmp <- as.vector(p1[1])
p2 <- c(a=log(tmp/(1-tmp)), b=0, p1[-1])

wait.mix2r <-
  ms(~ lmix2r(waiting[-1], duration[-299], a, b, u1, s1, u2, s2),
              start = p2, data = geyser, trace = tr.ms)
1154.238 : -0.811394 0 54.2026 4.952 80.3603 7.50764
   ....
985.319 : 16.14 -5.73629 55.1376 5.66314 81.0909 6.83759
```

There is clearly a very significant improvement in the fit. We can examine the fit by plotting the predictive densities for the next waiting time by (see Figure 8.7)

```
grid <- expand.grid(x = seq(1.5,5.5,0.1), y = seq(40,110,0.5))
grid$z <- exp(-lmix2r(grid$y, grid$x, 16.14, -5.74, 55.14,
                      5.663, 81.09, 6.838))
levelplot(z ~ x*y, grid, colorkey=F, at = seq(0, 0.075, 0.001),
        panel= function(...) {
          panel.levelplot(...)
          points(duration[-299], waiting[-1])
        }, xlab="previous duration", ylab="wait",
    col.regions = rev(trellis.par.get("regions")$col))
```

General facilities for minimization

The function `nlmin` finds an unconstrained local minimum of a function. For the mixture problem we can use

```
mix.f <- function(p) {
    e <- p[1]*dnorm((waiting-p[2])/p[3])/p[3] +
        (1-p[1])*dnorm((waiting-p[4])/p[5])/p[5]
    -sum(log(e)) }
waiting.init <- c(mean(waiting < 70), 50, 5, 80, 5)
nlmin(mix.f, waiting.init, print.level=1)
```

The argument `print.level=1` provides some tracing of the iterations (here 14, one less than the default limit). This fit took 1.2 seconds. There is no way to supply derivatives.

The most general minimization routine is `nlminb`, which can find a local minimum of a twice-differentiable function within a hypercube in parameter space. Either the gradient or gradient plus Hessian can be supplied; if no gradient is supplied it is approximated by finite differences. The underlying algorithm is a quasi-Newton optimizer, or a Newton optimizer if the Hessian is supplied. We can fit our mixture density (using zero, one and two derivatives) and giving the constraints that we have ignored hitherto.

```
attach(geyser)
mix.obj <- function(p, x) {
    e <- p[1]*dnorm((x-p[2])/p[3])/p[3] +
        (1-p[1])*dnorm((x-p[4])/p[5])/p[5]
    -sum(log(e)) }
mix.nl0 <- nlminb(waiting.init, mix.obj,
    scale = c(10, rep(1,4)), lower = c(0, -Inf, 0, -Inf, 0),
    upper = c(1, rep(Inf, 4)), x = waiting)

lmix2a <- deriv(
    ~ -log(p*dnorm((x-u1)/s1)/s1 + (1-p)*dnorm((x-u2)/s2)/s2),
    c("p", "u1", "s1", "u2", "s2"),
    function(x, p, u1, s1, u2, s2) NULL)
mix.gr <- function(p, x) {
    u1 <- p[2]; s1 <- p[3]; u2 <- p[4]; s2 <- p[5]; p <- p[1]
    e <- lmix2a(x, p, u1, s1, u2, s2)
    rep(1, length(x)) %*% attr(e, "gradient") }
mix.nl1 <- nlminb(waiting.init, mix.obj, mix.gr,
    scale = c(10, rep(1,4)), lower = c(0, -Inf, 0, -Inf, 0),
    upper = c(1, rep(Inf, 4)), x = waiting)
mix.grh <- function(p, x) {
    e <- lmix2(x, p[1], p[2], p[3], p[4], p[5])
    g <- attr(e, "gradient")
    g <- rep(1, length(x)) %*% g
    H <- apply(attr(e, "hessian"), c(2,3), sum)
    list(gradient=g, hessian=H[row(H) <= col(H)]) }
mix.nl2 <- nlminb(waiting.init, mix.obj, mix.grh, T,
```

```
      scale = c(10, rep(1,4)), lower = c(0, -Inf, 0, -Inf, 0),
      upper = c(1, rep(Inf, 4)), x = waiting)
```

These fits took 1.6, 0.8 and 1.1 seconds. (It is not unusual for `nlminb` fits to be comparatively slow, but here we have calculated the derivatives in an inelegant way.) We use a `scale` parameter to set the step length on the first parameter much smaller than the others, which can speed convergence. Generally it is helpful to have the scale set so that the range of uncertainty in (scale × parameter) is about one.

It is also possible to supply a separate function to calculate the Hessian; however it is supplied, it is a vector giving the lower triangle of the Hessian in row-first order (unlike S matrices).

`nlminb` only computes the Hessian at the solution if a means to compute Hessians is supplied (and even then it returns a scaled version of the Hessian). Thus it provides no help in using the observed information to provide approximate standard errors for the parameter estimates in a maximum likelihood estimation problem. For `nlminb` we use a finite-difference approximation to the Hessian if it is not available, via the function `vcov.nlminb` in our library.

```
> sqrt(diag(vcov.nlminb(mix.nl0)))
[1] 0.030746 0.676833 0.539178 0.619138 0.509087
> sqrt(diag(vcov.nlminb(mix.nl1)))
[1] 0.030438 0.683067 0.518231 0.633388 0.507096
> sqrt(diag(vcov.nlminb(mix.nl2)))
[1] 0.030438 0.683066 0.518231 0.633387 0.507095
```

The differences reflect a difference in philosophy: when no derivatives are available we seek a quadratic approximation to the negative log-likelihood over the scale of random variation in the parameter estimate rather than at the MLE.

Note that the theory used here is unreliable if the parameter estimate is close to or on the boundary; `vcov.nlminb` issues a warning.

Fitting GLM models

The Fisher scoring algorithm is not the only possible way to fit GLMs. Indeed, the function `glm` allows other methods to be used via its `method` argument, although we know of no such methods and the return values required of a method are geared to the IWLS algorithm.

The purpose of this section is to point out how easy it is to fit specific GLMs by direct maximization of the likelihood, and the code can easily be modified for other fit criteria.

Logistic regression

We consider maximum likelihood fitting of a binomial logistic regression. For other link functions just replace `plogis` and `dlogis`, for example, by `pnorm` and `dnorm` for a probit regression.

```
logitreg <- function(x, y, wt=rep(1, length(y)),
                intercept = T, start, trace = T, ...)
{
    fmin <- function(x, y, w, beta) {
        eta <- x %*% beta
        p <- plogis(eta)
        g <- -2 * dlogis(eta) * ifelse(y, 1/p, -1/(1-p))
        value <- -2 * w * ifelse(y, log(p), log(1-p))
        attr(value, "gradient") <- x*g*w
        value
    }
    if(is.null(dim(x))) dim(x) <- c(length(x), 1)
    dn <- dimnames(x)[[2]]
    if(!length(dn)) dn <- paste("Var", 1:ncol(x), sep="")
    p <- ncol(x) + intercept
    if(intercept) {
        x <- cbind(1, x)
        dn <- c("(Intercept)", dn)
    }
    if(missing(start)) start <- rep(0, p)
    fit <- ms(~ fmin(x, y, wt, beta), start=list(beta=start),
            trace=trace, ...)
    names(fit$par) <- dn
    cat("\nCoefficients:\n")
    print(fit$parameters)
    cat("Residual Deviance:", format(fit$value), "\n")
    invisible(fit)
}
```

This can easily be modified to allow constraints on the parameters (by using nlminb rather than ms; see the following), the handling of missing values by multiple imputation (Ripley, 1996), the possibility of errors in the recorded x or y values (Copas, 1988) and shrinkage and approximate Bayesian estimators, none of which can readily be handled by the glm algorithm.

Constrained Poisson regressions

Workers on the backcalculation of AIDS have used a Poisson regression model with identity link and non-negative parameters (for example, Rosenberg & Gail, 1991). The history of AIDS is divided into J time intervals $[T_{j-1}, T_j)$, and the data are counts Y_j of cases in each time interval. Then for a Poisson incidence process of rate ν, the counts are independent Poisson random variables with mean

$$EY_j = \mu_j = \int_0^{T_j} \left[F(T_j - s) - F(T_{j-1} - s) \right] \nu(s) \, ds$$

where F is the cumulative distribution of the incubation time. If we model ν by a step function

$$\nu(s) = \sum_i I(t_{i-1} \leqslant s < t_i)\beta_i$$

then $EY_j = \sum x_{ji}\beta_i$ for constants x_{ji} which are considered known.

This is a Poisson regression with identity link, no intercept and constraint $\beta_i \geqslant 0$. The deviance is $2 \sum_j \mu_j - y_j - y_j \log(\mu_j/y_j)$.

```
AIDSfit <- function(y, z, start=rep(mean(y), ncol(z)), ...)
{
    deviance <- function(beta, y, z) {
        mu <- z %*% beta
        2 * sum(mu - y - y*log(mu/y)) }
    grad <- function(beta, y, z) {
        mu <- z %*% beta
        2 * t(1 - y/mu) %*% z }
    nlminb(start, deviance, grad, lower = 0, y = y, z = z, ...)
}
```

As an example, we consider the history of AIDS in Belgium using the numbers of cases for 1981 to 1993 (as reported by WHO in March 1994). We use two-year intervals for the step function and do the integration crudely using a Weibull model for the incubation times.

```
Y <- scan()
12 14 33 50 67 74 123 141 165 204 253 246 240

library(nnet) # for class.ind
s <- seq(0, 13.999, 0.01); tint <- 1:14
X <- expand.grid(s, tint)
Z <- matrix(pweibull(pmax(X[,2] - X[,1],0), 2.5, 10),length(s))
Z <- Z[,2:14] - Z[,1:13]
Z <- t(Z) %*% class.ind(factor(floor(s/2))) * 0.01
round(AIDSfit(Y, Z)$param)
515    0 140 882    0    0    0
rm(s, X, Y, Z)
```

This has infections only in 1981–2 and 1985–8. We do not pursue the medical implications!

8.8 Non-linear mixed effects models

Non-linear mixed effects models are fitted by the function `nlme`, which exists in S-PLUS but we describe the enhanced version in version 3.0 of the `nlme` library included in S-PLUS 2000 (see page 471).

Most of its many arguments are specified in the same way as for `lme`, and the class of models that can be fitted is based on those described by equations (6.8) to (6.10) on page 205. The difference is that

$$y_{ij} = \mu_{ij} + \epsilon_{ij}, \qquad \mu_{ij} = f(\boldsymbol{x}_{ij}, \boldsymbol{\beta}, \boldsymbol{\eta}_i) \qquad (8.7)$$

so the conditional mean is a non-linear function specified by giving both fixed and random parameters. Which parameters are fixed and which are random is

specified by the arguments `fixed` and `random`, each of which is a set of formulae with left-hand side the parameter name. Parameters can be in both sets, and
random effects will have mean zero unless they are. For example, we can specify
a simple exponential growth curve for each tree in the `Sitka` data with random
intercept and asymptote by

```
library(nlme3, first=T)    # check the form needed on your system
options(contrasts = c("contr.treatment", "contr.poly"))
sitka.nlme <- nlme(size ~ A + B * (1 - exp(-(Time-100)/C)),
    fixed = list(A ~ treat, B ~ treat, C ~ 1),
    random =   A + B ~ 1 | tree, data = Sitka,
    start  = list(fixed = c(2, 0, 4, 0, 100)),
    method = "ML", verbose = T)
```

Here the shape of the curve is taken as common to all trees. It is necessary to
specify starting values for all the fixed parameters (in the order they occur) and
starting values can be specified for other parameters. Note the way the formula
for `random` is expressed: this is one of a variety of forms described on the help
page for `nlme`.

   It is not usually possible to find the exact likelihood of such a model, as
the non-linearity prevents integration over the random effects. Various linearization schemes have been proposed; `nlme` uses the strategy of Lindstrom & Bates
(1990), linearization about the conditional modes of the random effects η_i. As
`nlme` calls `lme` iteratively and `lme` fits can be slow, `nlme` fits can be very slow
(and also memory-intensive). Adding the argument `verbose = T` helps to mon-
itor progress. Finding good starting values can help a great deal. Our starting
values were chosen by inspection of the growth curves.

```
> summary(sitka.nlme)
Nonlinear mixed-effects model fit by maximum likelihood
  Model: size ~ A + B * (1 - exp( - (Time - 100)/C))
 Data: Sitka
       AIC     BIC logLik
  -96.275 -60.465 57.138

Random effects:
 Formula: list(A ~ 1, B ~ 1)
 Level: tree
 Structure: General positive-definite
                StdDev   Corr
A.(Intercept) 0.83561 A.(Int
B.(Intercept) 0.81954 -0.69
      Residual 0.10297

Fixed effects: list(A ~ treat, B ~ treat, C ~ 1)
               Value Std.Error  DF t-value p-value
A.(Intercept)  2.304    0.1995 312  11.547  <.0001
      A.treat  0.175    0.2117 312   0.826  0.4096
B.(Intercept)  3.921    0.1808 312  21.687  <.0001
```

```
      B.treat   -0.564    0.2156 312  -2.618  0.0093
            C   81.769    4.7270 312  17.299  <.0001
      ....
```

We can test for the difference in the treatment groups by

```
> sitka.nlme2 <- update(sitka.nlme,
      fixed = list(A ~ 1, B ~ 1, C ~ 1),
      start = list(fixed=c(2.4, 3.6, 82)))
> summary(sitka.nlme2)
Nonlinear mixed-effects model fit by maximum likelihood
  Model: size ~ A + B * (1 - exp( - (Time - 100)/C))
 Data: Sitka
      AIC     BIC logLik
   -91.587 -63.735 52.794
      ....
> anova(sitka.nlme2, sitka.nlme)
            Model df     AIC     BIC logLik    Test Lik.Ratio
sitka.nlme2     1  7 -91.588 -63.736 52.794
 sitka.nlme     2  9 -96.275 -60.465 57.138 1 vs. 2    8.6879
            p-value
sitka.nlme2
 sitka.nlme    0.013
```

so there is convincing evidence of a treatment difference.

For this model we can get a convincing fit with serial autocorrelation, using

```
> sitka.nlme3 <- update(sitka.nlme,
                        corr = corCAR1(0.95, ~Time | tree))
> summary(sitka.nlme3)
Nonlinear mixed-effects model fit by maximum likelihood
  Model: size ~ A + B * (1 - exp( - (Time - 100)/C))
 Data: Sitka
      AIC     BIC logLik
   -104.5 -64.715 62.252

Random effects:
 Formula: list(A ~ 1, B ~ 1)
 Level: tree
 Structure: General positive-definite
               StdDev   Corr
A.(Intercept) 0.81602 A.(Int
B.(Intercept) 0.76069 -0.674
    Residual 0.13068

Correlation Structure: Continuous AR(1)
 Parameter estimate(s):
     Phi
 0.96751
Fixed effects: list(A ~ treat, B ~ treat, C ~ 1)
                Value Std.Error  DF t-value p-value
```

Table 8.2: Data from a blood pressure experiment with five rabbits.

Treatment	Rabbit	6.25	12.5	25	50	100	200
Placebo	1	0.50	4.50	10.00	26.00	37.00	32.00
	2	1.00	1.25	4.00	12.00	27.00	29.00
	3	0.75	3.00	3.00	14.00	22.00	24.00
	4	1.25	1.50	6.00	19.00	33.00	33.00
	5	1.50	1.50	5.00	16.00	20.00	18.00
MDL 72222	1	1.25	0.75	4.00	9.00	25.00	37.00
	2	1.40	1.70	1.00	2.00	15.00	28.00
	3	0.75	2.30	3.00	5.00	26.00	25.00
	4	2.60	1.20	2.00	3.00	11.00	22.00
	5	2.40	2.50	1.50	2.00	9.00	19.00

Dose of Phenylbiguanide (μg)

```
A.(Intercept)    2.313   0.2052 312  11.271  <.0001
      A.treat    0.171   0.2144 312   0.796  0.4267
B.(Intercept)    3.892   0.1813 312  21.466  <.0001
      B.treat   -0.564   0.2162 312  -2.607  0.0096
            C   80.901   5.2920 312  15.288  <.0001
```

The correlation is in units of days; at the average spacing between observations of 26.5 days the estimated correlation is $0.9675^{26.5} \approx 0.42$. A good initial guess at the correlation is needed to persuade nlme to fit this parameter.

Blood pressure in rabbits

For a more extensive example, consider the data in Table 8.2 described in Ludbrook (1994).[4] To quote from the paper:

> Five rabbits were studied on two occasions, after treatment with saline (control) and after treatment with the 5-HT$_3$ antagonist MDL 72222. After each treatment ascending doses of phenylbiguanide (PBG) were injected intravenously at 10 minute intervals and the responses of mean blood pressure measured. The goal was to test whether the cardiogenic chemoreflex elicited by PBG depends on the activation of 5-HT$_3$ receptors.

The response is the *change* in blood pressure relative to the start of the experiment. The data set is a data frame Rabbit in our library MASS.

There are three strata of variation:

1. between animals,
2. within animals between occasions and
3. within animals within occasions.

[4] We are grateful to Professor Ludbrook for supplying us with the numerical data.

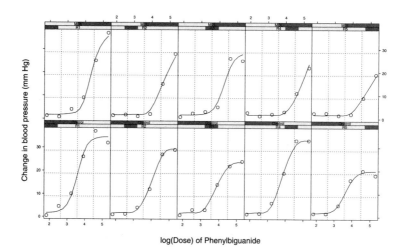

log(Dose) of Phenylbiguanide

Figure 8.8: Data from a cardiovascular experiment using five rabbits on two occasions: on control and with treatment. On each occasion the animals are given increasing doses of phenylbiguanide at about 10 minutes apart. The response is the change in blood pressure. The fitted curve is derived from a model on page 277.

We can specify this by having the random effects depend on `Animal/Run`. We take `Treatment` to be a factor on which the fixed effects in the model might depend.

A plot of the data shown in Figure 8.8 shows the blood pressure rising with the dose of PBG on each occasion generally according to a sigmoid curve. Evidently the main effect of the treatment is to suppress the change in blood pressure at lower doses of PBG, hence translating the response curve a distance to the right. There is a great deal of variation, though, so there may well be other effects as well.

Ludbrook suggests a four-parameter logistic response function in $\log(\text{dose})$ (as is commonly used in this context) and we adopt his suggestion. This function has the form

$$f(\alpha, \beta, \lambda, \theta, x) = \alpha + \frac{\beta - \alpha}{1 + \exp[(x - \lambda)/\theta]}$$

Notice that $f(\alpha, \beta, \lambda, \theta, x)$ and $f(\beta, \alpha, \lambda, -\theta, x)$ are identically equal in x; to resolve this lack of identification we require that θ be positive. For our example this makes α the right asymptote, β the left asymptote (or 'baseline'), λ the $\log(\text{dose})$ (LD50) leading to a response exactly halfway between asymptotes and θ an abscissa scale parameter determining the rapidity of ascent.

We start by considering the two treatment groups separately (when there are only two strata of variation). We guess some initial values then use an `nls` fit to refine them.

```
Fpl <- deriv(~ A + (B-A)/(1 + exp((log(d) - ld50)/th)),
       c("A","B","ld50","th"), function(d, A, B, ld50, th) {})
```

```
st <- nls(BPchange ~ Fpl(Dose, A, B, ld50, th),
          start = c(A=25, B=0, ld50=4, th=0.25),
          data = Rabbit)$par
Rc.nlme <- nlme(BPchange ~ Fpl(Dose, A, B, ld50, th),
      fixed = list(A ~ 1, B ~ 1, ld50 ~ 1, th ~ 1),
      random = A + ld50 ~ 1 | Animal, data = Rabbit,
      subset = Treatment=="Control",
      start = list(fixed=st))
Rm.nlme <- update(Rc.nlme, subset = Treatment=="MDL")
```

We may now look at the results of the separate analyses.

```
> Rc.nlme
Nonlinear mixed-effects model fit by maximum likelihood
  Model: BPchange ~ Fpl(Dose, A, B, ld50, th)
  Data: Rabbit
  Log-likelihood: -66.502
  Fixed: list(A ~ 1, B ~ 1, ld50 ~ 1, th ~ 1)
       A      B   ld50       th
 28.332 1.5134 3.7744 0.28957

Random effects:
 Formula: list(A ~ 1, ld50 ~ 1)
 Level: Animal
 Structure: General positive-definite
          StdDev  Corr
       A 5.76889 A
    ld50 0.17953 0.112
Residual 1.36735

> Rm.nlme
Nonlinear mixed-effects model fit by maximum likelihood
  Model: BPchange ~ Fpl(Dose, A, B, ld50, th)
  Data: Rabbit
  Log-likelihood: -65.422
  Fixed: list(A ~ 1, B ~ 1, ld50 ~ 1, th ~ 1)
       A      B   ld50       th
 27.521 1.7839 4.5257 0.24236

Random effects:
 Formula: list(A ~ 1, ld50 ~ 1)
 Level: Animal
 Structure: General positive-definite
          StdDev   Corr
       A 5.36549 A
    ld50 0.18999 -0.594
Residual 1.44172
```

This suggests that the random variation of ld50 between animals in the each group is small. The separate results suggest a combined model in which the distribution of the random effects does not depend on the treatment. Initially we allow all the parameter means to differ by group.

```
> options(contrasts=c("contr.treatment", "contr.poly"))
> c1 <- c(28, 1.6, 4.1, 0.27, 0)
> R.nlme1 <- nlme(BPchange ~ Fpl(Dose, A, B, ld50, th),
      fixed = list(A ~ Treatment, B ~ Treatment,
                   ld50 ~ Treatment, th ~ Treatment),
      random =  A + ld50 ~ 1 | Animal/Run, data = Rabbit,
      start = list(fixed=c1[c(1,5,2,5,3,5,4,5)]))
> summary(R.nlme1)
Nonlinear mixed-effects model fit by maximum likelihood
  Model: BPchange ~ Fpl(Dose, A, B, ld50, th)
 Data: Rabbit
      AIC    BIC  logLik
   292.63 324.04 -131.31

Random effects:
 Formula: list(A ~ 1, ld50 ~ 1)
 Level: Animal
 Structure: General positive-definite

                   StdDev   Corr
  A.(Intercept) 4.6063 A.(Int
ld50.(Intercept) 0.0626 -0.166

 Formula: list(A ~ 1, ld50 ~ 1)
 Level: Run %in% Animal
 Structure: General positive-definite
                   StdDev   Corr
  A.(Intercept) 3.2489 A.(Int
ld50.(Intercept) 0.1707 -0.348
        Residual 1.4113

Fixed effects:
                 Value Std.Error DF t-value p-value
  A.(Intercept) 28.326    2.7802 43  10.188  <.0001
    A.Treatment -0.727    2.5184 43  -0.288  0.7744
  B.(Intercept)  1.525    0.5155 43   2.958  0.0050
    B.Treatment  0.261    0.6460 43   0.405  0.6877
ld50.(Intercept) 3.778    0.0955 43  39.579  <.0001
  ld50.Treatment 0.747    0.1286 43   5.809  <.0001
   th.(Intercept) 0.290   0.0323 43   8.957  <.0001
     th.Treatment -0.047  0.0459 43  -1.020  0.3135
```

Note that most of the random variation in ld50 is between occasions rather than between animals wheras for A both strata of variation are important but the larger effect is that between animals.

This suggests that only ld50 depends on the treatment, confirmed by

```
> R.nlme2 <- update(R.nlme1,
      fixed = list(A ~ 1, B ~ 1, ld50 ~ Treatment, th ~ 1),
      start = list(fixed=c1[c(1:3,5,4)]))
> anova(R.nlme2, R.nlme1)
          Model df     AIC     BIC  logLik     Test Lik.Ratio
R.nlme2       1 12  287.29  312.43 -131.65
R.nlme1       2 15  292.63  324.04 -131.31 1 vs. 2     0.66896
> summary(R.nlme2)
    ....
                   Value Std.Error DF t-value p-value
              A   28.170    2.4909 46  11.309  <.0001
              B    1.667    0.3069 46   5.433  <.0001
ld50.(Intercept)  3.779    0.0921 46  41.036  <.0001
   ld50.Treatment  0.759    0.1217 46   6.233  <.0001
              th   0.271    0.0226 46  11.964  <.0001
    ....
```

Finally we may display the results and a spline approximation to the (BLUP) fitted curve.

```
xyplot(BPchange ~ log(Dose) | Animal * Treatment, Rabbit,
    xlab = "log(Dose) of Phenylbiguanide",
    ylab = "Change in blood pressure (mm Hg)",
    subscripts = T, aspect = "xy", panel =
        function(x, y, subscripts) {
            panel.grid()
            panel.xyplot(x, y)
            sp <- spline(x, fitted(R.nlme2)[subscripts])
            panel.xyplot(sp$x, sp$y, type="l")
    })
```

The result, shown in Figure 8.8 on page 274, seems to fit the data very well.

8.9   Exercises

8.1.   For the weight loss example compare the negative exponential model with quadratic and cubic polynomial regression alternative models, in particular check the behaviour of each model under extrapolation into the future.

8.2.   Fit the negative exponential weight loss model in the 'goal weight' form, equation (8.6), for the three goal weights, $w_0 = 110$, 100 and $90\,$kg. Plot the profiles.

8.3.   The model used in connection with the Stormer data may also be expressed as a generalized linear model. To do this we write

$$\frac{\beta_1 v}{w - \beta_2} = \frac{1}{\gamma_1 z_1 + \gamma_2 z_2}$$

where $\gamma_1 = 1/\beta_1$, $\gamma_2 = \beta_2/\beta_1$, $z_1 = w/v$ and $z_2 = -1/v$. This has the form of a generalized linear model with inverse link. Fit the model in this form using a quasi family with `inverse` link and `constant` variance function.

Back transform the estimated coefficients and show that they agree with the values obtained using the non-linear regression approach.

Also compute the estimated standard errors and verify that they also agree with the values obtained directly by the non-linear regression approach.

Finding the standard errors is more challenging. We need first to find the variance matrix from the generalized linear model. The large sample variance matrix for $\widehat{\beta}$ is related to that for $\widehat{\gamma}$ by $\text{var}\left(\widehat{\beta}\right) = J \text{var}\left(\widehat{\gamma}\right) J^T$ where J is the Jacobian matrix of the inverse of the parameter transformation:

$$J = \begin{bmatrix} \partial\beta_1/\partial\gamma_1 & \partial\beta_1/\partial\gamma_2 \\ \partial\beta_2/\partial\gamma_1 & \partial\beta_2/\partial\gamma_2 \end{bmatrix} = \begin{bmatrix} -1/\gamma_1^2 & 0 \\ -\gamma_2/\gamma_1^2 & 1/\gamma_1 \end{bmatrix}$$

(To achieve close agreement you may need to tighten the convergence criteria for the `glm` fit, for example, by setting `eps=1.0e-10`.)

8.4.   *Stable parameters*

Ross (1970) has suggested using *stable parameters* for non-linear regression, mainly to achieve estimates that are as near to uncorrelated as possible. It turns out that in many cases stable parameters also define a coordinate system within the solution locus with a small curvature.

The idea is to use the means at p well-separated points in sample space as the parameters. Writing the regression function in terms of the stable parameters is often intractable, but in the case of a negative exponential decay model of the type we considered for the weight loss data it is possible if the points are chosen equally spaced.

If the three mean parameters μ_i are chosen at x-points $x_0 + i\delta_x$, $i = 0, 1, 2$, show that the model may be written explicitly as:

$$\eta = \frac{\mu_0\mu_2 - \mu_1^2}{\mu_0 - 2\mu_1 + \mu_2} + \frac{(\mu_0 - \mu_1)^2}{\mu_0 - 2\mu_1 + \mu_2}\left(\frac{\mu_1 - \mu_2}{\mu_0 - \mu_1}\right)^{(x-x_0)/\delta_x}$$

Fit the negative exponential decay model to the weight loss data using this parametrization and choosing, say, $x_0 = 40$ days and $\delta_x = 80$ days. Look at the characteristics of the fit, including the correlations between the parameter estimates. Explain in heuristic terms why they are relatively low.

Examine the profiles of the fit and check for straightness. Can you give a possible statistical explanation for why they appear as straight as they do?

8.5.   *Heteroscedastic regression models*

A common heteroscedastic regression model specifies that the observations have constant coefficient of variation; that is, $Y \sim N(\mu, \theta\mu^2)$ where $\theta > 0$ and μ depends on regressor variables according to some linear model perhaps with a link function such as $\mu = \exp \eta$. Write a function to fit such models and try it out using the Quine data. Compare with the negative binomial models fitted in Section 7.4, page 234ff.

8.6. A deterministic relationship between pressure and temperature in saturated steam can be written as

$$\text{Pressure} = \alpha \exp \left(\frac{\beta T}{\gamma + T} \right)$$

where T is the temperature, considered the determining variable. Data collected to estimate the unknown parameters α, β and γ are contained in the data frame steam.

(a) Fit this model as a non-linear regression assuming additive errors in the pressure scale. Devise a suitable method for arriving at initial values.

(b) Fit the model again, this time taking logarithms of the relationship and assuming that the errors are additive in the log(pressure) scale, (and hence multiplicative on the original scale).

(c) Which model do you consider is better supported on the basis of model checks?

8.7. The data frame Puromycin supplied with S-PLUS contains data from a Michaelis–Menten experiment conducted at two levels of a factor state. The usual non-linear regression model proposed for such cases is

$$V_{ij} = \frac{K_j c_{ij}}{c_{ij} + \theta_j} + \varepsilon_{ij}, \qquad j = 1, 2$$

where V is the (initial) velocity of the reaction, c is the substrate concentration and j refers to the level of the factor.

Fit the model with separate asymptotes V_j and separate θ_js. Test the hypothesis that $\theta_1 = \theta_2$ and report. [Hint: use the anova method for nls objects.]

For the final model you adopt show the data and fitted curves in a two-panel Trellis display.

8.8. Fit the non-linear model (8.4) to the muscle data with (α_j) as a random effect; that is, a mean asymptote plus one variance component. Compare the predictions (both mean and BLUP curves) of this model with the fixed-effects model fitted in the text.

8.9. Sarah Hogan collected data on the 'binaural hearing' ability of children with a history of otitis media with effusion (OME). Some of the data (and a description of the problem) are in data frame OME. Fit a suitable non-linear model, and assess if there is a change in ability with age and OME status.

(a) The suggested model is a logistic curve that ranges from 0.5 at low noise levels (when the response is effectively a guess) to 1.0 at high noise levels.

Then the most important parameter will be the noise level L75 at which the child has a 75% success rate. The amount of data on each child is small, so fit a model with a common slope but a separate L75 for each child, and analyse the fitted parameters by age and group. [You may want to look up the function `nlsList`.]

(b) Consider a linear model for L75 on age, and differences between the OME groups, for each type of noise stimulus. Assess the significance of your results via standard errors and/or F-tests.

(c) The analysis thus far does not take into account the differences between individual subjects. Repeat the analysis using non-linear mixed-effects models.

8.10. Write a function to fit a gamma distribution to n observations by maximum likelihood.

8.11. McLachlan & Jones (1988) (see also McLachlan & Krishnan, 1997, pp. 73ff) give the following grouped data on red blood cell volume, in 18 equally spaced bins of width 7.2fl, starting at 21.6fl.

```
Set 1: 10 21 51 77 70 50 44 40 46 54 53 54 44 36 29 21 16 13
Set 2:  9 32 64 69 56 68 88 93 87 67 44 36 30 24 21 14  8  7
```

McLachlan and Jones fit a mixture of two normal densities on log scale by an involved method using the EM algorithm. Fit this model directly to each set of data by a small modification of the approach in Section 8.7.

Chapter 9

Smooth Regression

S-PLUS has a 'Modern Regression Module' which contains functions for a number of regression methods. These are not necessarily non-linear in the sense of Chapter 8, which refers to a non-linear parametrization, but they do allow non-linear functions of the independent variables to be chosen by the procedures. The methods are all fairly computer-intensive, and so are only feasible in the era of plentiful computing power (and hence are 'modern'). There are few texts covering this material. Although they do not cover all our topics in equal detail, for what they do cover Hastie & Tibshirani (1990), Simonoff (1996) and Bowman & Azzalini (1997) are good references.

9.1 Additive models and scatterplot smoothers

For linear regression we have a dependent variable Y and a set of predictor variables $X_1, \ldots, X_p$, and model

$$Y = \alpha + \sum_{j=1}^{p} \beta_j X_j + \epsilon$$

Additive models replace the linear function $\beta_j X_j$ by a non-linear function to get

$$Y = \alpha + \sum_{j=1}^{p} f_j(X_j) + \epsilon \qquad (9.1)$$

Since the functions f_j are rather general, they can subsume the α. Of course, it is not useful to allow an arbitrary function f_j, and it helps to think of it as a *smooth* function.

The classical way to introduce non-linear functions of dependent variables is to add a limited range of transformed variables to the model, for example, quadratic and cubic terms, or to split the range of the variable and use a piecewise constant function. (In S this can be achieved using the functions `poly` and `cut`.) A modern alternative is to use *spline* functions. (Green & Silverman, 1994, provide a gentle introduction to splines.) We only need cubic splines. Divide the real line by an ordered set of points $\{z_i\}$ known as *knots*. On the interval $[z_i, z_{i+1}]$ the spline is a cubic polynomial, and it is continuous and has continuous first and second derivatives, imposing 3 conditions at each knot. With n knots, $n + 4$ parameters are needed to represent the cubic spline (from $4(n + 1)$ for the cubic polynomials minus $3n$ continuity conditions). Of the many possible parametrizations, that of B-splines has desirable properties. The S function `bs` generates a matrix of B-splines, and so can be included in a linear-regression fit.

A restricted form of B-splines known as *natural splines* and implemented by the S function `ns` is linear on $(-\infty, z_1]$ and $[z_n, \infty)$ and thus would have n parameters. However, `ns` adds an extra knot at each of the maximum and minimum of the data points, and so has $n + 2$ parameters, dropping the requirement for the derivative to be continuous at z_1 and z_n. The functions `bs` and `ns` may have the knots specified or be allowed to choose the knots as quantiles of the empirical distribution of the variable to be transformed, by specifying the number `df` of parameters.

Prediction from models including splines (and indeed the orthogonal polynomials generated by `poly`) needs care, as the basis for the functions depends on the observed values of the independent variable. If `predict.lm` is used, it will form a new set of basis functions and then erroneously apply the fitted coefficients. The function `predict.gam` will work more nearly correctly. (It uses both the old and new data to choose the knots.)

Splines are by no means the only way to fit smooth functions in (9.1), and indeed as we show, fitting as part of the regression is not the only way to use splines to find a smooth term f_j. To consider these approaches we need to digress to consider the simplest case of just one independent variable, so the relationship between x and y can be shown on a scatterplot. The single-variable methods are used as building-blocks in fitting multi-variable additive models.

Scatterplot smoothing

These methods depend on being able to estimate 'smooth' functions. They make use of scatterplot smoothers, which, given a scatterplot of (x, y) values, draw a smooth curve against x. It is not necessary to understand the exact details, and there are several alternative methods, including splines, running means and running lines. However, some users will want to understand the smoother being used.

Our running example is a set of data on 133 observations of acceleration against time for a simulated motorcycle accident, taken from Silverman (1985). A series of smoothers is shown in Figure 9.1, obtained by the following code.

```
attach(mcycle)
par(mfrow = c(3,2))
plot(times, accel, main="Polynomial regression")
lines(times, fitted(lm(accel ~ poly(times, 3))))
lines(times, fitted(lm(accel ~ poly(times, 6))), lty=3)
legend(40, -100, c("degree=3", "degree=6"), lty=c(1,3),bty="n")
```

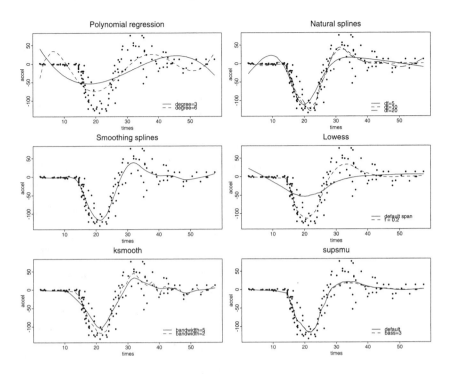

Figure 9.1: Scatterplot smoothers for the simulated motorcycle data given by Silverman (1985).

```
plot(times, accel, main="Natural splines")
lines(times, fitted(lm(accel ~ ns(times, df=5))))
lines(times, fitted(lm(accel ~ ns(times, df=10))), lty=3)
lines(times, fitted(lm(accel ~ ns(times, df=20))), lty=4)


legend(40, -100, c("df=5", "df=10", "df=20"), lty=c(1,3,4),
    bty="n")
plot(times, accel, main="Smoothing splines")
lines(smooth.spline(times, accel))
plot(times, accel, main="Lowess")
lines(lowess(times, accel))
lines(lowess(times, accel, 0.2), lty=3)
legend(40, -100, c("default span", "f = 0.2"), lty=c(1,3),
    bty="n")
plot(times, accel, main ="ksmooth")
lines(ksmooth(times, accel,"normal", bandwidth=5))
lines(ksmooth(times, accel,"normal", bandwidth=2), lty=3)
legend(40, -100, c("bandwidth=5", "bandwidth=2"), lty=c(1,3),
    bty="n")
plot(times, accel, main ="supsmu")
```

```
lines(supsmu(times, accel))
lines(supsmu(times, accel, bass=3), lty=3)
legend(40, -100, c("default", "bass=3"), lty=c(1,3), bty="n")
```

We have already discussed polynomials and regression splines, and Figure 9.1 shows how much better splines are than polynomials at adapting to general smooth curves. (The `df` parameter controls the number of terms in the regression spline.)

Suppose we have n pairs (x_i, y_i). A *smoothing spline* minimizes a compromise between the fit and the degree of smoothness of the form

$$\sum w_i [y_i - f(x_i)]^2 + \lambda \int (f''(x))^2 \, \mathrm{d}x$$

over all (measurably twice-differentiable) functions f. It is a cubic spline with knots at the x_i, but does not interpolate the data points for $\lambda > 0$ and the degree of fit is controlled by λ. The S function `smooth.spline` allows λ[1] or the (equivalent) degrees of freedom to be specified, otherwise it will choose the degree of smoothness automatically by cross-validation (see page 137). There are various definitions of equivalent degrees of freedom; see Green & Silverman (1994, pp. 27–8) and Hastie & Tibshirani (1990, Appendix B). That used in `smooth.spline` is the trace of the smoother matrix S; as fitting a smoothing spline is a linear operation, there is an $n \times n$ matrix S such $\hat{y} = Sy$. (In a regression fit S is the hat matrix (page 162; this has trace equal to the number of free parameters.)

For $\lambda = 0$ the smoothing spline will interpolate the data points if the x_i are distinct. There are simpler methods to fit interpolating cubic splines, implemented in the function `spline`. This can be useful to draw smooth curves thorough the result of some expensive smoothing algorithm: we could also use linear interpolation implemented by `approx`.

The algorithm used by `lowess` is quite complex; it uses robust locally linear fits. A window is placed about x; data points that lie inside the window are weighted so that nearby points get the most weight and a robust weighted regression is used to predict the value at x. The parameter f controls the window size and is the proportion of the data which is included. The default, `f=2/3`, is often too large for scatterplots with appreciable structure. The S function `loess` is an extension of the ideas of `lowess` which will work in one, two or more dimensions in a similar way. The function `scatter.smooth` plots a `loess` line on a scatter plot, using `loess.smooth`.

A *kernel smoother* is of the form

$$\hat{y}_i = \sum_{j=1}^{n} y_i K\left(\frac{x_i - x_j}{b}\right) \Big/ \sum_{j=1}^{n} K\left(\frac{x_i - x_j}{b}\right) \tag{9.2}$$

where b is a bandwidth parameter, and K a kernel function, as in density estimation. In our example we use the function `ksmooth` and take K to be a

---

[1] More precisely λ with the (x_i) scaled to $[0, 1]$ and the weights scaled to average 1.

standard normal density. The critical parameter is the bandwidth b. The function `ksmooth` seems rather slow, and the faster alternatives in Matt Wand's library `KernSmooth` are recommended for large problems. These also include ways to select the bandwidth.

The smoother `supsmu` is the one used by the S-PLUS functions `ace`, `avas` and `ppreg` and our function `ppr`. It is based on a symmetric k-nearest neighbour linear least squares procedure. (That is, $k/2$ data points on each side of x are used in a linear regression to predict the value at x.) This is run for three values of k, $n/2$, $n/5$ and $n/20$, and cross-validation is used to choose a value of k for each x that is approximated by interpolation between these three. Larger values of the parameter `bass` (up to 10) encourage smooth functions.

Kernel regression can be seen as local fitting of a constant. Theoretical work (see Wand & Jones, 1995) has shown advantages in local fitting of polynomials, especially those of odd order. (Of course, the advantages of local fitting of lines have been demonstrated by `lowess` and `supsmu` for many years and have been well known in the time-series literature.) Library `KernSmooth` contains code for local polynomial fitting and bandwidth selection.

There are problems in which the main interest is in estimating the first or second derivative of a smoother. One idea is to fit locally a polynomial of high enough order and report its derivative (implemented in library `KernSmooth`). We can differentiate a spline fit, although it may be desirable to fit splines of higher order than cubic: library `pspline` provides a suitable generalization of `smooth.spline`.

Fitting additive models

As we have seen, smooth terms parametrized by regression splines can be fitted by `lm`. For smoothing splines it would be possible to set up a penalized least-squares problem and minimize that, but there would be computational difficulties in choosing the smoothing parameters simultaneously (Wahba, 1990; Wahba *et al.*, 1995). Instead an iterative approach is often used.

The *backfitting* algorithm fits the smooth functions f_j in (9.1) one at a time by taking the residuals

$$Y - \sum_{k \neq j} f_k(X_k)$$

and smoothing them against X_j using one of the scatterplot smoothers of the previous subsection. The process is repeated until it converges. Linear terms in the model (including any linear terms in the smoother) are fitted by least squares.

This procedure is implemented in S by the function `gam`. The model formulae are extended to allow the terms `s(x)` and `lo(x)` which, respectively, specify a smoothing spline and a loess smoother. These have similar parameters to the scatterplot smoothers; for `s()` the default degrees of freedom is 4, and for `lo()` the window width is controlled by `span` with default 0.5. There is a plot method, `plot.gam`, which shows the smooth function fitted for each term in the additive model.

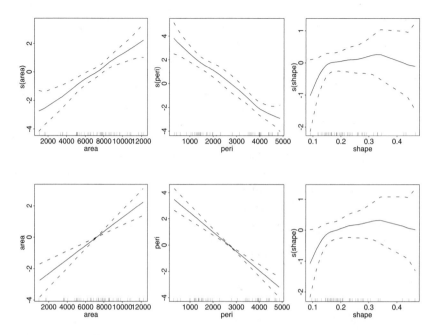

Figure 9.2: The results of fitting additive models to the dataset `rock` with smooth functions of all three predictors (top row) and linear functions of `area` and `perimeter` (bottom row). The dashed lines are approximate 95% pointwise confidence intervals. The tick marks show the locations of the observations on that variable.

Our dataset `rock` contains measurements on four cross-sections of each of 12 oil-bearing rocks; the aim is to predict permeability y (a property of fluid flow) from the other three measurements. As permeabilities vary greatly (6.3–1300), we use a log scale. The measurements are the end product of a complex image-analysis procedure and represent the total area, total perimeter and a measure of 'roundness' of the pores in the rock cross-section.

We first fit a linear model, then a full additive model. In this example convergence of the backfitting algorithm is unusually slow, so the `control` limits must be raised. The plots are shown in Figure 9.2.

```
rock.lm <- lm(log(perm) ~ area + peri + shape, data=rock)
summary(rock.lm)
rock.gam <- gam(log(perm) ~ s(area) + s(peri) + s(shape),
    control=gam.control(maxit=50, bf.maxit=50), data=rock)
summary(rock.gam)
anova(rock.lm, rock.gam)
par(mfrow=c(2,3), pty="s")
plot(rock.gam, se=T)
rock.gam1 <- gam(log(perm) ~ area + peri + s(shape), data=rock)
plot(rock.gam1, se=T)
anova(rock.lm, rock.gam1, rock.gam)
```

It is worth showing the output from summary.gam and the analysis of variance table from anova(rock.lm, rock.gam):

```
Deviance Residuals:
      Min         1Q  Median       3Q      Max
  -1.6855  -0.46962 0.12531  0.54248   1.2963


(Dispersion Parameter ... 0.7446 )

       Null Deviance: 126.93 on 47 degrees of freedom
   Residual Deviance: 26.065 on 35.006 degrees of freedom

Number of Local Scoring Iterations: 1

DF for Terms and F-values for Nonparametric Effects

             Df Npar Df Npar F   Pr(F)
(Intercept)   1
    s(area)   1       3 0.3417 0.79523
    s(peri)   1       3 0.9313 0.43583
   s(shape)   1       3 1.4331 0.24966


Analysis of Variance Table

                            Terms Resid. Df    RSS    Test
1             area + peri + shape    44.000 31.949
2 s(area) + s(peri) + s(shape)      35.006 26.065 1 vs. 2
       Df Sum of Sq F Value   Pr(F)
1
2 8.9943    5.8835  0.8785 0.55311
```

This shows that each smooth term has one linear and three non-linear degrees of freedom. The reduction of RSS from 31.95 (for the linear fit) to 26.07 is not significant with an extra nine degrees of freedom, but Figure 9.2 shows that only one of the functions appears non-linear, and even that is not totally convincing. Although suggestive, the non-linear term for peri is not significant. With just that non-linear term the RSS is 29.00.

For the cpus data we can experiment with step.gam. It fits a series of gam models, at each stage selecting one term from a list. Here we allow each variable to be dropped, entered linearly or taken as a smooth term with two or four (equivalent) degrees of freedom.

```
cpus0 <- cpus[, 2:8]
for(i in 1:3) cpus0[,i] <- log10(cpus0[,i])
cpus.gam <- gam(log10(perf) ~ ., data=cpus0[cpus.samp, ])
cpus.gam2 <- step.gam(cpus.gam, scope=list(
   "syct" = ~ 1 + syct + s(syct, 2) + s(syct),
   "mmin" = ~ 1 + mmin + s(mmin, 2) + s(mmin),
   "mmax" = ~ 1 + mmax + s(mmax, 2) + s(mmax),
   "cach" = ~ 1 + cach + s(cach, 2) + s(cach),
```

```
   "chmin" = ~ 1 + chmin + s(chmin, 2) + s(chmin),
    "chmax" = ~ 1 + chmax + s(chmax, 2) + s(chmax)
))
> print(cpus.gam2$anova, digits=3)

Initial Model:
log10(perf) ~ syct + mmin + mmax + cach + chmin + chmax

Final Model:
log10(perf) ~ s(mmin, 2) + mmax + s(cach, 2) + s(chmax, 2)

Scale: 0.034525

      From           To Df Deviance Resid. Df Resid. Dev  AIC
1                                         93        3.21 3.69
2  mmin  s(mmin, 2) -1   -0.160         92        3.05 3.60
3  syct               1    0.019         93        3.07 3.55
4  cach  s(cach, 2) -1   -0.115         92        2.95 3.51
5 chmax s(chmax, 2) -1   -0.095         91        2.86 3.48
6 chmin               1    0.055         92        2.91 3.47
> test.cpus(cpus.gam2)
[1] 0.2037684
```

The non-linear terms give only a very slight improvement.

Other ways to fit additive models in S-PLUS are available from the contributions of users. These are generally more ambitious than gam and step.gam in their choice of terms and the degree of smoothness of each term, and by relying heavily on compiled code can be very substantially faster. These are discussed in the on-line complements.

Generalized additive models

These stand in the same relationship to additive models as generalized linear models do to regression models. Consider, for example, a logistic regression model. There are n observational units, each of which records a random variable Y_i with a binomial (n_i, p_i) distribution, and (p_i) is determined by

$$\text{logit}(p) = \alpha + \sum_{j=1}^{p} \beta_j X_j \tag{9.3}$$

This is readily generalized to the logistic additive model:

$$\text{logit}(p) = \alpha + \sum_{j=1}^{p} f_j(X_j) \tag{9.4}$$

and the full form of a GAM is then obvious; replace the linear predictor in a GLM by an additive predictor. (Direct minimization with smoothing splines has been used here too: Wahba *et al.*, 1995.)

We return to the data on low birth weights in data frame `birthwt` studied in Section 7.2. We examine the linearity in age and mother's weight, using generalized additive models and including smooth terms.

```
> attach(bwt)
> age1 <- age*(ftv=="1"); age2 <- age*(ftv=="2+")
> birthwt.gam <- gam(low ~ s(age) + s(lwt) + smoke + ptd +
      ht + ui + ftv + s(age1) + s(age2) + smoke:ui, binomial,
      bwt, bf.maxit=25)
> summary(birthwt.gam)

Residual Deviance: 170.35 on 165.18 degrees of freedom

DF for Terms and Chi-squares for Nonparametric Effects

          Df Npar Df Npar Chisq  P(Chi)
  s(age)   1    3.0        3.1089 0.37230
  s(lwt)   1    2.9        2.3392 0.48532
  s(age1)  1    3.0        3.2504 0.34655
  s(age2)  1    3.0        3.1472 0.36829

> table(low, predict(birthwt.gam) > 0)
    FALSE TRUE
0    115   15
1     28   31
> plot(birthwt.gam, ask=T, se=T)
```

Creating the variables `age1` and `age2` allows us to fit smooth terms for the *difference* in having one or more visits in the first trimester. Both the summary and the plots show no evidence of non-linearity. Note that the convergence of the fitting algorithm is slow in this example, so we increased the control parameter `bf.maxit` from 10 to 25. The parameter `ask=T` allows us to choose plots from a menu. Our choice of plots is shown in Figure 9.3.

Chambers & Hastie (1992) cover `gam` in considerably greater detail than this chapter.

9.2 Projection-pursuit regression

Now suppose that the explanatory vector $X = (X_1, \ldots, X_p)$ is of high dimension. The additive model (9.1) may be too flexible as it allows a few degrees of freedom per X_j, yet it does not cover the effect of interactions between the independent variables. Projection pursuit regression (Friedman & Stuetzle, 1981) applies an additive model to projected variables. That is, it is of the form:

$$Y = \alpha_0 + \sum_{j=1}^{M} f_j(\alpha_j^T X) + \epsilon \qquad (9.5)$$

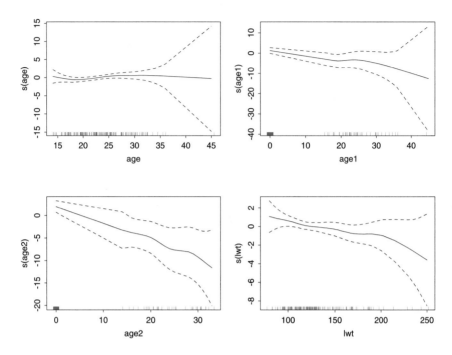

Figure 9.3: Plots of smooth terms in a generalized additive model for the data on low birth weight. The dashed lines indicate plus and minus two pointwise standard deviations.

for vectors α_j, and a dimension M to be chosen by the user. Thus it uses an additive model on predictor variables which are formed by projecting X in M carefully chosen directions. For large enough M such models can approximate (uniformly on compact sets and in many other senses) arbitrary continuous functions of X (for example, Diaconis & Shahshahani, 1984). The terms of (9.5) are called ridge functions, since they are constant in all but one direction.

The function ppr [2] in our library MASS fits (9.5) by least squares, and constrains the vectors α_k to be of unit length. It first fits $M_{\max}$ terms sequentially, then prunes back to M by at each stage dropping the least effective term and re-fitting. The function returns the proportion of the variance explained by all the fits from $M, \ldots, M_{\max}$.

For the rock example we can use[3]

```
> attach(rock)
> rock1 <- data.frame(area=area/10000, peri=peri/10000,
                      shape=shape, perm=perm)
> detach()
> rock.ppr <- ppr(log(perm) ~ area + peri + shape, data=rock1,
                  nterms=2, max.terms=5)
```

---

[2] Like the S-PLUS function ppreg based on the SMART program described in Friedman (1984), but using double precision internally.

[3] The exact results depend on the machine used.

```
> rock.ppr
Call:
ppr.formula(formula = log(perm) ~ area + peri +
        shape, data = rock1, nterms = 2, max.terms = 5)

Goodness of fit:
 2 terms 3 terms 4 terms 5 terms
 11.2317  7.3547  5.9445  3.1141
```

The summary method gives a little more information.

```
> summary(rock.ppr)
Call:
ppr.formula(formula = log(perm) ~ area + peri +
        shape, data = rock1, nterms = 2, max.terms = 5)


Goodness of fit:
 2 terms 3 terms 4 terms 5 terms
 11.2317  7.3547  5.9445  3.1141

Projection direction vectors:
          term 1     term 2
 area   0.314287   0.428802
 peri  -0.945525  -0.860929
shape   0.084893   0.273732

Coefficients of ridge terms:
   term 1   term 2
 0.93549 0.81952
```

The added information is the direction vectors $\boldsymbol{\alpha}_k$ and the coefficients β_{ij} in

$$Y_i = \alpha_{i0} + \sum_{j=1}^{M} \beta_{ij} f_j(\boldsymbol{\alpha}_j^T \boldsymbol{X}) + \epsilon \qquad (9.6)$$

Note that this is the extension of (9.5) to multiple responses, and so we separate the scalings from the smooth functions f_j (which are scaled to have zero mean and unit variance over the projections of the dataset).

We can examine the fitted functions f_j by (see Figure 9.4)

```
par(mfrow=c(3,2))
plot(rock.ppr)
plot(update(rock.ppr, bass=5))
plot(update(rock.ppr, sm.method="gcv", gcvpen=2))
```

We first increase the amount of smoothing in the 'super smoother' supsmu to fit a smoother function, and then change to using a smoothing spline with smoothness chosen by GCV (generalized cross-validation) with an increased complexity penalty. We can then examine the details of this fit by

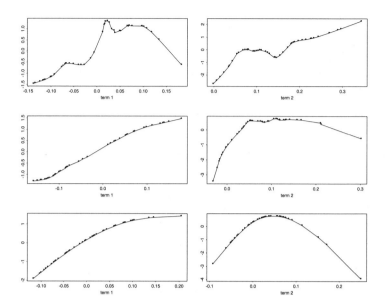

Figure 9.4: Plots of the ridge functions for three two-term projection pursuit regressions fitted to the `rock` dataset. The top two fits used `supsmu`, whereas the bottom fit used smoothing splines.

```
rock.ppr2 <- update(rock.ppr, sm.method="gcv", gcvpen=2)
summary(rock.ppr2)
    ....

Goodness of fit:
 2 terms 3 terms 4 terms 5 terms
 22.523  21.564  21.564   0.000

Projection direction vectors:
          term 1     term 2
 area   0.348531   0.442850
 peri  -0.936987  -0.856179
 shape  0.024124  -0.266162

Coefficients of ridge terms:
  term 1  term 2
 1.46479 0.20077

Equivalent df for ridge terms:
 term 1 term 2
   2.68      2
```

This fit is substantially slower since the effort put into choosing the amount of smoothing is much greater. Note that here only two effective terms could be found, and that `area` and `peri` dominate. We can arrange to view the surface

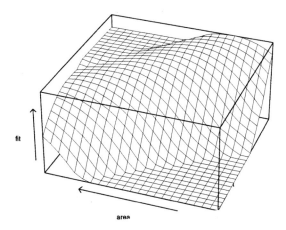

Figure 9.5: A two-dimensional fitted section of a projection pursuit regression surface fitted to the `rock` data. Note that the prediction extends the ridge functions as constant beyond the fitted functions, hence the planar regions shown. For display on paper we set `drape=F`.

for a typical value of `shape` (Figure 9.5). Users of **S-PLUS** 4.x can use the rotatable 3D-plots in the GUI graphics.

```
summary(rock1) # to find the ranges of the variables
Xp <- expand.grid(area=seq(0.1,1.2,0.05),
                  peri=seq(0,0.5,0.02), shape=0.2)
trellis.device()
rock.grid <- cbind(Xp,fit=predict(rock.ppr2, Xp))
wireframe(fit ~ area + peri, rock.grid, screen=list(z=160,x=-60),
          aspect=c(1,0.5), drape=T)
```

An example: The cpus data

We can also consider the `cpus` test problem. Our experience suggests that smoothing the terms rather more than the default for `supsmu` is a good idea.

```
> cpus.ppr <- ppr(log10(perf) ~ ., data=cpus0[cpus.samp,],
                  nterms=2, max.terms=10, bass=5)
> cpus.ppr
Call:
ppr.formula(formula = log10(perf) ~ ., data = cpus0[cpus.samp,],
        nterms = 2, max.terms = 10, bass = 5)

Goodness of fit:
 2 terms 3 terms 4 terms 5 terms 6 terms 7 terms 8 terms
 2.1371  2.4223  1.9865  1.8331  1.5806  1.5055  1.3962
 9 terms 10 terms
 1.2723  1.2338
```

```
> cpus.ppr <- ppr(log10(perf) ~ ., data=cpus0[cpus.samp,],
                  nterms=8, max.terms=10, bass=5)
> test.cpus(cpus.ppr)
[1] 0.18225
> ppr(log10(perf) ~ ., data=cpus0[cpus.samp,],
      nterms=2, max.terms=10, sm.method="spline")
Goodness of fit:
 2 terms 3 terms 4 terms 5 terms 6 terms 7 terms 8 terms
 2.6752  2.2854  2.0998  2.0562  1.6744  1.4438  1.3948
 9 terms 10 terms
 1.3843   1.3395
> cpus.ppr2 <- ppr(log10(perf) ~ ., data=cpus0[cpus.samp,],
      nterms=7, max.terms=10, sm.method="spline")
> test.cpus(cpus.ppr2)
[1] 0.18739
> res3 <- log10(cpus0[-cpus.samp, "perf"]) -
              predict(cpus.ppr, cpus0[-cpus.samp,])
> wilcox.test(res2^2, res3^2, paired=T, alternative="greater")
signed-rank normal statistic with correction Z = 0.6712,
  p-value = 0.2511
```

In these experiments projection pursuit regression outperformed all the additive models, but not by much, and is not significantly better than the best linear model with discretized variables.

9.3 Response transformation models

If we want to predict Y, it may be better to transform Y as well, so we have

$$\theta(Y) = \alpha + \sum_{j=1}^{p} f_j(X_j) + \epsilon \tag{9.7}$$

for an invertible smooth function $\theta()$, for example the log function we used for the rock dataset.

The ACE (alternating conditional expectation) algorithm of Breiman & Friedman (1985) chooses the functions θ and $f_1, \ldots, f_j$ to maximize the correlation between the predictor $\alpha + \sum_{j=1}^{p} f_j(X_j)$ and $\theta(Y)$. Tibshirani's (1988) procedure AVAS (additivity and variance stabilising transformation) fits the same model (9.7), but with the aim of achieving constant variance of the residuals for monotone θ.

The functions ace and avas fit (9.7). Both allow the functions f_j and θ to be constrained to be monotone or linear, and the functions f_j to be chosen for circular or categorical data. Thus these functions provide another way to fit additive models (with linear θ) but they do not provide measures of fit nor standard errors. They do, however, automatically choose the degree of smoothness. Our experience has been that AVAS is much more reliable than ACE.

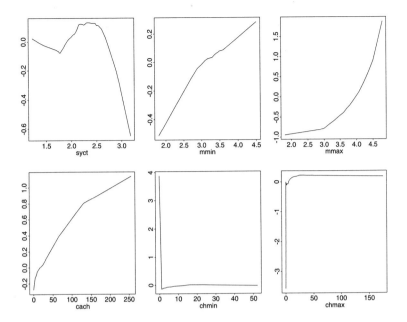

Figure 9.6: AVAS transformations of the regressors in the cpus dataset.

We can consider the cpus data: we have already log-transformed some of the variables.

```
attach(cpus0)
cpus.avas <- avas(cpus0[, 1:6], perf)
plot(log10(perf), cpus.avas$ty)
par(mfrow=c(2,3))
for(i in 1:6) {
  o <- order(cpus0[, i])
  plot(cpus0[o, i], cpus.avas$tx[o, i], type="l",
       xlab=names(cpus0[i]), ylab="")
}
detach()
```

This accepts the log-scale for the response, but has some interesting shapes for the regressors (Figure 9.6). The strange shape of the transformation for chmin and chmax is probably due to local collinearity as there are five machines without any channels.

For the rock dataset we can use the following code. The S function rug produces the ticks to indicate the locations of data points, as used by gam.plot.

```
attach(rock)
x <- cbind(area, peri, shape)
o1 <- order(area); o2 <- order(peri); o3 <- order(shape)
a <- avas(x, perm)
par(mfrow=c(2,2))
```

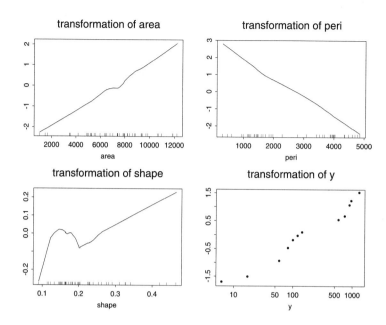

Figure 9.7: AVAS on permeabilities

```
plot(area[o1], a$tx[o1,1], type="l")          see Figure 9.7
rug(area)
plot(peri[o2], a$tx[o2,2], type="l")
rug(peri)
plot(shape[o3], a$tx[o3,3], type="l")
rug(shape)
plot(perm, a$ty, log="x")                     note log scale
a <- avas(x, log(perm))                       looks like $\log(y)$
a <- avas(x, log(perm), linear=0)             so force $\theta(y) = \log(y)$
# repeat plots
```

Here AVAS indicates a log transformation of permeabilities (expected on physical grounds) but little transformation of area or perimeter.

9.4   Neural networks

Feed-forward neural networks provide a flexible way to generalize linear regression functions. General references are Bishop (1995); Hertz, Krogh & Palmer (1991) and Ripley (1993, 1996).

We start with the simplest but most common form with one hidden layer as shown in Figure 9.8. The input units just provide a 'fan out' and distribute the inputs to the 'hidden' units in the second layer. These units sum their inputs, add

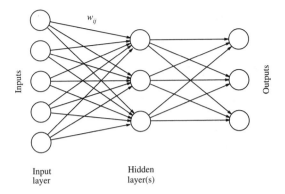

Figure 9.8: A generic feed-forward neural network.

a constant (the 'bias') and take a fixed function ϕ_h of the result. The output units are of the same form, but with output function ϕ_o. Thus

$$y_k = \phi_o \left(\alpha_k + \sum_h w_{hk} \, \phi_h \left(\alpha_h + \sum_i w_{ih} \, x_i \right) \right) \tag{9.8}$$

The 'activation function' ϕ_h of the hidden layer units is almost always taken to be the logistic function

$$\ell(z) = \frac{\exp(z)}{1 + \exp(z)}$$

and the output units are linear, logistic or threshold units. (The latter have $\phi_o(x) = I(x > 0)$.) Note the similarity to projection pursuit regression (*cf* (9.5)), which has linear output units but general smooth hidden units. (However, arbitrary smooth functions can be approximated by sums of rescaled logistics.)

The general definition allows more than one hidden layer, and also allows 'skip-layer' connections from input to output when we have

$$y_k = \phi_o \left(\alpha_k + \sum_{i \to k} w_{ik} x_i + \sum_{j \to k} w_{jk} \phi_h \left(\alpha_j + \sum_{i \to j} w_{ij} x_i \right) \right) \tag{9.9}$$

which allows the non-linear units to perturb a linear functional form.

We can eliminate the biases α_i by introducing an input unit 0 which is permanently at $+1$ and feeds every other unit. The regression function f is then parametrized by the set of weights w_{ij}, one for every link in the network (or zero for links that are absent).

The original biological motivation for such networks stems from McCulloch & Pitts (1943) who published a seminal model of a neuron as a binary thresholding device in discrete time, specifically that

$$n_i(t) = H \left(\sum_{j \to i} w_{ji} n_j(t-1) - \theta_i \right)$$

the sum being over neurons j connected to neuron i. Here H denotes the Heaviside or threshold function $H(x) = I(x > 0)$, $n_i(t)$ is the output of neuron i at time t and $0 < w_{ij} < 1$ are attenuation weights. Thus the effect is to threshold a weighted sum of the inputs at value θ_i. Real neurons are now known to be more complicated; they have a graded response rather than the simple thresholding of the McCulloch–Pitts model, work in continuous time, and can perform more general non-linear functions of their inputs, for example, logical functions. Nevertheless, the McCulloch–Pitts model has been extremely influential in the development of artificial neural networks.

Feed-forward neural networks can equally be seen as a way to parametrize a fairly general non-linear function. Such networks *are* rather general: Cybenko (1989), Funahashi (1989), Hornik, Stinchcombe & White (1989) and later authors have shown that neural networks with linear output units can approximate any continuous function f uniformly on compact sets, by increasing the size of the hidden layer.

The approximation results are non-constructive, and in practice the weights have to be chosen to minimize some fitting criterion, for example, least squares

$$E = \sum_{p} \|t^p - y^p\|^2$$

where t^p is the target and y^p the output for the pth example pattern. Other measures have been proposed, including for $y \in [0, 1]$ 'maximum likelihood' (in fact minus the logarithm of a conditional likelihood) or equivalently the Kullback–Leibler distance, which amount to minimizing

$$E = \sum_{p} \sum_{k} \left[t_k^p \log \frac{t_k^p}{y_k^p} + (1 - t_k^p) \log \frac{1 - t_k^p}{1 - y_k^p} \right] \tag{9.10}$$

This is half the deviance for a logistic model with linear predictor given by (9.8) or (9.9).

One way to ensure that f is smooth is to restrict the class of estimates, for example, by using a limited number of spline knots. Another way is *regularization* in which the fit criterion is altered to

$$E + \lambda C(f)$$

with a penalty C on the 'roughness' of f. *Weight decay*, specific to neural networks, uses as penalty the sum of squares of the weights w_{ij}. (This only makes sense if the inputs are rescaled to range about $[0, 1]$ to be comparable with the outputs of internal units.) The use of weight decay seems both to help the optimization process and to avoid over-fitting. Arguments in Ripley (1993, 1994) based on a Bayesian interpretation suggest $\lambda \approx 10^{-4}$–10^{-2} depending on the degree of fit expected, for least-squares fitting to variables of range one and $\lambda \approx 0.01$–0.1 for the entropy fit.

Software to fit feed-forward neural networks with a single hidden layer but allowing skip-layer connections (as in (9.9)) is provided in our library nnet. The format of the call to the fitting function nnet is

```
nnet(formula, data, weights, size, Wts, linout=F, entropy=F,
     softmax=F, skip=F, rang=0.7, decay=0, maxit=100, trace=T)
```

The non-standard arguments are as follows.

`size`	number of units in the hidden layer.
`Wts`	optional initial vector for w_{ij}.
`linout`	logical for linear output units.
`entropy`	logical for entropy rather than least-squares fit.
`softmax`	logical for log-probability models.
`skip`	logical for links from inputs to outputs.
`rang`	if `Wts` is missing, use random weights from `runif(n,-rang, rang)`.
`decay`	parameter λ.
`maxit`	maximum of iterations for the optimizer.
`Hess`	should the Hessian matrix at the solution be returned?
`trace`	logical for output from the optimizer. Very reassuring!

There are predict, print and summary methods for neural networks, and a function nnet.Hess to compute the Hessian with respect to the weight parameters and so check if a secure local minimum has been found. For our rock example we have

```
> library(nnet)
> attach(rock)
> area1 <- area/10000; peri1 <- peri/10000
> rock1 <- data.frame(perm, area=area1, peri=peri1, shape)
> rock.nn <- nnet(log(perm) ~ area + peri + shape, rock1,
       size=3, decay=1e-3, linout=T, skip=T, maxit=1000, Hess=T)
# weights:  19
initial  value 1092.816748
iter  10 value 32.272454
    ....
final  value 14.069537
converged
> summary(rock.nn)
a 3-3-1 network with 19 weights
options were - skip-layer connections  linear output units
  decay=0.001
  b->h1 i1->h1 i2->h1 i3->h1
   1.21   8.74 -15.00  -3.45
  b->h2 i1->h2 i2->h2 i3->h2
   9.50  -4.34 -12.66   2.48
  b->h3 i1->h3 i2->h3 i3->h3
   6.20  -7.63 -10.97   3.12
   b->o   h1->o  h2->o   h3->o   i1->o   i2->o   i3->o
   7.74   20.17  -7.68   -7.02  -12.95  -15.19    6.92
> sum((log(perm) - predict(rock.nn))^2)
[1] 12.231
> detach()
```

```
> eigen(rock.nn$Hess, T)$values      # rock.nn@Hess on 5.x
 [1] 9.1533e+02 1.6346e+02 1.3521e+02 3.0368e+01 7.3914e+00
 [6] 3.4012e+00 2.2879e+00 1.0917e+00 3.9823e-01 2.7867e-01
[11] 1.9953e-01 7.5159e-02 3.2513e-02 2.5950e-02 1.9077e-02
[16] 1.0834e-02 6.8937e-03 3.7671e-03 2.6974e-03
```

(There are several solutions and a random starting point, so your results may well differ.) The quoted values include the weight decay term. The eigenvalues of the Hessian suggest that a secure local minimum has been achieved. In the summary the b refers to the bias unit (input unit 0), and i, h and o to input, hidden and bias units.

To view the fitted surface for the rock dataset we can use essentially the same code as we used for the fits by ppr.

```
Xp <- expand.grid(area=seq(0.1,1.2,0.05),
                   peri=seq(0,0.5,0.02), shape=0.2)
trellis.device()
rock.grid <- cbind(Xp,fit=predict(rock.nn, Xp))
wireframe(fit ~ area + peri, rock.grid, screen=list(z=160,x=-60),
          aspect=c(1,0.5), drape=T)
```

An example: The cpus data

To use the nnet software effectively it is essential to scale the problem. A preliminary run with a linear model demonstrates that we get essentially the same results as the conventional approach to linear models.

```
attach(cpus0)
cpus1 <-
  data.frame(syct=syct-2, mmin=mmin-3, mmax=mmax-4, cach=cach/256,
            chmin=chmin/100, chmax=chmax/100, perf=perf)
detach()

test.cpus <- function(fit)
  sqrt(sum((log10(cpus1[-cpus.samp, "perf"]) -
          predict(fit, cpus1[-cpus.samp,]))^2)/109)
cpus.nn1 <- nnet(log10(perf) ~ ., cpus1[cpus.samp,], linout=T,
                skip=T, size=0)
test.cpus(cpus.nn1)
[1] 0.21295
```

We now consider adding non-linear terms to the model.

```
cpus.nn2 <- nnet(log10(perf) ~ ., cpus1[cpus.samp,], linout=T,
                skip=T, size=4, decay=0.01, maxit=1000)
final  value 2.369581
test.cpus(cpus.nn2)
[1] 0.21132
cpus.nn3 <- nnet(log10(perf) ~ ., cpus1[cpus.samp,], linout=T,
                skip=T, size=10, decay=0.01, maxit=1000)
```

```
final  value 2.338387
test.cpus(cpus.nn3)
[1] 0.21068
cpus.nn4 <- nnet(log10(perf) ~ ., cpus1[cpus.samp,], linout=T,
                skip=T, size=25, decay=0.01, maxit=1000)
final  value 2.339850
test.cpus(cpus.nn4)
[1] 0.23
```

This demonstrates that the degree of fit is almost completely controlled by the amount of weight decay rather than the number of hidden units (provided there are sufficient). We have to be able to choose the amount of weight decay *without* looking at the test set. To do so we borrow the ideas of Chapter 11.5, by using cross-validation and by averaging across multiple fits.

```
CVnn.cpus <- function(formula, data=cpus1[cpus.samp, ],
    size = c(0, 4, 4, 10, 10),
    lambda = c(0, rep(c(0.003, 0.01), 2)),
    nreps = 5, nifold = 10, ...)
{
  CVnn1 <- function(formula, data, nreps=1, ri,  ...)
  {
    truth <- log10(data$perf)
    res <- numeric(length(truth))
    cat("  fold")
    for (i in sort(unique(ri))) {
      cat(" ", i,  sep="")
      for(rep in 1:nreps) {
        learn <- nnet(formula, data[ri !=i,], trace=F, ...)
        res[ri == i] <- res[ri == i] +
                        predict(learn, data[ri == i,])
      }
    }
    cat("\n")
    sum((truth - res/nreps)^2)
  }
  choice <- numeric(length(lambda))
  ri <- sample(nifold, nrow(data), replace=T)
  for(j in seq(along=lambda)) {
    cat("  size =", size[j], "decay =", lambda[j], "\n")
    choice[j] <- CVnn1(formula, data, nreps=nreps, ri=ri,
                   size=size[j], decay=lambda[j], ...)
  }
  cbind(size=size, decay=lambda, fit=sqrt(choice/100))
}
CVnn.cpus(log10(perf) ~ ., data=cpus1[cpus.samp,],
          linout=T, skip=T, maxit=1000)
     size decay      fit
[1,]    0 0.000 0.19746
[2,]    4 0.003 0.23297
```

```
[3,]    4 0.010 0.20404
[4,]   10 0.003 0.22803
[5,]   10 0.010 0.20130
```

This took around 6 Mb and 5 minutes on the PC. The cross-validated results seem rather insensitive to the choice of model. The non-linearity does not seem justified.

More extensive examples of the use of neural networks are given in Section 11.5 and in the on-line complements.

9.5   Conclusions

We have considered a large, perhaps bewildering, variety of extensions to linear regression. These can be thought of as belonging to two classes, the 'black-box' fully automatic and maximally flexible routines represented by projection pursuit regression and neural networks, and the small steps under full user control of additive models. Although the latter may gain in interpretation, as we saw in the rock example they may not be general enough, and this is a common experience.

What is best for any particular problem depends on its aim, in particular whether prediction or interpretation is paramount. The methods of this chapter are powerful tools with very little distribution theory to support them so it is very easy to over-fit and over-explain features of the data. Be warned!

9.6   Exercises

9.1.   This data frame `gilgais` was collected on a line transect survey in gilgai territory in New South Wales, Australia. Gilgais are natural gentle depressions in otherwise flat land, and sometimes seem to be regularly distributed. The data collection was stimulated by the question: are these patterns reflected in soil properties? At each of 365 sampling locations on a linear grid of 4 metres spacing, samples were taken at depths 0–10cm, 30–40cm and 80–90cm below the surface. pH, electrical conductivity and chloride content were measured on a 1:5 soil:water extract from each sample.

Produce smoothed maps of the measurements.

9.2.   Exercise 6.5 considered linear regression for the GAG in urine data in data frame `GAGurine`. Consider using a non-linear or smooth regression for the same task.

9.3.   Use neural networks to fit a smooth curve to the `mcycle` data used in Figure 9.1. Investigate ways of choosing the degree of smoothness automatically.

Chapter 10

Tree-based Methods

The use of tree-based models will be relatively unfamiliar to statisticians, although researchers in other fields have found trees to be an attractive way to express knowledge and aid decision-making. Keys such as Figure 10.1 are common in botany and in medical decision-making, and provide a way to encapsulate and structure the knowledge of experts to be used by less-experienced users. Notice how this tree uses both categorical variables and splits on continuous variables. (It is a tree, and readers are encouraged to draw it.)

The automatic construction of decision trees dates from work in the social sciences by Morgan & Sonquist (1963) and Morgan & Messenger (1973). In statistics Breiman *et al.* (1984) had a seminal influence both in bringing the work to the attention of statisticians and in proposing new algorithms for constructing trees. At around the same time decision tree induction was beginning to be used in the field of *machine learning*, notably by Quinlan (1979, 1983, 1986, 1993), and in engineering (Henrichon & Fu, 1969; Sethi & Sarvarayudu, 1982). Whereas there is now an extensive literature in machine learning, further statistical contributions are still sparse. The introduction within S of tree-based models described by Clark & Pregibon (1992) made the methods much more freely available. Their methods are very much in the spirit of exploratory data analysis, with many functions to investigate trees. The library `rpart` (Therneau & Atkinson, 1997) provides a faster and more tightly-packaged set of S functions for fitting trees to data.

Ripley (1996, Chapter 7) gives a comprehensive survey of the subject, with proofs of the theoretical results.

Constructing trees may be seen as a type of variable selection. Questions of interaction between variables are handled automatically, and to a large extent so is monotonic transformation of both the x and y variables. These issues are reduced to which variables to divide on, and how to achieve the split.

Figure 10.1 is a *classification* tree since its endpoint is a factor giving the species. Although this is the most common use, it is also possible to have *regression* trees in which each terminal node gives a predicted value, as shown in Figure 10.2 for our dataset `cpus`.

Much of the machine learning literature is concerned with logical variables and correct decisions. The end point of a tree is a (labelled) partition of the space

1.	Leaves subterete to slightly flattened, plant with bulb	*2.*
	Leaves flat, plant with rhizome	*4.*
2.	Perianth-tube > 10 mm	**I. × hollandica**
	Perianth-tube < 10 mm	*3.*
3.	Leaves evergreen	**I. xiphium**
	Leaves dying in winter	**I. latifolia**
4.	Outer tepals bearded	**I. germanica**
	Outer tepals not bearded	*5.*
5.	Tepals predominately yellow	*6.*
	Tepals blue, purple, mauve or violet	*8.*
6.	Leaves evergreen	**I. foetidissima**
	Leaves dying in winter	*7.*
7.	Inner tepals white	**I. orientalis**
	Tepals yellow all over	**I. pseudocorus**
8.	Leaves evergreen	**I. foetidissima**
	Leaves dying in winter	*9.*
9.	Stems hollow, perianth-tube 4–7mm	**I. sibirica**
	Stems solid, perianth-tube 7–20mm	*10.*
10.	Upper part of ovary sterile	*11.*
	Ovary without sterile apical part	*12.*
11.	Capsule beak 5–8mm, 1 rib	**I. enstata**
	Capsule beak 8–16mm, 2 ridges	**I. spuria**
12.	Outer tepals glabrous, many seeds	**I. versicolor**
	Outer tepals pubescent, 0–few seeds	**I. × robusta**

Figure 10.1: Key to British species of the genus *Iris*. Simplified from Stace (1991, p. 1140), by omitting parts of his descriptions.

$\mathcal{X}$ of possible observations. In logical problems it is assumed that there *is* a partition of the space $\mathcal{X}$ that will correctly classify all observations, and the task is to find a tree to describe it succinctly. A famous example of Donald Michie (for example, Michie, 1989) is whether the space shuttle pilot should use the autolander or land manually (Table 10.1). Some enumeration will show that the decision has been specified for 253 out of the 256 possible observations. Some cases have been specified twice. This body of expert opinion needed to be reduced to a simple decision aid, as shown in Figure 10.3. (Table 10.1 appears to result from a decision tree that differs from Figure 10.3 in reversing the order of two pairs of splits.)

Note that the botanical problem is treated as if it were a logical problem, although there will be occasional specimens that do not meet the specification for their species.

10.1 Partitioning methods

The ideas for classification and regression trees are quite similar, but the terminology differs, so we consider classification first. Classification trees are more

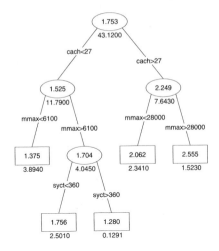

Figure 10.2: A regression tree for the cpu performance data on $\log_{10}$ scale. The value in each node is the prediction for the node; those underneath the nodes indicate the deviance contributions D_i.

Table 10.1: Example decisions for the space shuttle autolander problem.

stability	error	sign	wind	magnitude	visibility	decision
any	any	any	any	any	no	auto
xstab	any	any	any	any	yes	noauto
stab	LX	any	any	any	yes	noauto
stab	XL	any	any	any	yes	noauto
stab	MM	nn	tail	any	yes	noauto
any	any	any	any	Out of range	yes	noauto
stab	SS	any	any	Light	yes	auto
stab	SS	any	any	Medium	yes	auto
stab	SS	any	any	Strong	yes	auto
stab	MM	pp	head	Light	yes	auto
stab	MM	pp	head	Medium	yes	auto
stab	MM	pp	tail	Light	yes	auto
stab	MM	pp	tail	Medium	yes	auto
stab	MM	pp	head	Strong	yes	noauto
stab	MM	pp	tail	Strong	yes	auto

familiar and it is a little easier to justify the tree-construction procedure, so we consider them first.

Classification trees

We have already noted that the endpoint for a tree is a partition of the space $\mathcal{X}$, and we compare trees by how well that partition corresponds to the correct decision rule for the problem. In logical problems the easiest way to compare partitions is

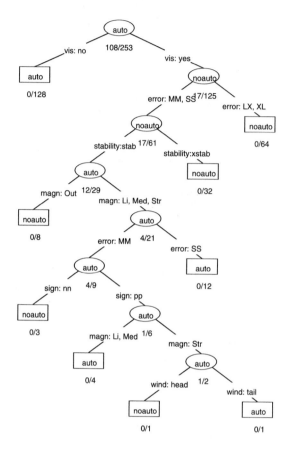

Figure 10.3: Decision tree for shuttle autolander problem. The numbers m/n denote the proportion of training cases reaching that node wrongly classified by the label.

to count the number of errors, or, if we have a prior over the space $\mathcal{X}$, to compute the probability of error.

In statistical problems the distributions of the classes over $\mathcal{X}$ usually overlap, so there is no partition that completely describes the classes. Then for each cell of the partition there will be a probability distribution over the classes, and the Bayes decision rule will choose the class with highest probability. This corresponds to assessing partitions by the overall probability of misclassification. Of course, in practice we do not have the whole probability structure, but a training set of n classified examples that we assume are an independent random sample. Then we can estimate the misclassification rate by the proportion of the training set that is misclassified.

Almost all current tree-construction methods, including those in S, use a one-step lookahead. That is, they choose the next split in an optimal way, without attempting to optimize the performance of the whole tree. (This avoids a combinatorial explosion over future choices, and is akin to a very simple strategy for

playing a game such as chess.) However, by choosing the right measure to optimize at each split, we can ease future splits. It does not seem appropriate to use
the misclassification rate to choose the splits.

What class of splits should we allow? Both Breiman *et al.*'s CART methodology and the S methods only allow binary splits, which avoids one difficulty in
comparing splits, that of normalization by size. For a continuous variable x_j the
allowed splits are of the form $x_j < t$ versus $x_j \geqslant t$. For ordered factors the
splits are of the same type. For general factors the levels are divided into two
classes. (Note that for L levels there are 2^L possible splits, and if we disallow
the empty split and ignore the order, there are still $2^{L-1} - 1$. For ordered factors there are only $L - 1$ possible splits.) Some algorithms, including CART
but excluding S, allow linear combination of continuous variables to be split, and
Boolean combinations to be formed of binary variables.

The justification for the S methodology is to view the tree as providing a
probability model (hence the title 'tree-based models' of Clark & Pregibon, 1992).
At each node i of a classification tree we have a probability distribution p_{ik} over
the classes. The partition is given by the *leaves* of the tree (also known as terminal
nodes). Each case in the training set is assigned to a leaf, and so at each leaf we
have a random sample n_{ik} from the multinomial distribution specified by p_{ik}.

We now condition on the observed variables x_i in the training set, and hence
we know the numbers n_i of cases assigned to every node of the tree, in particular
to the leaves. The conditional likelihood is then proportional to

$$\prod_{\text{cases } j} p_{[j]y_j} = \prod_{\text{leaves } i} \prod_{\text{classes } k} p_{ik}^{n_{ik}}$$

where $[j]$ denotes the leaf assigned to case j. This allows us to define a deviance
for the tree as

$$D = \sum_i D_i, \qquad D_i = -2 \sum_k n_{ik} \log p_{ik}$$

as a sum over leaves.

Now consider splitting node s into nodes t and u. This changes the probability model within node s, so the reduction in deviance for the tree is

$$D_s - D_t - D_u = 2 \sum_k \left[n_{tk} \log \frac{p_{tk}}{p_{sk}} + n_{uk} \log \frac{p_{uk}}{p_{sk}} \right]$$

Since we do not know the probabilities, we estimate them from the proportions in
the split node, obtaining

$$\hat{p}_{tk} = \frac{n_{tk}}{n_t}, \qquad \hat{p}_{uk} = \frac{n_{uk}}{n_u}, \qquad \hat{p}_{sk} = \frac{n_t \hat{p}_{tk} + n_u \hat{p}_{uk}}{n_s} = \frac{n_{sk}}{n_s}$$

so the reduction in deviance is

$$D_s - D_t - D_u = 2 \sum_k \left[n_{tk} \log \frac{n_{tk} n_s}{n_{sk} n_t} + n_{uk} \log \frac{n_{uk} n_s}{n_{sk} n_u} \right]$$

$$= 2\left[\sum_k n_{tk} \log n_{tk} + n_{uk} \log n_{uk} - n_{sk} \log n_{sk}\right.$$

$$\left. + n_s \log n_s - n_t \log n_t - n_u \log n_u\right]$$

This gives a measure of the value of a split. Note that it is size-biased; there is more value in splitting leaves with large numbers of cases.

The tree construction process takes the maximum reduction in deviance over all allowed splits of all leaves, to choose the next split. (Note that for continuous variates the value depends only on the split of the ranks of the observed values, so we may take a finite set of splits.) The tree construction continues until the number of cases reaching each leaf is small (by default $n_i < 10$ in S) or the leaf is homogeneous enough (by default its deviance is less than 1% of the deviance of the root node in S, which is a size-biased measure). Note that as all leaves not meeting the stopping criterion will eventually be split, an alternative view is to consider splitting any leaf and choose the best allowed split (if any) for that leaf, proceeding until no further splits are allowable.

This justification for the value of a split follows Ciampi *et al.* (1987) and Clark & Pregibon, but differs from most of the literature on tree construction. The more common approach is to define a measure of the impurity of the distribution at a node, and choose the split that most reduces the average impurity. Two common measures are the *entropy* or *information* $\sum p_{ik} \log p_{ik}$ and the *Gini index*

$$\sum_{j \neq k} p_{ij} p_{ik} = 1 - \sum_k p_{ik}^2$$

As the probabilities are unknown, they are estimated from the node proportions. With the entropy measure, the average impurity differs from D by a constant factor, so the tree construction process is the same, except perhaps for the stopping rule. Breiman *et al.* preferred the Gini index.

Regression trees

The prediction for a regression tree is constant over each cell of the partition of $\mathcal{X}$ induced by the leaves of the tree. The deviance is defined as

$$D = \sum_{\text{cases } j} (y_j - \mu_{[j]})^2$$

and so clearly we should estimate the constant μ_i for leaf i by the mean of the values of the training-set cases assigned to that node. Then the deviance is the sum over leaves of D_i, the corrected sum of squares for cases within that node, and the value of a split is the reduction in the residual sum of squares.

The obvious probability model (and that proposed by Clark & Pregibon) is to take a normal $N(\mu_i, \sigma^2)$ distribution within each leaf. Then D is the usual scaled deviance for a Gaussian GLM. However, the distribution at internal nodes of the tree is then a mixture of normal distributions, and so D_i is only appropriate at the leaves. The tree-construction process has to be seen as a hierarchical refinement of probability models, very similar to forward variable selection in regression. In contrast, for a classification tree, one probability model can be used throughout the tree-construction process.

Missing values

One attraction of tree-based methods is the ease with which missing values can be handled. Consider the botanical key of Figure 10.1. We only need to know about a small subset of the 10 observations to classify any case, and part of the art of constructing such trees is to avoid observations that will be difficult or missing in some of the species (or as in capsules, for some of the cases). A general strategy is to 'drop' a case down the tree as far as it will go. If it reaches a leaf we can predict y for it. Otherwise we use the distribution at the node reached to predict y, as shown in Figure 10.2, which has predictions at all nodes.

An alternative strategy is used by many botanical keys and can be seen at nodes 9 and 12 of Figure 10.1. A list of characteristics is given, the most important first, and a decision made from those observations that are available. This is codified in the method of *surrogate splits* in which surrogate rules are available at non-terminal nodes to be used if the splitting variable is unobserved. Another attractive strategy is to split cases with missing values, and pass part of the case down each branch of the tree (Ripley, 1996, p. 232).

Cutting trees down to size

With 'noisy' data, that is when the distributions for the classes overlap, it is quite possible to grow a tree which fits the training set well, but which has adapted too well to features of that subset of $\mathcal{X}$. Similarly, regression trees can be too elaborate and over-fit the training data. We need an analogue of variable selection in regression.

The established methodology is cost-complexity *pruning*, first introduced by Breiman *et al.* (1984). They considered rooted subtrees of the tree $\mathcal{T}$ grown by the construction algorithm, that is the possible results of snipping off terminal subtrees on $\mathcal{T}$. The pruning process chooses one of the rooted subtrees. Let R_i be a measure evaluated at the leaves, such as the deviance or the number of errors, and let R be the value for the tree, the sum over the leaves of R_i. Let the size of the tree be the number of leaves. Then Breiman *et al.* showed that the set of rooted subtrees of $\mathcal{T}$ which minimize the cost-complexity measure

$$R_\alpha = R + \alpha \, \text{size}$$

is itself nested. That is, as we increase α we can find the optimal trees by a sequence of snip operations on the current tree (just like pruning a real tree). This produces a sequence of trees from the size of $\mathcal{T}$ down to just the root node, but it may prune more than one node at a time. (Short proofs of these assertions are given by Ripley, 1996, Chapter 7. The tree $\mathcal{T}$ is not necessarily optimal for $\alpha = 0$, as we illustrate.)

We need a good way to choose the degree of pruning. If a separate validation set is available, we can predict on that set, and compute the deviance versus α for the pruned trees. This will often have a minimum, and we can choose the smallest tree whose deviance is close to the minimum.

If no validation set is available we can make one by splitting the training set. Suppose we split the training set into 10 (roughly) equally sized parts. We can then use 9 to grow the tree and test it on the tenth. This can be done in 10 ways, and we can average the results.

10.2 Implementation in `rpart`

The simplest way to use tree-based methods is via the library section `rpart` by Terry Therneau and Beth Atkinson (Therneau & Atkinson, 1997) available from `statlib`. The underlying philosophy is of one function, `rpart`, that both grows and computes where to prune a tree; although there is a function `prune.rpart` it merely further prunes the tree at points already determined by the call to `rpart`, which has itself done some pruning. It is also possible to print a pruned tree by giving a pruning parameter to `print.rpart`. By default `rpart` runs a 10-fold cross-validation and the results are stored in the `rpart` object to allow the user to choose the degree of pruning at a later stage. Since all the work is done in a C function the calculations are quite fast.

The `rpart` system was designed to be easily extended to new types of responses. We only consider the following types, selected by the argument `method`.

`"anova"` A regression tree, with the impurity criterion the reduction in sum of squares on creating a binary split of the data at that node. The criterion $R(T)$ used for pruning is the mean square error of the predictions of the tree on the current dataset (that is, the residual mean square).

`"class"` A classification tree, with a categorical or factor response and default impurity criterion the Gini index. The deviance-based approach corresponds to the entropy index, selected by the argument `parms = list(split="information")`. The pruning criterion $R(T)$ is the predicted loss, normally the error rate.

If the `method` argument is missing an appropriate type is inferred from the response variable in the formula.

It is helpful to consider a few examples. First we consider a regression tree for our `cpus` data discussed on page 188, then a classification tree for the `iris` data. The model is specified by a model formula with terms separated by `+`; interactions make no sense for trees, and `-` terms are ignored. The precise meaning of the argument `cp` is explained later; it is proportional to α in the cost-complexity measure.

```
> library(rpart)
> set.seed(123)
> cpus.rp <- rpart(log10(perf) ~ ., cpus[ ,2:8], cp=1e-3)
> cpus.rp  # gives a large tree not show here.
> print(cpus.rp, cp=0.01)
node), split, n, deviance, yval
      * denotes terminal node

  1) root 209 43.000 1.8
    2) cach<27 143 12.000 1.5
      4) mmax<6100 78   3.900 1.4
        8) mmax<1750 12  0.780 1.1 *
        9) mmax>1750 66  1.900 1.4 *
      5) mmax>6100 65   4.000 1.7
       10) syct>360 7  0.130 1.3 *
       11) syct<360 58  2.500 1.8
         22) chmin<5.5 46  1.200 1.7 *
         23) chmin>5.5 12  0.550 2.0 *
    3) cach>27 66   7.600 2.2
      6) mmax<28000 41   2.300 2.1
       12) cach<96.5 34  1.600 2.0
         24) mmax<11240 14  0.420 1.8 *
         25) mmax>11240 20  0.380 2.1 *
       13) cach>96.5 7  0.170 2.3 *
      7) mmax>28000 25   1.500 2.6
       14) cach<56 7   0.069 2.3 *
       15) cach>56 18  0.650 2.7 *
```

This shows the predicted value (yval) and deviance within each node. We can plot the full tree by

```
plot(cpus.rp, uniform=T); text(cpus.rp, digits=3)
```

There are lots of options to produce prettier trees; see help(plot.rpart) for details. In particular, a PostScript figure like Figure 10.2 can be obtained by a call to post.rpart.

```
post(cpus.rp, filename="CpusTree.eps", horizontal=F, pointsize=8)

> ird <- data.frame(rbind(iris[,,1], iris[,,2],iris[,,3]),
          Species=c(rep("s",50), rep("c",50), rep("v",50)))
> ir.rp <- rpart(Species ~ ., data=ird, method="class", cp=1e-3)
> ir.rp
node), split, n, loss, yval, (yprob)
      * denotes terminal node

  1) root 150 100 c ( 0.33 0.33 0.33 )
    2) Petal.L.>2.45 100  50 c ( 0.50 0.00 0.50 )
      4) Petal.W.<1.75 54   5 c ( 0.91 0.00 0.092 ) *
      5) Petal.W.>1.75 46   1 v ( 0.02 0.00 0.98 ) *
    3) Petal.L.<2.45 50   0 s ( 0.00 1.00 0.00 ) *
```

The (yprob) give the distribution by class within each node.

Note that neither tree has yet been pruned to final size. We can now consider pruning by using printcp to print out the information stored in the rpart object.

```
> printcp(cpus.rp)

Regression tree:
rpart(formula = log10(perf) ~ ., data = cpus[, 2:8], cp = 0.001)

Variables actually used in tree construction:
[1] cach  chmax chmin mmax  syct

Root node error: 43.1/209 = 0.206

         CP nsplit rel error xerror   xstd
1  0.54927      0     1.000  1.005 0.0972
2  0.08934      1     0.451  0.480 0.0487
3  0.03282      3     0.274  0.322 0.0322
4  0.02692      4     0.241  0.306 0.0306
5  0.01856      5     0.214  0.278 0.0294
6  0.00946      9     0.147  0.288 0.0323
7  0.00548     10     0.138  0.247 0.0289
8  0.00440     12     0.127  0.245 0.0287
9  0.00229     13     0.123  0.242 0.0284
10 0.00141     15     0.118  0.240 0.0282
11 0.00100     16     0.117  0.238 0.0279
```

The columns xerror and xstd are random, depending on the random partition used in the cross-validation. We can see the same output graphically (Figure 10.4) by a call to plotcp.

```
plotcp(cpus.rp)
```

We need to explain the *complexity parameter* cp; this is just the parameter α divided by the number $R(T_\emptyset)$ for the root tree.[1] A 10-fold cross-validation has been done within rpart to compute the entries[2] xerror and xstd; the complexity parameter may then be chosen to minimize xerror. An alternative procedure is to use the 1-SE rule, the largest value with xerror within one standard deviation of the minimum. In this case the 1-SE rule gives $0.238 + 0.0279$, so we choose line 7, a tree with 10 splits and hence 11 leaves.[3] We can examine this by

```
> print(cpus.rp, cp=0.006, digits=3)
node), split, n, deviance, yval
      * denotes terminal node
```

---

[1] Thus for most measures of fit the complexity parameter lies in $[0, 1]$.
[2] All the errors are scaled so the root tree has error $R(T_\emptyset)$ scaled to one.
[3] The number of leaves is always one more than the number of splits.

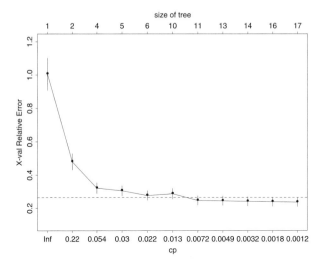

Figure 10.4: Plot by `plotcp` of the `rpart` object `cpus.rp1`.

```
1) root 209 43.1000 1.75
  2) cach<27 143 11.8000 1.52
    4) mmax<6100 78  3.8900 1.37
      8) mmax<1750 12  0.7840 1.09 *
      9) mmax>1750 66  1.9500 1.43 *
    5) mmax>6100 65  4.0500 1.70
     10) syct>360 7  0.1290 1.28 *
     11) syct<360 58  2.5000 1.76
       22) chmin<5.5 46  1.2300 1.70
         44) cach<0.5 11  0.2020 1.53 *
         45) cach>0.5 35  0.6160 1.75 *
       23) chmin>5.5 12  0.5510 1.97 *
  3) cach>27 66  7.6400 2.25
    6) mmax<28000 41  2.3400 2.06
     12) cach<96.5 34  1.5900 2.01
       24) mmax<11240 14  0.4250 1.83 *
       25) mmax>11240 20  0.3830 2.14 *
     13) cach>96.5 7  0.1720 2.32 *
    7) mmax>28000 25  1.5200 2.56
     14) cach<56 7  0.0693 2.27 *
     15) cach>56 18  0.6540 2.67 *
```

or

```
> cpus.rp1 <- prune(cpus.rp, cp=0.006)
> plot(cpus.rp1, branch=0.4, uniform=T)
> text(cpus.rp1, digits=3)
```

The plot is shown in Figure 10.5.

For the `iris` data we have

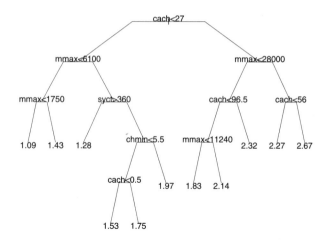

Figure 10.5: Plot of the rpart object cpus.rp1.

```
> printcp(ir.rp)
    ....
Variables actually used in tree construction:
[1] Petal.L. Petal.W.

Root node error: 100/150 = 0.66667

      CP nsplit rel error xerror   xstd
1 0.500      0     1.00   1.13 0.0528
2 0.440      1     0.50   0.58 0.0596
3 0.001      2     0.06   0.12 0.0332
```

which suggests no pruning, but that too small a tree has been grown since xerror
may not have reached its minimum.

The summary method, summary.rpart, produces voluminous output:

```
> summary(ir.rp)
Call:
rpart(formula = Species ~ ., data = ird, method = "class",
     cp = 0.001)

      CP nsplit rel error xerror   xstd
1 0.500      0     1.00   1.13 0.053
2 0.440      1     0.50   0.58 0.060
3 0.001      2     0.06   0.12 0.033

Node number 1: 150 observations,    complexity param=0.5
  predicted class= c  expected loss= 0.67
      class counts:    50 50 50
    probabilities:  0.33 0.33 0.33
  left son=2 (100 obs) right son=3 (50 obs)
```

```
    Primary splits:
        Petal.L. < 2.5 to the right, improve=50, (0 missing)
        Petal.W. < 0.8 to the right, improve=50, (0 missing)
        Sepal.L. < 5.4 to the left,  improve=34, (0 missing)
        Sepal.W. < 3.3 to the left,  improve=19, (0 missing)
    Surrogate splits:
        Petal.W. < 0.8 to the right, agree=1.00, (0 split)
        Sepal.L. < 5.4 to the right, agree=0.92, (0 split)
        Sepal.W. < 3.3 to the left,  agree=0.83, (0 split)

Node number 2: 100 observations,    complexity param=0.44
  predicted class= c  expected loss= 0.5
      class counts:   50   0  50
     probabilities:  0.5 0.0 0.5
  left son=4 (54 obs) right son=5 (46 obs)
  Primary splits:
        Petal.W. < 1.8 to the left,  improve=39.0, (0 missing)
        Petal.L. < 4.8 to the left,  improve=37.0, (0 missing)
        Sepal.L. < 6.1 to the left,  improve=11.0, (0 missing)
        Sepal.W. < 2.5 to the left,  improve= 3.6, (0 missing)
    Surrogate splits:
        Petal.L. < 4.8 to the left,  agree=0.91, (0 split)
        Sepal.L. < 6.1 to the left,  agree=0.73, (0 split)
        Sepal.W. < 3   to the left,  agree=0.67, (0 split)

Node number 3: 50 observations
  predicted class= s  expected loss= 0
      class counts:    0  50   0
     probabilities:   0  1  0

Node number 4: 54 observations
  predicted class= c  expected loss= 0.093
      class counts:   49   0   5
     probabilities:  0.91 0.00 0.09

Node number 5: 46 observations
  predicted class= v  expected loss= 0.022
      class counts:    1   0  45
     probabilities:  0.02 0.00 0.98
```

The initial table is that given by `printcp`. The summary method gives the top few (default up to five) splits and their reduction in impurity, plus up to five surrogates, splits on other variables with a high agreement with the chosen split. In this case the limit on tree growth is the restriction on the size of child nodes (which by default must cover at least seven cases).

Two arguments to `summary.rpart` can help with the volume of output: as with `print.rpart` the argument `cp` effectively prunes the tree before analysis, and the argument `file` allows the output to be redirected to a file (via `sink`).

Fine control

The function `rpart.control` is usually used to collect arguments for the `control` parameter of `rpart`. The parameter `minsplit` gives the smallest node that will be considered for a split: this defaults to 20. Parameter `minbucket` is the minimum number of observations in a daughter node, which defaults to 7 (`minsplit`/3, rounded up).

If a split does not result in a branch T_t with $R(T_t)$ at least $cp \times |T_t| \times R(T_\emptyset)$ it is not considered further. This is a form of 'pre-pruning'; the tree presented has been pruned to this value and the knowledge that this will happen can be used to stop tree growth.[4] In many of our examples the minimum of `xerror` occurs for values of `cp` less than 0.01 (the default), so we choose a smaller value.

Parameters `maxcompete` and `maxsurrogate` give the number of good attributes and surrogates that are retained. Set `maxsurrogate` to zero if it is known that missing values will not be encountered.

The number of cross-validations is controlled by parameter `xval`, default 10. This can be set to zero at early stages of exploration, since this will produce a very significant speedup.

Note that these parameters may also be passed directly to `rpart`. For example, for the `iris` data we can use

```
> ir.rp1 <- rpart(Species ~ ., ird, cp=0, minsplit=5,
                  maxsurrogate=0)
> printcp(ir.rp1)
  ....
Root node error: 100/150 = 0.667

    CP nsplit rel error xerror    xstd
1 0.50      0      1.00   1.18 0.0502
2 0.44      1      0.50   0.63 0.0604
3 0.02      2      0.06   0.08 0.0275
4 0.01      3      0.04   0.08 0.0275
5 0.00      4      0.03   0.07 0.0258

> print(ir.rp1, cp=0.015)
node), split, n, loss, yval, (yprob)
      * denotes terminal node

1) root 150 100 c ( 0.33 0.33 0.33 )
   2) Petal.L.>2.45 100  50 c ( 0.50 0.00 0.50 )
     4) Petal.W.<1.75 54   5 c ( 0.91 0.00 0.09 )
       8) Petal.L.<4.95 48   1 c ( 0.98 0.00 0.02 ) *
       9) Petal.L.>4.95 6    2 v ( 0.33 0.00 0.67 ) *
     5) Petal.W.>1.75 46   1 v ( 0.02 0.00 0.98 ) *
   3) Petal.L.<2.45 50    0 s ( 0.00 1.00 0.00 ) *
```

which suggests a tree with four leaves using only two of the variables.

---

[4] If $R(T_t) \geqslant 0$, splits of nodes with $R(t) < cpR(T_\emptyset)$ will always be pruned.

Plots

There are plot methods for use on a standard **S-PLUS** graphics device
(plot.rpart and text.rpart), plus a method for post for plots in **PostScript**.
Note that unlike post.tree, post.rpart is just a wrapper for calls to
plot.rpart and text.rpart on a postscript device.

The function plot.rpart has a wide range of options to choose the layout
of the plotted tree. Let us consider some examples

```
plot(cpus.rp, branch=0.6, compress=T, uniform=T)
text(cpus.rp, digits=3, all=T, use.n=T)
```

The argument branch controls the slope of the branches. Arguments uniform
and compress control whether the spacing reflects the importance of the fits (by
default it does) and whether a compact style is used. The call to text.rpart
may have additional arguments all which gives the value at all nodes (not just
the leaves) and use.n which if true gives the numbers of cases reaching the node
(and for classification trees the number of errors).

Forensic glass

For the forensic glass dataset fgl which has six classes we can use

```
> set.seed(123)
> fgl.rp <- rpart(type ~ ., fgl, cp=0.001)
> plotcp(fgl.rp)
> printcp(fgl.rp)

Classification tree:
rpart(formula = type ~ ., data = fgl, cp = 0.001)

Variables actually used in tree construction:
[1] Al Ba Ca Fe Mg Na RI

Root node error: 138/214 = 0.645

      CP nsplit rel error xerror   xstd
1 0.2065      0     1.000  1.000 0.0507
2 0.0725      2     0.587  0.594 0.0515
3 0.0580      3     0.514  0.587 0.0514
4 0.0362      4     0.457  0.551 0.0507
5 0.0326      5     0.420  0.536 0.0504
6 0.0109      7     0.355  0.478 0.0490
7 0.0010      9     0.333  0.500 0.0495

> print(fgl.rp, cp=0.02)
node), split, n, loss, yval, (yprob)
      * denotes terminal node
```

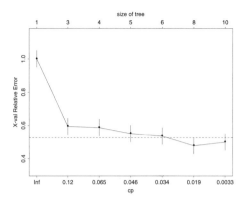

Figure 10.6: Plot by `plotcp` of the `rpart` object `fgl.rp`.

```
 1) root 214 140 WinNF ( 0.33 0.36 0.07 0.06 0.04 0.14 )
   2) Ba<0.335 185 110 WinNF ( 0.37 0.41 0.09 0.06 0.04 0.01 )
     4) Al<1.42 113   50 WinF ( 0.56 0.27 0.12 0.00 0.02 0.01 )
       8) Ca<10.48 101   38 WinF ( 0.62 0.21 0.13 0.00 0.02 0.02 )
        16) RI>-0.93 85   25 WinF ( 0.71 0.20 0.07 0.00 0.01 0.01)
          32) Mg<3.865 77   18 WinF ( 0.77 0.14 0.06 0.00 0.01 0.0 1 )
          33) Mg>3.865 8    2 WinNF ( 0.12 0.75 0.12 0.00 0.00 0.0 0 )
        17) RI<-0.93 16    9 Veh ( 0.19 0.25 0.44 0.00 0.06 0.06 ) *
       9) Ca>10.48 12    2 WinNF ( 0.00 0.83 0.00 0.08 0.08 0.00 ) *
     5) Al>1.42 72   28 WinNF ( 0.08 0.61 0.05 0.15 0.08 0.01 )
      10) Mg>2.26 52   11 WinNF ( 0.12 0.79 0.07 0.00 0.01 0.00 ) *
      11) Mg<2.26 20    9 Con ( 0.00 0.15 0.00 0.55 0.25 0.05 )
        22) Na<13.495 12    1 Con ( 0.00 0.08 0.00 0.92 0.00 0.00) *
        23) Na>13.495 8    3 Tabl ( 0.00 0.25 0.00 0.00 0.62 0.12) *
   3) Ba>0.335 29    3 Head ( 0.03 0.03 0.00 0.03 0.00 0.90 ) *
```

This suggests (Figure 10.6) a tree of size 8, plotted in Figure 10.7 by

```
fgl.rp2 <- prune(fgl.rp, cp=0.02)
plot(fgl.rp2, uniform=T); text(fgl.rp2, use.n=T)
```

Missing data

If the control parameter `maxsurrogate` is positive (without altering the parameter `usesurrogate`), the surrogates are used to handle missing cases both in training and in prediction (including cross-validation to choose the complexity). Each of the surrogate splits is examined in turn, and if the variable is available that split is used to decide whether to send the case left or right. If no surrogate is available or none can be used, the case is sent with the majority unless `usesurrogate` < 2 when it is left at the node.

The default `na.action` during training is `na.rpart`, which excludes cases only if the response or *all* the explanatory variables are missing. (This looks

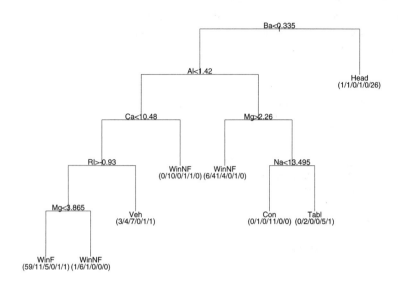

Figure 10.7: Plot of the `rpart` object `fgl.rp`.

like a sub-optimal choice, as cases with missing response are useful for finding surrogate variables.)

When missing values are encountered in considering a split they are ignored and the probabilities and impurity measures are calculated from the non-missing values of that variable. Surrogate splits are then used to allocate the missing cases to the daughter nodes.

Surrogate splits are chosen to match as well as possible the primary split (viewed as a binary classification), and retained provided they send at least two cases down each branch, and agree as well as the rule of following the majority. The measure of agreement is the number of cases that are sent the same way, possibly after swapping 'left' and 'right' for the surrogate. (As far as we can tell, missing values on the surrogate are ignored, so this measure is biased towards surrogate variables with few missing values.)

10.3 Implementation in `tree`

The S-PLUS implementation is based on a class `tree`. The tree is a regression tree unless the response variable in the model formula is a factor, and the internal structure differs in the two cases. Beware: it is very easy to code the classes numerically and then fit a regression tree by mistake. The right-hand side of the formula is a series of terms separated by + just as for `rpart`.

The function `tree` constructs trees, and there are `print`, `summary` and `plot` methods for trees. As our first example consider the computer performance data in our data frame `cpus`. We first consider a tree for the performance and then for log-performance.

```
> cpus.tr <- tree(perf ~ syct+mmin+mmax+cach+chmin+chmax, cpus)
> summary(cpus.tr)
Regression tree:
tree(formula = perf ~ syct + mmin + mmax + cach + chmin + chmax,
        data = cpus)
Variables actually used in tree construction:
[1] "mmax"  "cach"  "chmax" "mmin"
Number of terminal nodes:  9
Residual mean deviance:  4523 = 904600 / 200
     ....
> print(cpus.tr)
node), split, n, deviance, yval
        * denotes terminal node

  1) root 209 5380000 105.60
    2) mmax<28000 182   585900   60.72
      4) cach<27 141     97850   39.64
        8) mmax<10000 113    36000   32.21 *
        9) mmax>10000 28     30470   69.61 *
      5) cach>27 41   209900  133.20
       10) cach<96.5 34     96490  114.40
         20) mmax<11240 14     12950   72.79 *
         21) mmax>11240 20     42240  143.60 *
       11) cach>96.5 7     43150  224.40 *
    3) mmax>28000 27 1954000  408.30
      6) chmax<59 22   436600  323.20
       12) mmin<12000 15   106500  244.50
         24) cach<56 9     26650  191.60 *
         25) cach>56 6     16700  324.00 *
       13) mmin>12000 7     38480  491.70 *
      7) chmax>59 5   658000  782.60 *
> plot(cpus.tr, type="u");  text(cpus.tr, srt=90)

> cpus.ltr <- update(cpus.tr, log10(perf) ~ .)
> summary(cpus.ltr)
Number of terminal nodes:  19
Residual mean deviance:  0.0239 = 4.55 / 190
     ....
> plot(cpus.ltr, type="u");  text(cpus.ltr, srt=90)
```

This output needs some explanation. The first tree does not use all of the variables, so the summary informs us. (The second tree does use all six.) Next we have a compact printout of the tree including the number, sum of squares and mean at the node. The plots are shown in Figure 10.8. The `type` parameter is used to reduce overcrowding of the labels; by default the depths reflect the values

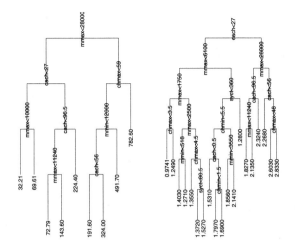

Figure 10.8: Regression trees for the cpu performance data on linear (left) and $\log_{10}$ scale (right).

of the splits. Plotting uses no labels and allows the tree topology to be studied: split conditions and the values at the leaves are added by `text` (which has a number of other options). The elegant plots such as Figure 10.2 are produced directly in PostScript by `post.tree`.

For examples of classification trees we use the `iris` data. We have

```
> ir.tr <- tree(Species ~., ird)
> summary(ir.tr)

Classification tree:
tree(formula = Species ~ ., data = ird)
Variables actually used in tree construction:
[1] "Petal.L." "Petal.W." "Sepal.L."
Number of terminal nodes:  6
Residual mean deviance:  0.125 = 18 / 144
Misclassification error rate: 0.0267 = 4 / 150

> ir.tr
node), split, n, deviance, yval, (yprob)
      * denotes terminal node

 1) root 150 330.0 c ( 0.330 0.33 0.330 )
   2) Petal.L.<2.45 50    0.0 s ( 0.000 1.00 0.000 ) *
   3) Petal.L.>2.45 100 140.0 c ( 0.500 0.00 0.500 )
     6) Petal.W.<1.75 54  33.0 c ( 0.910 0.00 0.093 )
      12) Petal.L.<4.95 48   9.7 c ( 0.980 0.00 0.021 )
        24) Sepal.L.<5.15 5   5.0 c ( 0.800 0.00 0.200 ) *
        25) Sepal.L.>5.15 43   0.0 c ( 1.000 0.00 0.000 ) *
      13) Petal.L.>4.95 6   7.6 v ( 0.330 0.00 0.670 ) *
```

```
   7) Petal.W.>1.75 46    9.6 v ( 0.022 0.00 0.980 )
     14) Petal.L.<4.95 6    5.4 v ( 0.170 0.00 0.830 ) *
     15) Petal.L.>4.95 40   0.0 v ( 0.000 0.00 1.000 ) *
```

The (yprob) give the distribution by class within the node. Note how the second split on Petal length occurs twice, and that splitting node 12 is attempting to classify one case of *I. virginica* without success. Thus viewed as a decision tree we would want to snip off nodes 24, 25, 14, 15, and this corresponds to error-rate pruning at $\alpha = 0$. We can do so interactively:

```
> plot(ir.tr)
> text(ir.tr, all=T)
> ir.tr1 <- snip.tree(ir.tr)
node number:  12
   tree deviance =  18.05
   subtree deviance =  22.77
node number:  7
   tree deviance =  22.77
   subtree deviance =  26.99
> ir.tr1
node), split, n, deviance, yval, (yprob)
      * denotes terminal node

 1) root 150 330.0 c ( 0.330 0.33 0.330 )
   2) Petal.L.<2.45 50    0.0 s ( 0.000 1.00 0.000 ) *
   3) Petal.L.>2.45 100 140.0 c ( 0.500 0.00 0.500 )
     6) Petal.W.<1.75 54  33.0 c ( 0.910 0.00 0.093 )
       12) Petal.L.<4.95 48   9.7 c ( 0.980 0.00 0.021 ) *
       13) Petal.L.>4.95 6    7.6 v ( 0.330 0.00 0.670 ) *
     7) Petal.W.>1.75 46   9.6 v ( 0.022 0.00 0.980 ) *
> summary(ir.tr1)

Classification tree:
   ....
Number of terminal nodes:  4
Residual mean deviance:  0.185 = 27 / 146
Misclassification error rate: 0.0267 = 4 / 150

par(pty="s")
plot(ird[, 3],ird[, 4], type="n",
   xlab="petal length", ylab="petal width")
text(ird[, 3], ird[, 4], labels=as.character(ird$Species))
par(cex=2)
partition.tree(ir.tr1, add=T)
par(cex=1)
```

where we clicked twice in succession with mouse button 1 on each of nodes 12 and 7 to remove their subtrees, then with button 2 to quit.

The decision region is now entirely in the petal length–petal width space, so we can show it by partition.tree. This example shows the limitations of

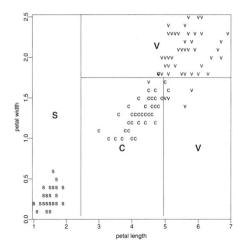

Figure 10.9: Partition for the `iris` data induced by the snipped tree.

the one-step-ahead tree construction, for Weiss & Kapouleas (1989) used a rule-induction program to find the set of rules

> If `Petal length` < 3 then *I. setosa.*
> If `Petal length` > 4.9 or `Petal width` > 1.6 then *I. virginica.*
> Otherwise *I. versicolor.*

Compare this with Figure 10.9, which makes one more error.

For the forensic glass dataset `fgl` which has six classes we can use

```
> fgl.tr <- tree(type ~ ., fgl)
> summary(fgl.tr)
Classification tree:
tree(formula = type ~ ., data = fgl)
Number of terminal nodes:   24
Residual mean deviance:   0.649 = 123 / 190
Misclassification error rate: 0.15 = 32 / 214
> plot(fgl.tr);   text(fgl.tr, all=T, cex=0.5)
> fgl.tr1 <- snip.tree(fgl.tr)
> tree.screens()
> plot(fgl.tr1)
> tile.tree(fgl.tr1, fgl$type)
> close.screen(all = T)
```

Once more snipping is required.

Fine control

Missing values are handled by the argument `na.action` of `tree`. The default is `na.fail`, to abort if missing values are found; an alternative is `na.omit` which omits all cases with a missing observation. The `weights` and `subset` arguments are available as in all model-fitting functions.

The `predict` method `predict.tree` can be used to predict future cases. This automatically handles missing values by dropping cases down the tree until a NA is encountered or a leaf is reached. With option `split=T` it splits cases as described in Ripley (1996, p. 232).

There are three control arguments specified under `tree.control`, but which may be passed directly to `tree`. The `mindev` controls the threshold for splitting a node. The parameters `minsize` and `mincut` control the size thresholds; `minsize` is the threshold for a node size, so nodes of size `minsize` or larger are candidates for a split. Daughter nodes must exceed `mincut` for a split to be allowed. The defaults are `minsize` = 10 and `mincut` = 5.

To make a tree fit exactly, the help page recommends the use of `mindev` = 0 and `minsize` = 2. However, for the shuttle data this will split an already pure node. A small value such as `mindev` = 1e-6 is safer. For example, for the shuttle data:

```
> shuttle.tr <- tree(use ~ ., shuttle, subset=1:253,
                       mindev=1e-6, minsize=2)
> shuttle.tr
node), split, n, deviance, yval, (yprob)
      * denotes terminal node

  1) root 253 350.0  auto ( 0.57 0.43 )
    2) vis: no 128    0.0  auto ( 1.00 0.00 ) *
    3) vis: yes 125  99.0  noauto ( 0.14 0.86 )
      6) error: MM, SS 61  72.0  noauto ( 0.28 0.72 )
       12) stability:stab 29  39.0  auto ( 0.59 0.41 )
         24) magn: Out 8    0.0  noauto ( 0.00 1.00 ) *
         25) magn: Light, Med, Str 21  20.0  auto ( 0.81 0.19 )
           50) error: MM 9  12.0  auto ( 0.56 0.44 )
            100) sign: nn 3    0.0  noauto ( 0.00 1.00 ) *
            101) sign: pp 6    5.4  auto ( 0.83 0.17 )
              202) magn: Light, Med 4  0.0  auto ( 1.00 0.00 ) *
              203) magn: Strong 2    2.8  auto ( 0.50 0.50 )
                406) wind: head 1    0.0  noauto ( 0.00 1.00 ) *
                407) wind: tail 1    0.0  auto ( 1.00 0.00 ) *
           51) error: SS 12    0.0  auto ( 1.00 0.00 ) *
       13) stability:xstab 32    0.0  noauto ( 0.00 1.00 ) *
      7) error: LX, XL 64    0.0  noauto ( 0.00 1.00 ) *
> post.tree(shuttle.tr)
> shuttle1 <- shuttle[254:256, ]  # 3 missing cases
  stability error sign  wind    magn  vis
1      stab    MM   nn  head   Light  yes
2      stab    MM   nn  head  Medium  yes
3      stab    MM   nn  head  Strong  yes
> predict(shuttle.tr, shuttle1)
   auto   noauto
```

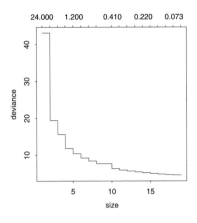
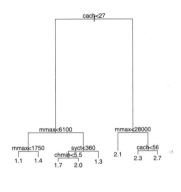

Figure 10.10: Pruning `cpus.ltr` : (left) deviance versus size. The top axis is of α; (right) the pruned tree of size 8.

```
254    0       1
255    0       1
256    0       1
```

Let us prune the cpus tree. Plotting the output of `prune.tree`, the deviance against size, allows us to choose a likely breakpoint (Figure 10.10):

```
> plot(prune.tree(cpus.ltr))
> cpus.ltr1 <- prune.tree(cpus.ltr, best=8)
> plot(cpus.ltr1);    text(cpus.ltr1)
```

Note that size 9 does not appear in the tree sequence. We can test the prediction accuracy by

```
sqrt(sum((log10(cpus[-cpus.samp, "perf"]) -
          predict(cpus.ltr1, cpus[-cpus.samp,]))^2)/109)
[1] 0.17692
```

which beats all the linear and smooth non-linear methods.

One way to choose the parameter α is to consider pruning as a method of variable selection. Akaike's information criterion (AIC) penalizes minus twice log-likelihood by twice the number of parameters. For classification trees with K classes choosing $\alpha = 2(K-1)$ will find the rooted subtree with minimum AIC. For regression trees one approximation (Mallows' C_p) to AIC is to replace the minus twice log-likelihood by the residual sum of squares divided by $\hat{\sigma}^2$ (our best estimate of σ^2), so we could take $\alpha = 2\hat{\sigma}^2$. One way to select $\hat{\sigma}^2$ would be from the fit of the full tree model. In our example this suggests $\alpha = 2 \times 0.02394$, which makes no change. Other criteria suggest that AIC and C_p tend to over-fit and choose larger constants in the range 2–6. (Note that counting parameters in this way does not take the selection of the splits into account.)

For the classification tree on the `fgl` data this suggests $\alpha = 2 \times (6-1) = 10$, and we have

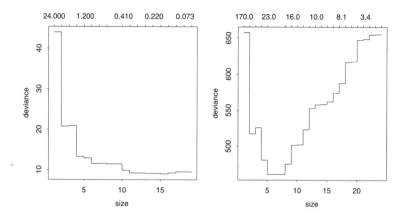

Figure 10.11: Cross-validation plots for pruning: (left) cpus.ltr and (right) fgl.tr.

```
> summary(prune.tree(fgl.tr, k=10))
Classification tree:
snip.tree(tree = fgl.tr, nodes = c(11, 10, 15, 108, 109, 12, 26))
Variables actually used in tree construction:
[1] "Mg" "Na" "Al" "Fe" "Ba" "RI" "K"  "Si"
Number of terminal nodes:  13
Residual mean deviance:  0.961 = 193 / 201
Misclassification error rate: 0.182 = 39 / 214
```

This is done by the function cv.tree. Note that as 10 trees must be grown, the process can be slow, and that the averaging is done for fixed α and not fixed tree size.

```
set.seed(123)
plot(cv.tree(cpus.ltr,, prune.tree))
post.tree(prune.tree(cpus.ltr, best=4))
```

The answers can be far from convincing as Figure 10.11 shows. The algorithm randomly divides the training set, and sometimes a second attempt will give a better answer. It may be worthwhile if CPU time allows us to average over several random divisions.

Let us prune the tree grown on the fgl dataset.

```
set.seed(123)
fgl.cv <- cv.tree(fgl.tr,, prune.tree)
for(i in 2:5)  fgl.cv$dev <- fgl.cv$dev +
    cv.tree(fgl.tr,, prune.tree)$dev
fgl.cv$dev <- fgl.cv$dev/5
plot(fgl.cv)
misclass.tree(fgl.tr)
[1] 32
misclass.tree(prune.tree(fgl.tr, best=5))
[1] 65
```

These resubstitution error counts are optimistically biased.

There is a difficulty with cross-validating the deviance; if at some leaf a class occurs in the test set but not in the training set the deviance will be infinity. (If this occurs at the root node, the deviance will be infinite for all prunings.) To avoid this, zero probabilities are replaced by a very small probability (10^{-3} by default).

Error-rate pruning

Pruning on error-rate is attractive, as it will remove terminal splits that have the same class for each leaf, and allows the use of cross-validation to estimate the error rate. Note that cv.tree passes its ... argument only to some of its calls to the pruning function. We can work around this by replacing prune.tree by the function prune.misclass:

```
> set.seed(123)
> fgl.cv <- cv.tree(fgl.tr,, prune.misclass)
> for(i in 2:5)  fgl.cv$dev <- fgl.cv$dev +
        cv.tree(fgl.tr,, prune.misclass)$dev
> fgl.cv$dev <- fgl.cv$dev/5
> fgl.cv
$size:
 [1] 24 18 16 12 11  9  6  5  4  3  1
$dev:
 [1]  73.6  74.2  74.0  72.6  73.0  73.4  78.8  89.0  90.4  92.6
[11] 146.4
> plot(fgl.cv)
> prune.misclass(fgl.tr)
$size:
 [1] 24 18 16 12 11  9  6  5  4  3  1
$dev:
 [1]  32  32  33  37  39  44  58  65  73  84 138
    ....
```

which suggests that a pruned tree of size 5 suffices. Note that in this example cross-validation gives a 35% higher estimate of the error rate.

We said on page 310 that we should take the smallest tree whose performance is close to the minimum. Breiman *et al.* (1984) suggest the *one s.e. rule* by taking the smallest pruned tree whose error rate is within one standard deviation of the minimum. Under cross-validation the number of errors made is approximately Poisson, so in this example we would seek an answer within nine of the minimum.

Chapter 11

Multivariate Analysis and Pattern Recognition

Multivariate analysis is concerned with datasets that have more than one response variable for each observational or experimental unit. The datasets can be summarized by data matrices X with n rows and p columns, the rows representing the observations or cases, and the columns the variables. The matrix can be viewed either way, depending on whether the main interest is in the relationships between the cases or between the variables. Note that for consistency we represent the variables of a case by the *row* vector x.

The main division in multivariate methods is between those methods that assume a given structure, for example, dividing the cases into groups, and those that seek to discover structure from the evidence of the data matrix alone (nowadays often called *data mining*). In pattern-recognition terminology the distinction is between *supervised* and *unsupervised* methods. One of our examples is the (in)famous iris data collected by Anderson (1935) and given and analysed by Fisher (1936). This has 150 cases, which are stated to be 50 of each of the three species *Iris setosa*, *I. virginica* and *I. versicolor*. Each case has four measurements on the length and width of its petals and sepals. *A priori* this seems a supervised problem, and the obvious questions are to use measurements on a future case to classify it, and perhaps to ask how the variables vary among the species. (In fact, Fisher used these data to test a genetic hypothesis which placed *I. versicolor* as a hybrid two-thirds of the way from *I. setosa* to *I. virginica*.) However, the classification of species is uncertain, and similar data have been used to identify species by grouping the cases. (Indeed, Wilson (1982) and McLachlan (1992, §6.9) consider whether the iris data can be split into subspecies.)

Classification is an increasingly important application of modern methods in statistics. In the statistical literature the word is used in two distinct senses. The entry (Hartigan, 1982) in the original *Encyclopedia of Statistical Sciences* uses the sense of *cluster analysis* discussed in Section 11.2. Modern usage is leaning to the other meaning (Ripley, 1997) of allocating future cases to one of g prespecified classes. Medical diagnosis is an archetypal classification problem in the modern sense. (The older statistical literature sometimes refers to this as *allocation*.)

Krzanowski (1988) and Mardia, Kent & Bibby (1979) are two general references on multivariate analysis. For pattern recognition we follow Ripley (1996), which also has a computationally-informed account of multivariate analysis.

Colour can be used very effectively to differentiate groups in the plots of this chapter, on screen if not on paper. The code given here uses both colours and symbols, but you may prefer to use only one of these to differentiate groups. (The colours used are chosen for use on a trellis device.)

11.1   Graphical methods

The simplest way to examine multivariate data is via a `splom` plot, enhanced to show the groups as discussed in Chapter 3. A similar effect can be obtained using the S-PLUS function `brush`

```
ir <- rbind(iris[,,1], iris[,,2], iris[,,3])
ir.species <- factor(c(rep("s",50), rep("c",50), rep("v",50)))
brush(ir)
```

by marking the cases with different symbols. (See Section 3.2 for further details.)

However, pairs plots can easily miss interesting structure in the data that depends on three or more of the variables, and genuinely multivariate methods explore the data in a less coordinate-dependent way. Many of the most effective routes to explore multivariate data use dynamic graphics such as exploratory projection pursuit (for example, Huber, 1985, Friedman, 1987, Jones & Sibson, 1987 and Ripley, 1996) which chooses 'interesting' rotations of the point cloud. These are not included in S-PLUS, but are available through an S interface to the package XGobi available from `statlib` (see Appendix C.2) for machines running X11.[1] A foretaste is given in Figure 11.17.

Principal component analysis

Linear methods are the heart of classical multivariate analysis, and depend on seeking linear combinations of the variables with desirable properties. For the unsupervised case the main method is *principal component* analysis, which seeks linear combinations of the columns of X with maximal (or minimal) variance. Because the variance can be scaled by rescaling the combination, we constrain the combinations to have unit length.

Let S denote the covariance matrix of the data X, which is defined[2] by

$$nS = (X - n^{-1}\mathbf{1}\mathbf{1}^T X)^T (X - n^{-1}\mathbf{1}\mathbf{1}^T X) = (X^T X - n\overline{x}\,\overline{x}^T)$$

where $\overline{x} = \mathbf{1}^T X/n$ is the row vector of means of the variables. Then the sample variance of a linear combination xa of a row vector x is $a^T \Sigma a$ and this is

---

[1] On UNIX *and* on Windows: a Windows port of XGobi is available at the sites on page 471.

[2] A divisor of $n - 1$ is more conventional, but `princomp` calls `cov.wt`, which uses n.

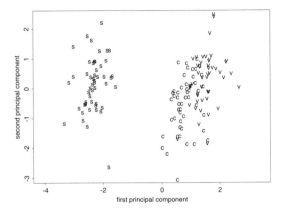

Figure 11.1: First two principal components for the log-transformed `iris` data.

to be maximized (or minimized) subject to $\|a\|^2 = a^T a = 1$. Since Σ is a non-negative definite matrix, it has an eigendecomposition

$$\Sigma = C^T \Lambda C$$

where Λ is a diagonal matrix of (non-negative) eigenvalues in decreasing order. Let $b = Ca$, which has the same length as a (since C is orthogonal). The problem is then equivalent to maximizing $b^T \Lambda b = \sum \lambda_i b_i^2$ subject to $\sum b_i^2 = 1$. Clearly the variance is maximized by taking b to be the first unit vector, or equivalently taking a to be the column eigenvector corresponding to the largest eigenvalue of Σ. Taking subsequent eigenvectors gives combinations with as large as possible variance which are uncorrelated with those that have been taken earlier. The ith principal component is then the ith linear combination picked by this procedure. (It is only determined up to a change of sign; you may get different signs in different implementations of S-PLUS.)

Another point of view is to seek new variables y_j which are rotations of the old variables to explain best the variation in the dataset. Clearly these new variables should be taken to be the principal components, in order. Suppose we use the first k principal components. Then the subspace they span contains the 'best' k-dimensional view of the data. It has both a maximal covariance matrix (both in trace and determinant) and best approximates the original points in the sense of minimizing the sum of squared distance from the points to their projections. The first few principal components are often useful to reveal structure in the data. The principal components corresponding to the smallest eigenvalues are the most nearly constant combinations of the variables, and can also be of interest.

Note that the principal components depend on the scaling of the original variables, and this will be undesirable except perhaps if (as in the `iris` data) they are in comparable units. (Even in this case, correlations would often be used.) Otherwise it is conventional to take the principal components of the *correlation* matrix, implicitly rescaling all the variables to have unit sample variance.

The function `princomp` computes principal components. The argument `cor` controls whether the covariance or correlation matrix is used (via rescaling the variables).

```
> ir.pca <- princomp(log(ir), cor=T)
> ir.pca
Standard deviations:
 Comp. 1 Comp. 2 Comp. 3 Comp. 4
   1.7125 0.95238  0.3647 0.16568
   . . . .
> summary(ir.pca)
Importance of components:
                        Comp. 1 Comp. 2  Comp. 3    Comp. 4
      Standard deviation 1.71246 0.95238 0.364703 0.1656840
Proportion of Variance 0.73313 0.22676 0.033252 0.0068628
 Cumulative Proportion 0.73313 0.95989 0.993137 1.0000000
> plot(ir.pca)
> loadings(ir.pca)
          Comp. 1 Comp. 2 Comp. 3 Comp. 4
Sepal L.   0.504   0.455   0.709   0.191
Sepal W.  -0.302   0.889  -0.331
Petal L.   0.577          -0.219  -0.786
Petal W.   0.567          -0.583   0.580
> ir.pc <- predict(ir.pca)
> eqscplot(ir.pc[,1:2], type="n",
      xlab = "first principal component",
      ylab = "second principal component")
> text(ir.pc[,1:2], labels = as.character(ir.species),
      col = 3+codes(ir.species))
```

In the terminology of this function, the *loadings* are columns giving the linear combinations a for each principal component, and the *scores* are the data on the principal components. The plot shows the `screeplot`, a barplot of the variances of the principal components labelled by $\sum_{i=1}^{j} \lambda_i/\text{trace}(\Sigma)$. The result of `loadings` is rather deceptive, as small entries are suppressed in printing but will be insignificant only if the correlation matrix is used, and that is *not* the default. The `predict` method rotates to the principal components.

As well as a data matrix x, the function `princomp` can accept data via a model formula with an empty left-hand side or as a variance or correlation matrix specified by argument `covlist`, of the form output by `cov.wt` and `cov.rob`. Using the latter is one way to robustify principal component analysis.

Figure 11.1 shows the first two principal components for the `iris` data based on the covariance matrix, revealing the group structure if it had not already been known. *A warning:* principal component analysis will reveal the gross features of the data, which may already be known, and is often best applied to residuals after the known structure has been removed.

There are two books devoted solely to principal components, Jackson (1991) and Jolliffe (1986), which we think overstates its value as a technique.

Distance methods

This is a class of methods based on representing the cases in a low-dimensional Euclidean space so that their proximity reflects the similarity of their variables. To do so we have to produce a measure of similarity. The S function `dist` uses one of four distance measures between the points in the p-dimensional space of variables; the default is Euclidean distance. Distances are often called *dissimilarities*. Jardine & Sibson (1971) discuss several families of similarity and dissimilarity measures. For categorical variables most dissimilarities are measures of agreement. The *simple matching coefficient* is the proportion of categorical variables on which the cases differ. The *Jaccard coefficient* applies to categorical variables with a preferred level. It is the proportion of such variables with one of the cases at the preferred level in which the cases differ. The `binary` method of `dist` is of this family, being the Jaccard coefficient if all non-zero levels are preferred. Applied to logical variables on two cases it gives the proportion of variables in which only one is true among those which are true on at least one case. There are many variants of these coefficients; Kaufman & Rousseeuw (1990, §2.5) provide a readable summary and recommendations.The function `daisy` in library `cluster` provides a more general way to compute dissimilarity matrices. The main extension is to variables that are not on interval scale, for example, ordinal, log-ratio and asymmetric binary variables.

The most obvious of the distance methods is *multidimensional scaling*, which seeks a configuration in $\mathbb{R}^d$ such that distances between the points best match (in a sense to be defined) those of the distance matrix. Only the classical or metric form of multidimensional scaling is implemented in S-PLUS, which is also known as *principal coordinate analysis*. For the `iris` data we can use:

```
ir.scal <- cmdscale(dist(ir), k = 2, eig = T)
ir.scal$points[, 2] <- -ir.scal$points[, 2]
eqscplot(ir.scal$points, type="n")
text(ir.scal$points, labels = as.character(ir.species),
     col = 3+codes(ir.species), cex = 0.8)
```

where care is taken to ensure correct scaling of the axes (see the top left plot of Figure 11.2). Note that a configuration can be determined only up to translation, rotation and reflection, since Euclidean distance is invariant under the group of rigid motions and reflections. (We chose to reflect this plot to match later ones.) An idea of how good the fit is can be obtained by calculating a measure of 'stress':

```
> distp <- dist(ir)
> dist2 <- dist(ir.scal$points)
> sum((distp - dist2)^2)/sum(distp^2)
[1] 0.001747
```

which shows the fit is good. Using classical multidimensional scaling with a Euclidean distance as here is precisely equivalent to plotting the first k principal components (without rescaling to correlations).

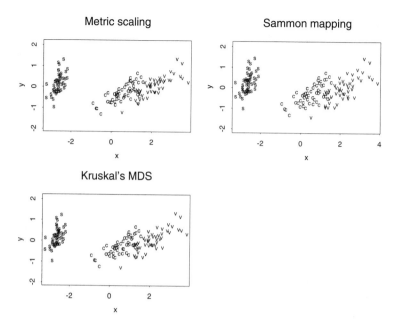

Figure 11.2: Distance-based representations of the `iris` data. The top left plot is by multidimensional scaling, the top right by Sammon's non-linear mapping, the bottom left by Kruskal's isotonic multidimensional scaling. Note that each is defined up to shifts, rotations and reflections.

A non-metric form of multidimensional scaling is Sammon's (1969) non-linear mapping, which given a distance d on n points constructs a k-dimensional configuration with distances $\tilde{d}$ to minimize a weighted 'stress'

$$E(d, \tilde{d}) = \frac{1}{\sum_{i \neq j} d_{ij}} \sum_{i \neq j} \frac{(d_{ij} - \tilde{d}_{ij})^2}{d_{ij}}$$

by an iterative algorithm implemented in our function `sammon`. We have to drop duplicate observations to make sense of $E(d, \tilde{d})$; running `sammon` will report which observations are duplicates.[3]

```
ir.sam <- sammon(dist(ir[-143,]))
eqscplot(ir.sam$points, type="n")
text(ir.sam$points, labels = as.character(ir.species[-143]),
     col = 3+codes(ir.species), cex = 0.8)
```

A more thoroughly non-metric version of multidimensional scaling goes back to Kruskal and Shepard in the 1960s (see Cox & Cox, 1994 and Ripley, 1996.

---

[3] In S we would use `(1:150)[duplicated(do.call("paste", data.frame(ir)))]`.

Figure 11.3: Isotonic multidimensional scaling representation of the `fgl` data. The groups are plotted by the initial letter, except `F` for window float glass, and `N` for window non-float glass. Small dissimilarities correspond to small distances on the plot and conversely.

The idea is to choose a configuration to minimize

$$STRESS^2 = \frac{\sum_{i \neq j}\left[\theta(d_{ij}) - \tilde{d}_{ij}\right]^2}{\sum_{i \neq j} \tilde{d}_{ij}^2}$$

over both the configuration of points and an increasing function θ. Now the location, rotation, reflection and scale of the configuration are all indeterminate. This is implemented in function `isoMDS` which we can use by

```
ir.iso <- isoMDS(dist(ir[-143,]))
eqscplot(ir.iso$points, type="n")
text(ir.iso$points, labels = as.character(ir.species[-143]),
        col = 3+codes(ir.species), cex = 0.8)
```

The optimization task is difficult and this can be quite slow.

These representations can be quite different in examples such as our dataset `fgl` that do not project well into a small number of dimensions; Figure 11.3 shows a non-metric MDS plot. (We omit one of an identical pair of fragments.)

```
fgl.iso <- isoMDS(dist(as.matrix(fgl[-40, -10])))
eqscplot(fgl.iso$points, type="n", xlab="", ylab="")
# either
for(i in seq(along=levels(fgl$type))) {
   set <- fgl$type[-40] == levels(fgl$type)[i]
   points(fgl.iso$points[set,], pch=18, cex=0.6, col=2+i)}
key(text=list(levels(fgl$type), col=3:8))
# or
text(fgl.iso$points, labels = c("F", "N", "V", "C", "T", "H")
     [fgl$type[-40]], cex=0.6)
```

Biplots

The biplot (Gabriel, 1971) is another method to represent both the cases and variables. We suppose that X has been centred to remove column means. The biplot represents X by two sets of vectors of dimensions n and p producing a rank-2

approximation to X. The best (in the sense of least squares) such approximation is given by replacing Λ in the singular value decomposition of X by D, a
diagonal matrix setting $\lambda_3, \ldots$ to zero, so

$$X \approx \widetilde{X} = [\boldsymbol{u}_1\, \boldsymbol{u}_2] \begin{bmatrix} \lambda_1 & 0 \\ 0 & \lambda_2 \end{bmatrix} \begin{bmatrix} \boldsymbol{v}_1^T \\ \boldsymbol{v}_2^T \end{bmatrix} = GH^T$$

where the diagonal scaling factors can be absorbed into G and H in a number
of ways. For example, we could take

$$G = n^{a/2}\, [\boldsymbol{u}_1\, \boldsymbol{u}_2] \begin{bmatrix} \lambda_1 & 0 \\ 0 & \lambda_2 \end{bmatrix}^{1-\lambda}, \qquad H = n^{-a/2}\, [\boldsymbol{v}_1\, \boldsymbol{v}_2] \begin{bmatrix} \lambda_1 & 0 \\ 0 & \lambda_2 \end{bmatrix}^{\lambda}$$

The biplot then consists of plotting the $n + p$ two-dimensional vectors that form
the rows of G and H. The interpretation is based on inner products between
vectors from the two sets, which give the elements of $\widetilde{X}$. For $\lambda = a = 0$ this is
just a plot of the first two principal components and the projections of the variable
axes.

The most popular choice is $\lambda = a = 1$ (which Gabriel, 1971, calls the *principal component biplot*). Then G contains the first two principal components
scaled to unit variance, so the Euclidean distances between the rows of G represent the Mahalanobis distances (page 347) between the observations and the inner
products between the rows of H represent the covariances between the (possibly
scaled) variables (Jolliffe, 1986, pp. 77–8); thus the lengths of the vectors represent the standard deviations.

Figure 11.4 shows a biplot with $\lambda = 1$, obtained by[4]

```
library(MASS, first=T)        # enhanced biplot.princomp
state <- state.x77[,2:7]; row.names(state) <- state.abb
biplot(princomp(state, cor=T), pc.biplot=T, cex = 0.7, ex=0.8)
```

We specified a rescaling of the original variables to unit variance. (There are
additional arguments `scale` which specifies λ and `expand` which specifies a
scaling of the rows of H relative to the rows of G, both of which default to 1.)

Gower & Hand (1996) in a book-length discussion of biplots criticize conventional plots such as Figure 11.4. In particular they point out that the axis scales
are not at all helpful. Notice the two sets of scales. That on the lower and left
axes refers to the values of the rows of G. The upper/right scale is for the values
of the rows of H which are shown as arrows.

11.2   Cluster analysis

Cluster analysis is concerned with discovering groupings among the cases of our
n by p matrix. Two general references are Gordon (1981) and Hartigan (1975).

---

[4] An enhanced version of `biplot.princomp` from our library is used.

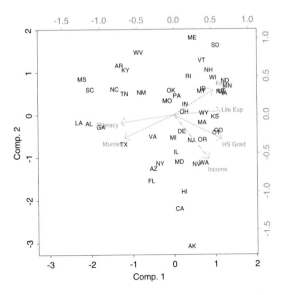

Figure 11.4: Principal component biplot of the part of the `state.x77` data. Distances between states represent Mahalanobis distance, and inner products between variables represent correlations. (The arrows extend 80% of the way along the variable's vector.)

Almost all methods are based on a measure of the similarity or dissimilarity between cases. A *dissimilarity coefficient* d is symmetric ($d(A, B) = d(B, A)$), non-negative and $d(A, A)$ is zero. A similarity coefficient has the scale reversed. Dissimilarities may be *metric*

$$d(A, C) \leqslant d(A, B) + d(B, C)$$

or *ultrametric*

$$d(A, B) \leqslant \max\bigl(d(A, C), d(B, C)\bigr)$$

but need not be either. We have already seen several dissimilarities calculated by `dist` and `daisy`.

Ultrametric dissimilarities have the appealing property that they can be represented by a *dendrogram* such as that shown in Figure 11.5, in which the dissimilarity between two cases can be read from the height at which they join a single group. Hierarchical clustering methods can be thought of as approximating a dissimilarity by an ultrametric dissimilarity. Jardine & Sibson argue that one method, single-link clustering, uniquely has all the desirable properties of a clustering method. This measures distances between clusters by the dissimilarity of the closest pair, and agglomerates by adding the shortest possible link (that is, joining the two closest clusters). Other authors disagree, and Kaufman & Rousseeuw (1990, §5.2) give a different set of desirable properties leading uniquely to their preferred method, which views the dissimilarity between clusters as the average of the dissimilarities between members of those clusters. An-

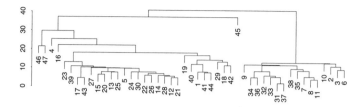

Figure 11.5: Dendrogram for the socio-economic data on Swiss provinces computed by single-link clustering.

other popular method is complete-linkage, which views the dissimilarity between clusters as the maximum of the dissimilarities between members.

The S function `hclust` implements these three metrics, selected by its `method` argument which takes values `compact` (the default, for complete-linkage), `average` and `connected` (for single-linkage).

The S dataset `swiss.x` gives five measures of socio-economic data on Swiss provinces about 1888, given by Mosteller & Tukey (1977, pp. 549–551). The data are percentages, so Euclidean distance is a reasonable choice. We use single-link clustering:

```
h <- hclust(dist(swiss.x), method="connected")
plclust(h)
cutree(h, 3)
plclust( clorder(h, cutree(h, 3) ))
```

The first plot suggests three main clusters, and the remaining code re-orders the dendrogram to display (see Figure 11.5) those clusters more clearly. Note that there are two main groups, with the point 45 well separated from them.

K-means

The K-means clustering algorithm (MacQueen, 1967; Hartigan, 1975; Hartigan & Wong, 1979) chooses a prespecified number of cluster centres to minimize the within-class sum of squares from those centres. As such it is most appropriate to continuous variables, suitably scaled. The algorithm needs a starting point, so we choose the means of the clusters identified by group-average clustering. The clusters *are* altered (cluster 3 contained just point 45), and are shown in principal-component space in Figure 11.6. (Its standard deviations show that a two-dimensional representation is reasonable.)

```
h <- hclust(dist(swiss.x), method="average")
initial <- tapply(swiss.x, list(rep(cutree(h, 3),
   ncol(swiss.x)), col(swiss.x)), mean)
dimnames(initial) <- list(NULL, dimnames(swiss.x)[[2]])
km <- kmeans(swiss.x, initial)
swiss.pca <- princomp(swiss.x)
swiss.pca
```

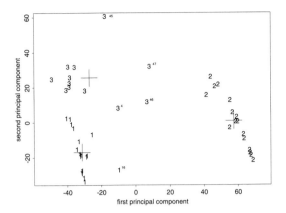

Figure 11.6: The Swiss provinces data plotted on its first two principal components. The labels are the groups assigned by K-means; the crosses denote the group means. Five points are labelled with smaller symbols.

```
Standard deviations:
  Comp. 1 Comp. 2 Comp. 3 Comp. 4 Comp. 5
   42.903   21.202   7.588   3.6879   2.7211
   ....
swiss.px <- predict(swiss.pca)
dimnames(km$centers)[[2]] <- dimnames(swiss.x)[[2]]
swiss.centers <- predict(swiss.pca, km$centers)
eqscplot(swiss.px[, 1:2], type="n",
   xlab="first principal component",
   ylab="second principal component")
text(swiss.px[,1:2], labels = km$cluster)
points(swiss.centers[,1:2], pch=3, cex=3)
identify(swiss.px[, 1:2], cex=0.5)
```

Model-based clustering

S-PLUS has functions for 'model-based' clustering (Banfield & Raftery, 1993) implemented by the functions `mclust`, `mclass` and `mreloc`. For Figure 11.7 we used

```
h <- mclust(swiss.x, method = "S*")$tree
plclust( clorder(h, cutree(h, 3) ))
```

Note that this works with a data matrix and not with a dissimilarity matrix.

The idea of 'model-based' clustering is that the data are independent samples from a series of group populations, but the group labels have been lost. If we knew that the vector γ gave the group labels, and each group had class-conditional pdf $f_i(x; \theta)$, then the likelihood would be

$$\prod_{i=1}^{n} f_{\gamma_i}(x_i; \theta) \qquad (11.1)$$

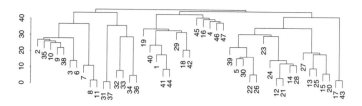

Figure 11.7: Dendrogram for the socio-economic data on Swiss provinces computed by "model-based" clustering.

Since the labels are unknown, these are regarded as parameters in (11.1), and the likelihood maximized over (θ, γ).

Choosing the class-conditional pdfs to be multivariate normal with different means but a common covariance matrix $\Sigma = \sigma^2 I$ leads to the criterion of minimizing the sum of squares to the cluster centre, that is, K-means. The other options available are based on multivariate normals with other constraints on the covariance matrices. For example, the default method S* allows clusters of different sizes and orientations but the same prespecified 'shape' (the ratio of axes of the ellipsoid), and S is similar but constrains to equal size.

The code returns an 'approximate weight of evidence' for the number of clusters, which in this case suggests two or three clusters. A further option is 'noise' to allow some of the points to come from a homogeneous Poisson process rather than from one of the cluster groups. (It may help to think of this as a $(k+1)$st cluster with a very diffuse distribution.)

Note that the hierarchical clustering given by mclust only optimizes the fit criterion in a very crude way. Once the number of clusters is chosen, mreloc can be used to optimize further, although in our example no change occurs. If we allow 'noise' two points are reallocated:

```
h <- mclust(swiss.x, method = "S*", noise=T)
hclass <- mclass(h, 3)
hclass$class
 [1]  1 2 2 4 1 2 2 2 2 2 2 1 1 1 1 1 1 1 1 1 1 1 1 1 1 1 1 1
[29]  1 1 2 2 2 2 2 2 2 2 1 1 1 1 1 1 4 4 4
mreloc(hclass, swiss.x, method = "S*", noise=T)
 [1]  1 2 2 4 1 2 2 2 2 2 2 1 1 1 1 4 1 1 1 1 1 1 1 1 1 1 1 1
[29]  1 1 2 2 2 2 2 2 2 2 1 1 1 4 1 1 4 4 4
```

Note that as this is a random algorithm the results are not repeatable, and that the class number is the lowest-numbered object in the cluster.

Library cluster

The library cluster provides an S interface to the FORTRAN clustering routines described in Kaufman & Rousseeuw (1990). These rejoice in the acronymic names of agnes, clara, daisy, diana, fanny, mona and pam, and all have print and summary methods.

The functions `pam`, `clara` and `fanny` are all partitioning methods like K-means; they are described in Kaufman & Rousseeuw (1990) and Ripley (1996, §9.3). Both `pam` and `clara` use the k-medoids criterion of Vinod (1969), which differs from K-means in requiring the cluster centre to be a data point and in using unsquared distances. As this criterion only uses distances between data points, it can also be applied to general dissimilarities. The function `pam` allows a general dissimilarity whereas `clara` works on Euclidean or Manhattan distance between rows of a data matrix, and uses a faster optimization procedure for large datasets.

```
> library(cluster)            # needed in S-PLUS 3.4 and 4.0 only
> swiss.pam <- pam(swiss.px, 3)
> summary(swiss.pam)
Medoids:
       Comp. 1   Comp. 2 Comp. 3 Comp. 4  Comp. 5
[1,] -29.716  18.22162  1.4265 -1.3206  0.95201
[2,]  58.609   0.56211  2.2320 -4.1778  4.22828
[3,] -28.844 -19.54901  3.1506  2.3870 -2.46842
Clustering vector:
 [1] 1 2 2 1 3 2 2 2 2 2 2 3 3 3 3 3 1 1 1 3 3 3 3 3 3 3 3 3
[29] 1 3 2 2 2 2 2 2 2 2 1 1 1 1 1 1 1 1 1
    . . . .
> eqscplot(swiss.px[, 1:2], type="n",
    xlab="first principal component",
    ylab="second principal component")
> text(swiss.px[,1:2], labels = swiss.pam$clustering)
> points(swiss.pam$medoid[,1:2], pch=3, cex=3)
```

The function `fanny` implements a 'fuzzy' version of the k-medoids criterion. Rather than point i having a membership of just one cluster v, its membership is partitioned among clusters as positive weights u_{iv} summing to one. The criterion then is

$$\min_{(u_{iv})} \sum_v \frac{\sum_{i,j} u_{iv}^2 u_{jv}^2 d_{ij}}{2 \sum_i u_{iv}^2}.$$

For our running example we find

```
> fanny(swiss.px, 3)
 iterations objective
        16    354.01
Membership coefficients:
         [,1]     [,2]     [,3]
[1,] 0.725016 0.075485 0.199499
[2,] 0.189978 0.643928 0.166094
[3,] 0.191282 0.643596 0.165123
    . . . .
Closest hard clustering:
 [1] 1 2 2 1 3 2 2 2 2 2 2 3 3 3 3 3 1 1 1 3 3 3 3 3 3 3 3 3
[29] 1 3 2 2 2 2 2 2 2 2 1 1 1 1 1 1 1 1 1
```

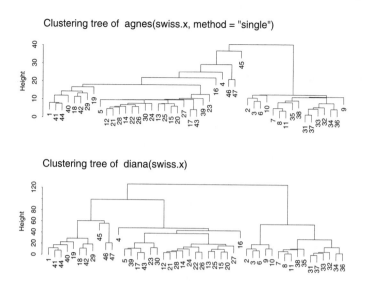

Figure 11.8: Hierarchical clustering of `swiss.x` using library `cluster`.

The remaining methods, `agnes`, `diana` and `mona`, are hierarchical cluster-ing methods; `mona` is a specialized method for binary observations only. The function `agnes` is very similar to `hclust` and also implements `average` (its default), `single` (which means `connected`) and `compact` methods of agglom-erative clustering. Thus we can use

```
pltree(agnes(swiss.x, method="single"))
```

to obtain a dendrogram (Figure 11.8) similar to Figure 11.5. The generic function `pltree` converts the output of `agnes` into a call to `plclust`.

Function `diana` performs *divisive* clustering, in which the clusters are repeat-edly subdivided rather than joined, using the algorithm of Macnaughton-Smith *et al.* (1964). Divisive clustering is an attractive option when a grouping into a few large clusters is of interest. The lower panel of Figure 11.8 was produced by `pltree(diana(swiss.x))`.

11.3   Correspondence analysis

Correspondence analysis is applied to two-way tables of counts. Suppose we have an $r \times c$ table N of counts. For example, consider Fisher's (1940) example on colours of eyes and hair of people in Caithness, Scotland:

	fair	red	medium	dark	black
blue	326	38	241	110	3
light	688	116	584	188	4
medium	343	84	909	412	26
dark	98	48	403	681	85

in our dataset `caith`. Correspondence analysis seeks 'scores' f and g for the rows and columns which are maximally correlated. Clearly the maximum correlation is one, attained by constant scores, so we seek the largest non-trivial solution. Let R and C be matrices of the group indicators of the rows and columns, so $R^T C = N$. Consider the singular value decomposition of their correlation matrix

$$X_{ij} = \frac{n_{ij}/n - (n_{i\cdot}/n)(n_{\cdot j}/n)}{\sqrt{(n_{i\cdot}/n)(n_{\cdot j}/n)}} = \frac{n_{ij} - n\, r_i\, c_j}{n\sqrt{r_i\, c_j}}$$

where $r_i = n_{i\cdot}/n$ and $c_j = n_{\cdot j}/n$ are the proportions in each row and column. Let D_r and D_c be the diagonal matrices of r and c. Correspondence analysis corresponds to selecting the first singular value and left and right singular vectors of X_{ij} and rescaling by $D_r^{-1/2}$ and $D_c^{-1/2}$, respectively. This is done by our function `corresp`:

```
> corresp(caith)
First canonical correlation: 0.44637

Row scores:
      blue    light   medium    dark
  -0.89679 -0.98732 0.075306  1.5743

Column scores:
     fair      red   medium    dark   black
  -1.2187 -0.52258 -0.094147  1.3189  2.4518
```

Can we make use of the subsequent singular values? In what Gower & Hand call 'classical CA' we consider $A = D_r^{-1/2}U\Lambda$ and $B = D_c^{-1/2}V\Lambda$. Then the first columns of A and B are what we have termed the row and column scores *scaled by* ρ, the first canonical correlation. More generally, we can see distances between the rows of A as approximating the distances between the row profiles (rows rescaled to unit sum) of the table N, and analogously for the rows of B and the column profiles.

Classical CA plots the first two columns of A and B on the same figure. This is a form of a biplot and is obtained with our software by plotting a correspondence analysis object with `nf` $\geqslant 2$ or as the default for the method `biplot.correspondence`. This is sometimes known as a 'symmetric' plot. Other authors (for example, Greenacre, 1992) advocate 'asymmetric' plots. The asymmetric plot for the rows is a plot of the first two columns of A with the column labels plotted at the first two columns of $\Gamma = D_c^{-1/2}V$; the corresponding plot for the columns has columns plotted at B and row labels at $\Phi = D_r^{-1/2}U$. The most direct interpretation for the row plot is that

$$A = D_r^{-1}N\Gamma$$

so A is a plot of the *row profiles* (the rows normalized to sum to one) as convex combinations of the column vertices given by Γ.

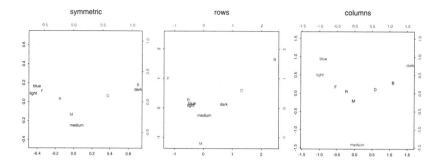

Figure 11.9: Three variants of correspondence analysis plots from Fisher's data on people in Caithness: (left) 'symmetric", (middle) 'row asymmetric' and (right) 'column asymmetric'.

By default `corresp` only retains one-dimensional row and column scores; then `plot.corresp` plots these scores and indicates the size of the entries in the table by the area of circles. The two-dimensional forms of the plot are shown in Figure 11.9 for Fisher's data on people from Caithness. These were produced by

```
caith <- caith
dimnames(caith)[[2]] <- c("F", "R", "M", "D", "B")
par(mfcol=c(1,3))
plot(corresp(caith, nf=2)); title("symmetric")
plot(corresp(caith, nf=2), type="rows"); title("rows")
plot(corresp(caith, nf=2), type="col"); title("columns")
```

Note that the symmetric plot (left) has the row points from the asymmetric row plot (middle) and the column points from the asymmetric column plot (right) superimposed on the same plot (but with different scales).

11.4   Discriminant analysis

Now suppose that we have a set of g classes, and for each case we know the class (assumed correctly). We can then use the class information to help reveal the structure. Let W denote the within-class covariance matrix, that is the covariance matrix of the variables centred on the class mean, and B denote the between-classes covariance matrix, that is, of the predictions by the class means. Let M be the $g \times p$ matrix of class means, and G be the $n \times g$ matrix of class indicator variables (so $g_{ij} = 1$ if and only if case i is assigned to class j). Then the predictions are GM. Let $\bar{x}$ be the means of the variables over the whole sample. Then the sample covariance matrices are

$$W = \frac{(X - GM)^T (X - GM)}{n - g}, \qquad B = \frac{(GM - 1\bar{x})^T (GM - 1\bar{x})}{g - 1} \quad (11.2)$$

Note that B has rank at most $\min(p, g - 1)$.

Fisher (1936) introduced a linear discriminant analysis seeking a linear combination xa of the variables which has a maximal ratio of the separation of the class means to the within-class variance, that is, maximizing the ratio $a^T Ba/a^T Wa$. To compute this, choose a scaling xS of the variables so that they have the identity as their within-group correlation matrix. (One such scaling is take the principal components with respect to W, and rescale each to unit variance.) On the rescaled variables the problem is to maximize $a^T Ba$ subject to $\|a\| = 1$, and as we saw before, this is solved by taking a to be the eigenvector of B corresponding to the largest eigenvalue. (Note that rescaling changes B to $S^T BS$.) The linear combination a is unique up to a change of sign (unless there are multiple eigenvalues, an event of probability zero). The exact multiple of a returned by a program will depend on its definition of the within-class variance matrix. We use the conventional divisor of $n - g$, but divisors of n and $n - 1$ have been used.

As for principal components, we can take further linear components corresponding to the next largest eigenvalues. There will be at most $r = \min(p, g-1)$ positive eigenvalues. Note that the eigenvalues are the proportions of the between-classes variance explained by the linear combinations, which may help us to choose how many to use. The corresponding transformed variables are called the *linear discriminants* or *canonical variates*. It is often useful to plot the data on the first few linear discriminants (Figure 11.10). Since the within-group covariances should be the identity, we chose an equal-scaled plot. (Using plot(ir.lda) will give this plot without the colours.) The linear discriminants are conventionally centered to have mean zero on dataset.

```
> ir.lda <- lda(log(ir), ir.species)
> ir.lda
Prior probabilities of groups:
      c       s       v
 0.33333 0.33333 0.33333

Group means:
   Sepal L. Sepal W. Petal L. Petal W.
 c   1.7773   1.0123  1.44293   0.27093
 s   1.6082   1.2259  0.37276  -1.48465
 v   1.8807   1.0842  1.70943   0.69675

Coefficients of linear discriminants:
               LD1       LD2
 Sepal L.   3.7798   4.27690
 Sepal W.   3.9405   6.59422
 Petal L.  -9.0240   0.30952
 Petal W.  -1.5328  -0.13605

Proportion of trace:
    LD1     LD2
 0.9965  0.0035
> ir.ld <- predict(ir.lda, dimen=2)$x
> eqscplot(ir.ld, type="n", xlab = "first linear discriminant",
```

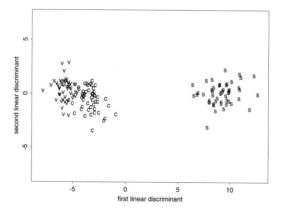

Figure 11.10: The log `iris` data on the first two discriminant axes.

```
      ylab = "second linear discriminant")
> text(ir.ld, labels = as.character(ir.species[-143]),
         col = 3+codes(ir.species), cex = 0.8)
```

This shows that 99.65% of the between-group variance is on the first discriminant axis. Using

```
plot(ir.lda, dimen=1)
plot(ir.lda, type="density", dimen=1)
```

will examine the distributions of the groups on the first linear discriminant.

The approach we have illustrated is the conventional one, following Bryan (1951), but it is not the only one. The definition of B at (11.2) weights the groups by their size in the dataset. Rao (1948) used the unweighted covariance matrix of the group means, and our software uses a covariance matrix weighted by the prior probabilities of the classes if these are specified.

Discrimination for normal populations

An alternative approach to discrimination is *via* probability models. Let π_c denote the prior probabilities of the classes, and $p(x \mid c)$ the densities of distributions of the observations for each class. Then the posterior distribution of the classes after observing x is

$$p(c \mid x) = \frac{\pi_c p(x \mid c)}{p(x)} \propto \pi_c p(x \mid c)$$

and it is fairly simple to show that the allocation rule which makes the smallest expected number of errors chooses the class with maximal $p(c \mid x)$; this is known as the *Bayes rule*. (We consider a more general version in Section 11.5.)

Now suppose the distribution for class c is multivariate normal with mean μ_c and covariance Σ_c. Then the Bayes rule minimizes

$$\begin{aligned}
Q_c &= -2 \log p(\boldsymbol{x}\,|\,c) - 2 \log \pi_c \\
&= (\boldsymbol{x} - \boldsymbol{\mu}_c)\Sigma_c^{-1}(\boldsymbol{x} - \boldsymbol{\mu}_c)^T + \log |\Sigma_c| - 2 \log \pi_c \qquad (11.3)
\end{aligned}$$

The first term of (11.3) is the squared *Mahalanobis distance* to the class centre, and can be calculated by the S function `mahalanobis`. The difference between the Q_c for two classes is a quadratic function of $\boldsymbol{x}$, so the method is known as *quadratic discriminant analysis* and the boundaries of the decision regions are quadratic surfaces in $\boldsymbol{x}$ space. This is implemented by our function `qda`.

Further suppose that the classes have a common covariance matrix Σ. Differences in the Q_c are then *linear* functions of $\boldsymbol{x}$, and we can maximize $-Q_c/2$ or

$$L_c = \boldsymbol{x}\Sigma^{-1}\boldsymbol{\mu}_c^T - \boldsymbol{\mu}_c\Sigma^{-1}\boldsymbol{\mu}_c^T/2 + \log \pi_c \qquad (11.4)$$

To use (11.3) or (11.4) we have to estimate $\boldsymbol{\mu}_c$ and Σ_c or Σ. The obvious estimates are used, the sample mean and covariance matrix within each class, and W for Σ.

How does this relate to Fisher's linear discrimination? The latter gives new variables, the linear discriminants, with unit within-class sample variance, and the differences between the group means lie entirely in the first r variables. Thus on these variables the Mahalanobis distance (with respect to $\widehat{\Sigma} = W$) is just

$$\|\boldsymbol{x} - \boldsymbol{\mu}_c\|^2$$

and only the first r components of the vector depend on c. Similarly, on these variables

$$L_c = \boldsymbol{x}\boldsymbol{\mu}_c^T - \|\boldsymbol{\mu}_c\|^2/2 + \log \pi_c$$

and we can work in r dimensions. If there are just two classes, there is a single linear discriminant, and

$$L_2 - L_1 = \boldsymbol{x}(\boldsymbol{\mu}_2 - \boldsymbol{\mu}_1)^T + \text{const}$$

This is an affine function of the linear discriminant, which has coefficient $(\boldsymbol{\mu}_2 - \boldsymbol{\mu}_1)^T$ rescaled to unit length.

We can use linear discriminant analysis on more than two classes, and illustrate this with the forensic glass dataset `fgl`.

```
fgl.lda <- lda(type ~ ., fgl)
fgl.ld <- predict(fgl.lda, dimen=2)$x
eqscplot(fgl.ld, type="n", xlab="LD2", ylab="LD1")
# either
for(i in seq(along=levels(fgl$type))) {
  set <- fgl$type[-40] == levels(fgl$type)[i]
  points(fgl.ld[set,], pch=18, cex=0.6, col=2+i)}
key(text=list(levels(fgl$type), col=3:8))
# or
text(fgl.ld, labels = c("F", "N", "V", "C", "T", "H")
  [fgl$type[-40]], cex=0.6)
```

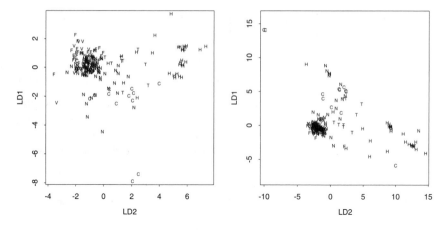

Figure 11.11: The `fgl` data on the first two discriminant axes. The right-hand plot used robust estimation of the common covariance matrix.

Robust estimation of multivariate location and scale

We may wish to consider more robust estimates of W (but not B). Somewhat counter-intuitively, it does not suffice to apply a robust location estimator to each component of a multivariate mean (Rousseeuw & Leroy, 1987, p. 250), and it is easier to consider the estimation of mean and variance simultaneously.

Multivariate variances are very sensitive to outliers. Two methods for robust covariance estimation are available via our function `cov.rob` and the S-PLUS functions `cov.mve` and `cov.mcd`[5] (Rousseeuw, 1984; Rousseeuw & Leroy, 1987). Suppose there are n observations of p variables. The *minimum volume ellipsoid* method seeks an ellipsoid containing $h = \lfloor (n+p+1)/2 \rfloor$ points which is of minimum volume, and the *minimum covariance determinant* method seeks h points whose covariance has minimum determinant (so the conventional confidence ellipsoid for the mean of those points has minimum volume). MCD is to be preferred for its higher statistical efficiency. Our function `cov.rob` implements both.

The search for an MVE or MCD provides h points whose mean and variance matrix (adjusted for selection) give an initial estimate. This is refined by selecting those points whose Mahalanobis distance from the initial mean using the initial covariance is not too large (specifically within the 97.5% point under normality), and returning their mean and variance matrix).

An alternative approach is to extend the idea of M-estimation to this setting, fitting a multivariate t_ν distribution for a small number ν of degrees of freedom. This is implemented in our function `cov.trob`; the theory behind the algorithm used is given in Kent, Tyler & Vardi (1994) and Ripley (1996). Normally `cov.trob` is faster than `cov.rob`, but it lacks the latter's extreme resistance.

---

[5] In S-PLUS 4.0 and later.

Our function `lda` has an argument `method = "mve"` to use the minimum volume ellipsoid estimate (but without robust estimation of the group centres) or the multivariate t_ν distribution by setting `method="t"`. This makes a considerable difference for the `fgl` forensic glass data, as Figure 11.11 shows. We use the default $\nu = 5$.

```
fgl.rlda <- lda(type ~ ., fgl, method="t")
fgl.rld <- predict(fgl.rlda, dimen=2)$x
eqscplot(fgl.rld, type="n", xlab="LD2", ylab="LD1")
# either
for(i in seq(along=levels(fgl$type))) {
   set <- fgl$type[-40] == levels(fgl$type)[i]
   points(fgl.rld[set,], pch=18, cex=0.6, col=2+i)}
key(text=list(levels(fgl$type), col=3:8))
# or
text(fgl.rld, labels = c("F", "N", "V", "C", "T", "H")
   [fgl$type[-40]], cex=0.6)
```

Try `method = "mve"` which gives an almost linear plot.

11.5 Classification theory

In the terminology of pattern recognition the given examples together with their classifications are known as the *training set*, and future cases form the *test set*. Our primary measure of success is the error (or misclassification) rate. Note that we would obtain (possibly seriously) biased estimates by re-classifying the training set, but that the error rate on a test set randomly chosen from the whole population will be an unbiased estimator.

It may be helpful to know the type of errors made. A *confusion matrix* gives the number of cases with true class i classified as of class j. In some problems some errors are considered to be worse than others, so we assign costs L_{ij} to allocating a case of class i to class j. Then we will be interested in the average error cost rather than the error rate.

It is fairly easy to show (Ripley, 1996, p. 19) that the average error cost is minimized by the *Bayes rule* which is to allocate to the class c minimizing $\sum_i L_{ic} p(i \mid \boldsymbol{x})$ where $p(i \mid \boldsymbol{x})$ is the posterior distribution of the classes after observing $\boldsymbol{x}$. If the costs of all errors are the same, this rule amounts to choosing the class c with the largest posterior probability $p(c \mid \boldsymbol{x})$. The minimum average cost is known as the *Bayes risk*.

We saw in Section 11.4 how $p(c \mid \boldsymbol{x})$ can be computed for normal populations, and how estimating the Bayes rule with equal error costs leads to linear and quadratic discriminant analysis. As our functions `predict.lda` and `predict.qda` return posterior probabilities, they can also be used for classification with error costs.

The posterior probabilities $p(c \mid \boldsymbol{x})$ may also be estimated directly. For just two classes we can model $p(1 \mid \boldsymbol{x})$ using a logistic regression, fitted by `glm`. For more than two classes we need a multiple logistic model: it may be possible to fit this using a surrogate log-linear Poisson GLM model, but using the `multinom` function in library `nnet` will usually be faster and easier.

Classification trees model the $p(c \mid x)$ directly by a special multiple logistic model, one in which the right-hand side is a single factor specifying which leaf the case will be assigned to by the tree. Again, since the posterior probabilities are given by the `predict` method it is easy to estimate the Bayes rule for unequal error costs.

Predictive and 'plug-in' rules

In the last few paragraphs we skated over an important point. To find the Bayes rule we need to know the posterior probabilities $p(c \mid x)$. Since these are unknown we use an explicit or implicit parametric family $p(c \mid x; \theta)$. In the methods considered so far we act as if $p(c \mid x; \hat{\theta})$ were the actual posterior probabilities, where $\hat{\theta}$ is an estimate computed from the training set $\mathcal{T}$, often by maximizing some appropriate likelihood. This is known as the 'plug-in' rule. However, the 'correct' estimate of $p(c \mid x)$ is (Ripley, 1996, §2.4) to use the *predictive* estimates

$$\tilde{p}(c \mid x) = P(c = c \mid X = x, \mathcal{T}) = \int p(c \mid x; \theta) p(\theta \mid \mathcal{T}) \, d\theta \qquad (11.5)$$

If we are very sure of our estimate $\hat{\theta}$ there will be little difference between $p(c \mid x; \hat{\theta})$ and $\tilde{p}(c \mid x)$; otherwise the predictive estimate will normally be less extreme (not as near 0 or 1). The 'plug-in' estimate ignores the uncertainty in the parameter estimate $\hat{\theta}$ which the predictive estimate takes into account.

It is not often possible to perform the integration in (11.5) analytically, but it *is* possible for linear and quadratic discrimination with appropriate 'vague' priors on θ (Aitchison & Dunsmore, 1975; Geisser, 1993; Ripley, 1996). This estimate is implemented by `method = "predictive"` of the `predict` methods for our functions `lda` and `qda`. Often the differences are small, especially for linear discrimination, *provided* there are enough data for a good estimate of the variance matrices. When there are not, Moran & Murphy (1979) argue that considerable improvement can be obtained by using an unbiased estimator of $\log \acute{p}(x \mid c)$, implemented by the argument `method = "debiased"`.

A simple example: Cushing's syndrome

We illustrate these methods by a small example taken from Aitchison & Dunsmore (1975, Tables 11.1–3) and used for the same purpose by Ripley (1996). The data are on diagnostic tests on patients with Cushing's syndrome, a hypersensitive disorder associated with over-secretion of cortisol by the adrenal gland. This dataset has three recognized types of the syndrome represented as a, b, c. (These encode 'adenoma', 'bilateral hyperplasia' and 'carcinoma', and represent the underlying cause of over-secretion. This can only be determined histopathologically.) The observations are urinary excretion rates (mg/24h) of the

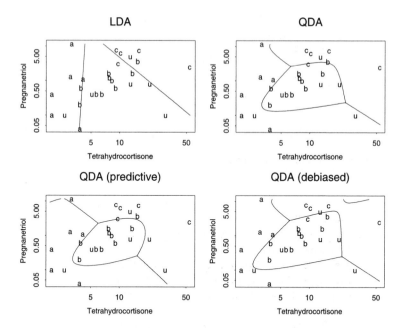

Figure 11.12: Linear and quadratic discriminant analysis applied to the Cushing's syndrome data.

steroid metabolites tetrahydrocortisone and pregnanetriol, and are considered on log scale.

There are six patients of unknown type (marked u), one of whom was later found to be of a fourth type, and another was measured faultily.

Figure 11.12 shows the classifications produced by lda and the various options of quadratic discriminant analysis. This was produced by

```
cush <- log(as.matrix(Cushings[, -3]))
tp <- factor(Cushings$Type[1:21])
cush.lda <- lda(cush[1:21,], tp); predplot(cush.lda, "LDA")
cush.qda <- qda(cush[1:21,], tp); predplot(cush.qda, "QDA")
predplot(cush.qda, "QDA (predictive)", method = "predictive")
predplot(cush.qda, "QDA (debiased)", method = "debiased")
```

(Function predplot is given in the scripts.)

We can contrast these with logistic discrimination performed by

```
library(nnet)
Cf <- data.frame(tp = tp,
    Tetrahydrocortisone = log(Cushings[1:21,1]),
    Pregnanetriol = log(Cushings[1:21,2]) )
cush.multinom <- multinom(tp ~ Tetrahydrocortisone
    + Pregnanetriol, Cf, maxit=250)
xp <- seq(0.6, 4.0, length=100); np <- length(xp)
```

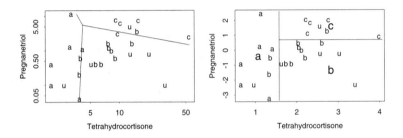

Figure 11.13: Logistic regression and classification trees applied to the Cushing's syndrome data.

```
yp <- seq(-3.25, 2.45, length=100)
cushT <- expand.grid(Tetrahydrocortisone=xp,
    Pregnanetriol=yp)
Z <- predict(cush.multinom, cushT, type="probs")
plot(Cushings[,1], Cushings[,2], log="xy", type="n",
 xlab="Tetrahydrocortisone", ylab = "Pregnanetriol")
for(il in 1:4) {
  set <- Cushings$Type==levels(Cushings$Type)[il]
  text(Cushings[set, 1], Cushings[set, 2],
      labels=as.character(Cushings$Type[set]), col = 2 + il) }
zp <- Z[,3] - pmax(Z[,2], Z[,1])
contour(xp/log(10), yp/log(10), matrix(zp, np),
    add=T, levels=0, labex=0)
zp <- Z[,1] - pmax(Z[,2], Z[,3])
contour(xp/log(10), yp/log(10), matrix(zp, np),
    add=T, levels=0, labex=0)
```

When, as here, the classes have quite different variance matrices, linear and logistic discrimination can give quite different answers (compare Figures 11.12 and 11.13).

For classification trees we can use

```
cush.tr <- tree(tp ~ Tetrahydrocortisone + Pregnanetriol, Cf)
plot(cush[,1], cush[,2], type="n",
    xlab="Tetrahydrocortisone", ylab = "Pregnanetriol")
for(il in 1:4) {
  set <- Cushings$Type==levels(Cushings$Type)[il]
  text(cush[set, 1], cush[set, 2],
      labels= as.character(Cushings$Type[set]), col = 2 + il) }
par(cex=1.5); partition.tree(cush.tr, add=T); par(cex=1)
```

With such a small dataset we make no attempt to refine the size of the tree, shown in Figure 11.13.

11.6   Other classification methods

Neural networks

Neural networks provide a flexible non-linear extension of multiple logistic regression, as we saw in Section 9.4. We can consider them for this example by the following code.[6]

```
library(nnet)
cush <- cush[1:21,]; tpi <- class.ind(tp)
# functions pltnn and plt.bndry given in the scripts
par(mfrow=c(2,2))
pltnn("Size = 2")
set.seed(1); plt.bndry(size=2, col=2)
set.seed(3); plt.bndry(size=2, col=3); plt.bndry(size=2, col=4)

pltnn("Size = 2, lambda = 0.001")
set.seed(1); plt.bndry(size=2, decay=0.001, col=2)
set.seed(2); plt.bndry(size=0, decay=0.001, col=4)

pltnn("Size = 2, lambda = 0.01")
set.seed(1); plt.bndry(size=2, decay=0.01, col=2)
set.seed(2); plt.bndry(size=2, decay=0.01, col=4)

pltnn("Size = 5, 20  lambda = 0.01")
set.seed(2); plt.bndry(size=5, decay=0.01, col=1)
set.seed(2); plt.bndry(size=20, decay=0.01, col=2)
```

The results are shown in Figure 11.14. We see that in all cases there are multiple local maxima of the likelihood.

Once we have a penalty, the choice of the number of hidden units is often not critical (see Figure 11.14). The spirit of the predictive approach is to average the predicted $p(c \mid x)$ over the local maxima. A simple average will often suffice:

```
# functions pltnn and b1 are in the scripts
pltnn("Many local maxima")
Z <- matrix(0, nrow(cushT), ncol(tpi))
for(iter in 1:20) {
    set.seed(iter)
    cush.nn <- nnet(cush, tpi,  skip=T, softmax=T, size=3,
        decay=0.01, maxit=1000, trace=F)
    Z <- Z + predict(cush.nn, cushT)
# In 5.x replace $ by @ in next line.
    cat("final value", format(round(cush.nn$value,3)), "\n")
    b1(predict(cush.nn, cushT), col=2, lwd=0.5)
}
pltnn("Averaged")
b1(Z, lwd=3)
```

Note that there are two quite different types of local maxima occurring here, and some local maxima occur several times (up to convergence tolerances).

---

[6] The colours are set for a Trellis device.

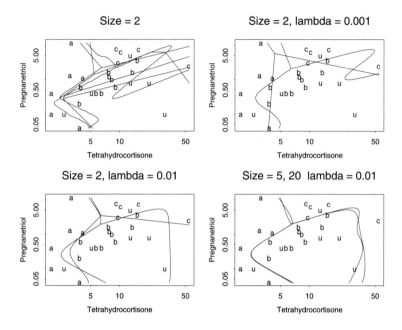

Figure 11.14: Neural networks applied to the Cushing's syndrome data. Each panel shows the fits from two or three local maxima of the (penalized) log-likelihood.

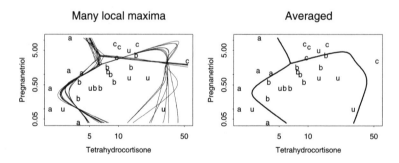

Figure 11.15: Neural networks with three hidden units and $\lambda = 0.01$ applied to the Cushing's syndrome data.

Non-parametric rules

There are a number of non-parametric classifiers based on non-parametric estimates of the class densities or of the log posterior. Library `class` implements the k-nearest neighbour classifier and related methods (Devijver & Kittler, 1982; Ripley, 1996) and learning vector quantization (Kohonen, 1990, 1995; Ripley, 1996). These are all based on finding the k nearest examples in some reference set, and taking a majority vote among the classes of these k examples, or, equivalently, estimating the posterior probabilities $p(c \,|\, x)$ by the proportions of the classes among the k examples.

   The methods differ in their choice of reference set. The k-nearest neighbour
methods use the whole training set or an edited subset. Learning vector quantization is similar to K-means in selecting points in the space other than the training
set examples to summarize the training set, but unlike K-means it takes the classes
of the examples into account.

   These methods almost always measure 'nearest' by Euclidean distance. For
the Cushing's syndrome data we use Euclidean distance on the logged covariates,
rather arbitrarily scaling them equally.

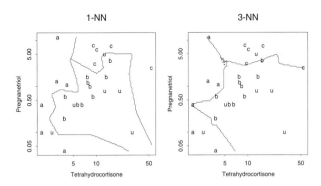

Figure 11.16: k-nearest neighbours applied to the Cushing's syndrome data.

```
library(class)
par(pty="s", mfrow=c(1,2))
plot(Cushings[,1], Cushings[,2], log="xy", type="n",
     xlab = "Tetrahydrocortisone", ylab = "Pregnanetriol",
     main = "1-NN")
for(il in 1:4) {
  set <- Cushings$Type==levels(Cushings$Type)[il]
  text(Cushings[set, 1], Cushings[set, 2],
       as.character(Cushings$Type[set]), col = 2 + il) }
Z <- knn(scale(cush, F, c(3.4, 5.7)),
         scale(cushT, F, c(3.4, 5.7)), tp)
contour(xp/log(10), yp/log(10), matrix(as.numeric(Z=="a"), np),
        add=T, levels=0.5, labex=0)
contour(xp/log(10), yp/log(10), matrix(as.numeric(Z=="c"), np),
        add=T, levels=0.5, labex=0)
plot(Cushings[,1], Cushings[,2], log="xy", type="n",
     xlab="Tetrahydrocortisone", ylab = "Pregnanetriol",
     main = "3-NN")
for(il in 1:4) {
  set <- Cushings$Type==levels(Cushings$Type)[il]
  text(Cushings[set, 1], Cushings[set, 2],
       as.character(Cushings$Type[set]), col = 2 + il) }
Z <- knn(scale(cush, F, c(3.4, 5.7)),
         scale(cushT, F, c(3.4, 5.7)), tp, k=3)
contour(xp/log(10), yp/log(10), matrix(as.numeric(Z=="a"), np),
```

```
             add=T, levels=0.5, labex=0)
    contour(xp/log(10), yp/log(10), matrix(as.numeric(Z=="c"), np),
             add=T, levels=0.5, labex=0)
```

This dataset is too small to try the editing and LVQ methods in library `class`.

11.7   Two extended examples

Leptograpsus variegatus crabs

Mahon (see Campbell & Mahon, 1974) recorded data on 200 specimens of *Leptograpsus variegatus* crabs on the shore in Western Australia. This occurs in two colour forms, blue and orange, and he collected 50 of each form of each sex and made five physical measurements. These were the carapace (shell) length CL and width CW, the size of the frontal lobe FL and rear width RW, and the body depth BD. Part of the authors' thesis was to establish that the two colour forms were clearly differentiated morphologically, to support classification as two separate species.

We consider two (fairly realistic) questions:

1. Is there evidence from these morphological data alone of a division into two forms?

2. Can we construct a rule to predict the sex of a future crab of unknown colour form (species)? How accurate do we expect the rule to be?

On the second question, the body depth was measured somewhat differently for the females, so should be excluded from the analysis.

The data are physical measurements, so a sound initial strategy is to work on log scale. This has been done throughout. The data are very highly correlated, and scatterplot matrices and brush plots are none too revealing (try them for yourself).

```
> lcrabs <- log(crabs[,4:8])
> crabs.grp <- factor(c("B", "b", "O", "o")[rep(1:4, rep(50,4))])
> lcrabs.pca <- princomp(lcrabs)
> lcrabs.pc <- predict(lcrabs.pca)
> dimnames(lcrabs.pc) <- list(NULL, paste("PC", 1:5, sep=""))
> lcrabs.pca
Standard deviations:
 Comp. 1  Comp. 2  Comp. 3  Comp. 4   Comp. 5
 0.51664 0.074654 0.047914 0.024804 0.0090522
> loadings(lcrabs.pca)
     Comp. 1 Comp. 2 Comp. 3 Comp. 4 Comp. 5
FL    0.452   0.157   0.438  -0.752   0.114
RW    0.387  -0.911
CL    0.453   0.204  -0.371          -0.784
CW    0.440          -0.672           0.591
BD    0.497   0.315   0.458   0.652   0.136
```

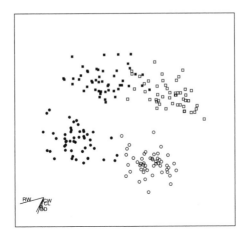

Figure 11.17: Projection pursuit view of the `crabs` data. Males are coded as filled symbols, females as open symbols, the blue colour form as squares and the orange form as circles.

We started by looking at the principal components. (As the data on log scale *are* very comparable, we did not rescale the variables to unit variance.) The first principal component had by far the largest standard deviation (0.52), with coefficients that show it to be a 'size' effect. A plot of the second and third principal components shows an almost total separation into forms (Figure 3.14 and 3.15 on pages 82 and 83) on the third PC, the second PC distinguishing sex. The coefficients of the third PC show that it is contrasting overall size with FL and BD.

Using projection pursuit in XGobi[7] allows us to visualize the data much more thoroughly. Try one of

```
library(xgobi)
xgobi(lcrabs, colors=c("SkyBlue", "SlateBlue", "Orange",
      "Red")[rep(1:4, rep(50, 4))])
xgobi(lcrabs, glyphs=12 + 5*rep(0:3, rep(50, 4)))
```

A result of optimizing by the 'holes' index is shown in Figure 11.17.

To proceed further, for example, to do a cluster analysis, we have to remove the dominant effect of size. We used the carapace area as a good measure of size, and divided all measurements by the square root of the area. It is also necessary to account for the sex differences, which we can do by analysing each sex separately, or by subtracting the mean for each sex, which we did:

```
cr.scale <- 0.5 * log(crabs$CL * crabs$CW)
slcrabs <- lcrabs - cr.scale
cr.means <- matrix(0, 2, 5)
cr.means[1,] <- apply(slcrabs[crabs$sex=="F",], 2, mean)
cr.means[2,] <- apply(slcrabs[crabs$sex=="M",], 2, mean)
```

---

[7] See page 330.

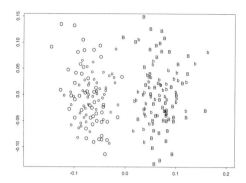

Figure 11.18: Sammon mapping of `crabs` data adjusted for size and sex. Males are coded as capitals, females as lower case, colours as the initial letter of blue or orange.

```
dslcrabs <- slcrabs - cr.means[as.numeric(crabs$sex),]
lcrabs.sam <- sammon(dist(dslcrabs))
eqscplot(lcrabs.sam$points, type="n", xlab="", ylab="")
text(lcrabs.sam$points, labels = as.character(crabs.grp))
```

As the Sammon mapping shows (Figure 11.18), Euclidean distance with this set of variables will be sensible. For this set of data, complete-link clustering makes three errors, and K-means one.

```
> crabs.h <- cutree(hclust(dist(dslcrabs)),2)
> table(crabs$sp, crabs.h)
    1   2
B 100   0
O   3  97
> cr.means[1,] <- apply(dslcrabs[crabs.h==1,], 2, mean)
> cr.means[2,] <- apply(dslcrabs[crabs.h==2,], 2, mean)
> crabs.km <- kmeans(dslcrabs, cr.means)
> table(crabs$sp, crabs.km$cluster)
    1   2
B  99   1
O   0 100
```

Discriminant analysis for sex

We noted that BD is measured differently for males and females, so it seemed prudent to omit it from the analysis. To start with, we ignore the differences between the forms. Linear discriminant analysis, for what are highly non-normal populations, finds a variable that is essentially $CL^3 RW^{-2} CW^{-1}$, a dimensionally neutral quantity. Six errors are made, all for the blue form:

```
> dcrabs.lda <- lda(crabs$sex ~ FL + RW + CL + CW, lcrabs)
> dcrabs.lda
Coefficients of linear discriminants:
        LD1
```

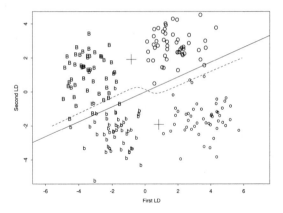

Figure 11.19: Linear discriminants for the crabs data. Males are coded as capitals, females as lower case, colours as the initial letter of blue or orange. The crosses are the group means for a linear discriminant for sex (solid line) and the dashed line is the decision boundary for sex based on four groups.

```
FL  -2.8896
RW -25.5176
CL  36.3169
CW -11.8280
> dcrabs.pred <- predict(dcrabs.lda)
> table(crabs$sex, dcrabs.pred$class)
    F  M
F 97  3
M  3 97
```

It does make sense to take the colour forms into account, especially as the within-group distributions look close to joint normality (look at the data on the linear discriminants). The first two linear discriminants dominate the between-group variation; Figure 11.19 shows the data on those variables.

```
> dcrabs.lda4 <- lda(crabs.grp ~ FL + RW + CL + CW, lcrabs)
> dcrabs.lda4
Proportion of trace:
    LD1    LD2    LD3
 0.6422 0.3491 0.0087
> dcrabs.pr4 <- predict(dcrabs.lda4, dimen=2)
> dcrabs.pr2 <- dcrabs.pr4$post[, c("B","O")] %*% c(1,1)
> table(crabs$sex, dcrabs.pr2 > 0.5)
   FALSE TRUE
F     96    4
M      3   97
```

We cannot represent all the decision surfaces exactly on a plot. However, using the first two linear discriminants as the data will provide a very good approximation; see Figure 11.19.

```
cr.t <- dcrabs.pr4$x[,1:2]
eqscplot(cr.t, type="n", xlab="First LD", ylab="Second LD")
text(cr.t, labels = as.character(crabs.grp))
perp <- function(x, y) {
    m <- (x+y)/2
    s <- - (x[1] - y[1])/(x[2] - y[2])
    abline(c(m[2] - s*m[1], s))
    invisible()
}
# For 5.x replace $means by @means
cr.m <- lda(cr.t, crabs$sex)$means
points(cr.m, pch=3, mkh=0.3)
perp(cr.m[1,], cr.m[2,])

cr.lda <- lda(cr.t, crabs.grp)
x <- seq(-6, 6, 0.25)
y <- seq(-2, 2, 0.25)
Xcon <- matrix(c(rep(x,length(y)),
            rep(y, rep(length(x),length(y)))),,2)
cr.pr <- predict(cr.lda, Xcon)$post[, c("B","O")] %*% c(1,1)
contour(x, y, matrix(cr.pr, length(x), length(y)),
    levels=0.5, labex=0, add=T, lty=3)
```

The reader is invited to try quadratic discrimination on this problem. It performs very marginally better than linear discrimination, not surprisingly since the covariances of the groups appear so similar, as can be seen from the result of

```
for(i in c("O", "o",  "B", "b"))
  print(var(lcrabs[crabs.grp==i, ]))
```

Forensic glass

The forensic glass dataset `fgl` has 214 points from six classes with nine measurements, and provides a fairly stiff test of classification methods. As we have seen (Figures 3.18 on page 86, 5.4 on page 123, 11.3 on page 335 and 11.11 on page 348) the types of glass do not form compact well-separated groupings, and the marginal distributions are far from normal. There are some small classes (with 9, 13 and 17 examples), so we cannot use quadratic discriminant analysis.

We assess their performance by 10-fold cross-validation, using the same random partition for all the methods. Logistic regression provides a suitable benchmark (as is often the case).

```
set.seed(123); rand <- sample (10, 214, replace=T)
con <- function(x,y)
{
    tab <- table(x,y)
    print(tab)
    diag(tab) <- 0
    cat("error rate = ", round(100*sum(tab)/length(x),2),"%\n")
```

```
      invisible()
}
CVtest <- function(fitfn, predfn, ...)
{
  res <- fgl$type
  for (i in sort(unique(rand))) {
    cat("fold ",i,"\n", sep="")
    learn <- fitfn(rand != i, ...)
    res[rand == i] <- predfn(learn, rand==i)
  }
  res
}
res.multinom <- CVtest(
   function(x, ...) multinom(type ~ ., fgl[x,], ...),
   function(obj, x) predict(obj, fgl[x, ],type="class"),
   maxit=1000, trace=F )

> con(fgl$type, res.multinom)
    ....
error rate =  38.79 %

res.lda <- CVtest(
   function(x, ...) lda(type ~ ., fgl[x, ], ...),
   function(obj, x) predict(obj, fgl[x, ])$class )
> con(fgl$type, res.lda)
    ....
error rate =  37.38 %
```

Using predictive LDA gives the same error rate.

Figure 11.3 on page 335 suggests that nearest neighbour methods might work well, and the 1-nearest neighbour classifier is (to date) unbeatable in this problem. We can estimate a lower bound for the Bayes risk as 10% by the method of Ripley (1996, pp. 196–7).

```
library(class)
fgl0 <- fgl[ ,-10] # drop type
{ res <- fgl$type
  for (i in sort(unique(rand))) {
     cat("fold ",i,"\n", sep="")
     sub <- rand == i
     res[sub] <- knn(fgl0[!sub, ], fgl0[sub,], fgl$type[!sub],
                     k=1)
  }
  res } -> res.knn1
> con(fgl$type, res.knn1)
      WinF WinNF Veh Con Tabl Head
    ....
error rate =  23.83 %
> res.lb <- knn(fgl0, fgl0, fgl$type, k=3, prob=T, use.all=F)
> table(attr(res.lb, "prob"))
```

```
    0.333333 0.666667    1
          10        64 140
1/3 * (64/214) = 0.099688
```

We saw in Chapter 10 that we could fit a classification tree of size about six to this dataset. We need to cross-validate over the choice of tree size, which does vary by group from five to eight.

```
library(rpart)
res.rpart <- CVtest(
  function(x, ...) {
    tr <- rpart(type ~ ., fgl[x,], ...)
    cp <- tr$cptable
    r <- cp[, 4] + cp[, 5]
    rmin <- min(seq(along=r)[cp[, 4] < min(r)])
    cp0 <- cp[rmin, 1]
    cat("size chosen was", cp[rmin, 2] + 1, "\n")
    prune(tr, cp=1.01*cp0)
  },
  function(obj, x)
    levels(fgl$type)[apply(predict(obj, fgl[x, ]), 1, which.is.max)],
  cp = 0.001
)
con(fgl$type, res.rpart)
    ....
error rate =  32.24 %
```

Neural networks

We write some general functions for testing neural network models by *V*-fold cross-validation. First we rescale the dataset so the inputs have range $[0, 1]$.

```
fgl1 <- lapply(fgl[, 1:9], function(x)
                {r <- range(x); (x-r[1])/diff(r)})
fgl1 <- data.frame(fgl1, type=fgl$type)
```

Then we can experiment with subset selection in multiple logistic regressions.

```
res.mult3 <- CVtest(
  function(xsamp, ...) {
    assign("xsamp", xsamp, frame=1)
    obj <- multinom(type ~ ., fgl1[xsamp,], trace=F, ...)
    stepAIC(obj)
    },
  function(obj, x) predict(obj, fgl1[x, ],type="class"),
  maxit=1000, decay=1e-3)
> con(fgl$type, res.mult3)
    ....
error rate =  41.12 %
```

In our runs the variables RI, Na, Mg, Al and Si were retained in all 10 folds, and Ca in all but one, the only one in which K and Ba were retained.

It is straightforward to fit a fully specified neural network in the same way. We, however, want to average across several fits and to choose the number of hidden units and the amount of weight decay by an inner cross-validation. To do so we wrote a fairly general function that can easily be used or modified to suit other problems. (See the scripts for the code.)

```
> res.nn2 <- CVnn2(type ~ ., fgl1, skip=T, maxit=500, nreps=10)
> con(fgl$type, res.nn2)
   ....
error rate =  28.5 %
```

This fits a neural network 1 000 times, and so is fairly slow (six hours on the PC) and memory-intensive (about 20 Mb).

This code chooses between neural nets on the basis of their cross-validated error rate. An alternative is to use logarithmic scoring, which is equivalent to finding the deviance on the validation set. Rather than count 0 if the predicted class is correct and 1 otherwise, we count $-\log p(c \mid x)$ for the true class c. We can easily code this variant by replacing the line

```
sum(as.numeric(truth) != max.col(res/nreps))
```

by

```
sum(-log(res[cbind(seq(along=truth),as.numeric(truth))]/nreps))
```

in CVnn2.

Learning vector quantization

For LVQ as for k-nearest neighbour methods we have to select a suitable metric. The following experiments used Euclidean distance on the original variables, but the rescaled variables or Mahalanobis distance could also be tried.

```
cd0 <- lvqinit(fgl0, fgl$type, prior=rep(1,6)/6,k=3)
cd1 <- olvq1(fgl0, fgl$type, cd0)
con(fgl$type, lvqtest(cd1, fgl0))
```

We set an even prior over the classes as otherwise there are too few representatives of the smaller classes. Our initialization code in lvqinit follows Kohonen's in selecting the number of representatives: in this problem 24 points are selected, four from each class.

```
CV.lvq <- function()
{
  res <- fgl$type
  for(i in sort(unique(rand))) {
    cat("doing fold",i,"\n")
    cd0 <- lvqinit(fgl0[rand != i,], fgl$type[rand != i],
             prior=rep(1,6)/6, k=3)
```

```
    cd1 <- olvq1(fgl0[rand != i,], fgl$type[rand != i], cd0)
    cd1 <- lvq3(fgl0[rand != i,], fgl$type[rand != i],
               cd1, niter=10000)
    res[rand == i] <- lvqtest(cd1, fgl0[rand == i,])
  }
  res
}
con(fgl$type, CV.lvq())
  ....
error rate =  26.17 %

# Try Mahalanobis distance
fgl0 <- scale(princomp(fgl[,-10])$scores)
con(fgl$type, CV.lvq())
  ....
error rate =  35.05 %
```

The initialization is random, so your results are likely to differ.

11.8 Calibration plots

One measure that a suitable model for $p(c \mid x)$ has been found is that the predicted probabilities are *well calibrated*; that is, that a fraction of about p of the events we predict with probability p actually occur. Methods for testing calibration of probability forecasts have been developed in connection with weather forecasts (Dawid, 1982, 1986).

For the forensic glass example we are making six probability forecasts for each case, one for each class. To ensure that they are genuine forecasts, we should use the cross-validation procedure. A minor change to the code gives the probability predictions:

```
CVprobs <- function(fitfn, predfn, ...)
{
  res <- matrix(, 214, 6)
  for (i in sort(unique(rand))) {
    cat("fold ",i,"\n", sep="")
    learn <- fitfn(rand != i, ...)
    res[rand == i,] <- predfn(learn, rand==i)
  }
  res
}
probs.multinom <- CVprobs(
   function(x, ...) multinom(type ~ ., fgl[x,], ...),
   function(obj, x) predict(obj, fgl[x, ],type="probs"),
   maxit=1000, trace=F )
```

We can plot these and smooth them by

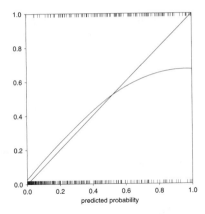

Figure 11.20: Calibration plot for multiple logistic fit to the `fgl` data.

```
probs.yes <- as.vector(class.ind(fgl$type))
probs <- as.vector(probs.multinom)
par(pty="s")
plot(c(0,1), c(0,1), type="n", xlab="predicted probability",
    ylab="", xaxs="i", yaxs="i", las=1)
rug(probs[probs.yes==0], 0.02, side=1, lwd=0.5)
rug(probs[probs.yes==1], 0.02, side=3, lwd=0.5)
abline(0,1)
newp <- seq(0, 1, length=100)
lines(newp, predict(loess(probs.yes ~ probs, span=1), newp))
```

A method with an adaptive bandwidth such as `loess` is needed here, as the distribution of points along the x-axis can be very much more uneven than in this example. The result is shown in Figure 11.20. This plot does show substantial over-confidence in the predictions, especially at probabilities close to one. Indeed, only 22/64 of the events predicted with probability greater than 0.9 occurred. (The underlying cause is the multimodal nature of some of the underlying class distributions.)

Where calibration plots are not straight, the best solution is to find a better model. Sometimes the over-confidence is minor, and mainly attributable to the use of plug-in rather than predictive estimates. Then the plot can be used to adjust the probabilities (which may need further adjustment to sum to one for more than two classes).

11.9 Exercises

11.1. Data frame `biopsy` contains data on 699 biopsies of breast tumours, which have been classified as benign or malignant (Mangasarian & Wolberg, 1990). The nine variables on each biopsy are a rating (1 to 10) by the coordinating physician; ratings on one variable are missing for some biopsies.

Analyse these data. In particular, investigate the differences in the two types of tumour, find a rule to classify tumours based solely on the biopsy variables and assess the accuracy of your rule.

11.2. Data frame `UScereals` describes 65 commonly available breakfast cereals in the USA, based on the information available on the mandatory food label on the packet. The measurements are normalized to a serving size of one American cup.

(i) Is there any way to discriminate among the major manufacturers by cereal characteristics, or do they each have a balanced portfolio of cereals?

(ii) Are there interpretable clusters of cereals?

(iii) Can you describe why cereals are displayed on high, low or middle shelves?

Chapter 12

Survival Analysis

S-PLUS contains extensive survival analysis facilities written by Terry Therneau (Mayo Foundation). There are several versions of his code in different versions of S-PLUS; the examples here were computed with `survival5` which is available from `http://www.mayo.edu/hsr/biostat.html` and is similar to the code included in S-PLUS 2000.

Survival analysis is concerned with the distribution of lifetimes, often of humans but also of components and machines. There are two distinct levels of mathematical treatment in the literature. Cox & Oakes (1984), Kalbfleisch & Prentice (1980), Lawless (1982), Miller (1981b) and Collett (1994) take a traditional and mathematically non-rigorous approach. The modern mathematical approach based on continuous-parameter martingales is given by Fleming & Harrington (1991) and Andersen *et al.* (1993). (The latter is needed here only to justify some of the distribution theory and for the concept of martingale residuals.)

Let T denote a lifetime random variable. It will take values in $(0, \infty)$, and its continuous distribution may be specified by a cumulative distribution function F with a density f. (Mixed distributions can be considered, but many of the formulae used by the software need modification.) For lifetimes it is more usual to work with the *survivor function* $S(t) = 1 - F(t) = P(T > t)$, the *hazard function* $h(t) = \lim_{\Delta t \to 0} P(t \leqslant T < t + \Delta t \mid T \geqslant t)/\Delta t$ and the *cumulative hazard function* $H(t) = \int_0^t h(s)\, ds$. These are all related; we have

$$h(t) = \frac{f(t)}{S(t)}, \qquad H(t) = -\log S(t)$$

Common parametric distributions for lifetimes are (Kalbfleisch & Prentice, 1980) the exponential, with $S(t) = \exp -\lambda t$ and hazard λ, the Weibull with

$$S(t) = \exp -(\lambda t)^{\alpha}, \qquad h(t) = \lambda \alpha (\lambda t)^{\alpha - 1}$$

the log-normal, the gamma and the log-logistic which has

$$S(t) = \frac{1}{1 + (\lambda t)^{\tau}}, \qquad h(t) = \frac{\lambda \tau (\lambda t)^{\tau - 1}}{1 + (\lambda t)^{\tau}}$$

The major distinguishing feature of survival analysis is *censoring*. An individual case may not be observed on the whole of its lifetime, so that, for example, we may only know that it survived to the end of the trial. More general patterns of censoring are possible, but all lead to data for each case of the form either of a precise lifetime or the information that the lifetime fell in some interval (possibly extending to infinity).

Clearly we must place some restrictions on the censoring mechanism, for if cases were removed from the trial just before death we would be misled. Consider right censoring, in which the case leaves the trial at time C_i, and we know either T_i if $T_i \leqslant C_i$ or that $T_i > C_i$. *Random censoring* assumes that T_i and C_i are independent random variables, and therefore in a strong sense that censoring is uninformative. This includes the special case of *type I* censoring, in which the censoring time is fixed in advance, as well as trials in which the patients enter at random times but the trial is reviewed at a fixed time. It excludes *type II* censoring in which the trial is concluded after a fixed number of failures. Most analyses (including all those based solely on likelihoods) are valid under a weaker assumption that Kalbfleisch & Prentice (1980, §5.2) call *independent* censoring in which the hazard at time t conditional on the whole history of the process only depends on the survival of that individual to time t. (Independent censoring does cover type II censoring.) Conventionally the time recorded is $\min(T_i, C_i)$ together with the indicator variable for observed death $\delta_i = I(T_i \leqslant C_i)$. Then under independent right censoring the likelihood for parameters in the lifetime distribution is

$$L = \prod_{\delta_i=1} f(t_i) \prod_{\delta_i=0} S(t_i) = \prod_{\delta_i=1} h(t_i)S(t_i) \prod_{\delta_i=0} S(t_i) = \prod_{i=1}^{n} h(t_i)^{\delta_i} S(t_i)$$

$$(12.1)$$

Usually we are not primarily interested in the lifetime distribution *per se*, but how it varies between groups (usually called *strata* in the survival context) or on measurements on the cases, called *covariates*. In the more complicated problems the hazard will depend on covariates that vary with time, such as blood pressure measurements or changes of treatments.

The function Surv(times, status) is used to describe the censored survival data to the S functions, and always appears on the left side of a model formula. In the simplest case of right censoring the variables are $\min(T_i, C_i)$ and δ_i (logical or 0/1 or 1/2). Further forms allow left and interval censoring. The results of printing the object returned by Surv are the vector of the information available, either the lifetime or an interval.

We consider three small running examples. Uncensored data on survival times for leukaemia (Feigl & Zelen, 1965; Cox & Oakes, 1984, p. 9) are in data frame leuk. This has two covariates, the white blood count wbc, and ag a test result that returns 'present' or 'absent'. Two-sample data (Gehan, 1965; Cox & Oakes, 1984, p. 7) on remission times for leukaemia are given in data frame gehan. This trial has 42 individuals in matched pairs, and no covariates (other than the treatment group).[1] Data frame `motors` contains the results of an accelerated life test experiment with 10 replicates at each of four temperatures reported by Nelson & Hahn (1972) and Kalbfleisch & Prentice (1980, pp. 4–5). The times are given in hours, but all but one is a multiple of 12, and only 14 values occur

17 21 22 56 60 70 73.5 115.5 143.5 147.5833 157.5 202.5 216.5 227

in days, which suggests that observation was not continuous. Thus this is a good example to test the handling of ties.

12.1   Estimators of survivor curves

The estimate of the survivor curve for uncensored data is easy; just take one minus the empirical distribution function. For the leukaemia data we have

```
attach(leuk)
plot(survfit(Surv(time) ~ ag, data=leuk), lty=2:3, col=2:3)
legend(80, 0.8, c("ag absent", "ag present"), lty=2:3, col=2:3)
```

and confidence intervals are obtained easily from the binomial distribution of $\widehat{S}(t)$. For example, the estimated variance is

$$\widehat{S}(t)[1 - \widehat{S}(t)]/n = r(t)[n - r(t)]/n^3 \qquad (12.2)$$

when $r(t)$ is the number of cases still alive (and hence 'at risk') at time t.

This computation introduces the function `survfit` and its associated `plot`, `print` and `summary` methods. It takes a model formula, and if there are factors on the right-hand side, splits the data on those factors, and plots a survivor curve for each factor combination, here just presence or absence of ag. (Although the factors can be specified additively, the computation effectively uses their interaction.)

For censored data we have to allow for the decline in the number of cases at risk over time. Let $r(t)$ be the number of cases at risk just before time t, that is, those which are in the trial and not yet dead. If we consider a set of intervals $I_i = [t_i, t_{i+1})$ covering $[0, \infty)$, we can estimate the probability p_i of surviving interval I_i as $[r(t_i) - d_i]/r(t_i)$ where d_i is the number of deaths in interval I_i. Then the probability of surviving until t_i is

$$P(T > t_i) = S(t_i) \approx \prod_0^{i-1} p_j \approx \prod_0^{i-1} \frac{r(t_i) - d_i}{r(t_i)}$$

Now let us refine the grid of intervals. Non-unity terms in the product will only appear for intervals in which deaths occur, so the limit becomes

$$\widehat{S}(t) = \prod \frac{r(t_i) - d_i}{r(t_i)}$$

---

[1] Andersen *et al.* (1993, p. 22) indicate that this trial had a sequential stopping rule which invalidates most of the methods used here; it should be seen as illustrative only.

the product being over times at which deaths occur before t (but they could occur simultaneously). This is the Kaplan–Meier estimator. Note that this becomes constant after the largest observed t_i, and for this reason the estimate is only plotted up to the largest t_i. However, the points at the right-hand end of the plot will be very variable, and it may be better to stop plotting when there are still a few individuals at risk.

We can apply similar reasoning to the cumulative hazard

$$H(t_i) \approx \sum_{j \leqslant i} h(t_j)(t_{j+1} - t_j) \approx \sum_{j \leqslant i} \frac{d_j}{r(t_j)}$$

with limit

$$\widehat{H}(t) = \sum \frac{d_j}{r(t_j)} \tag{12.3}$$

again over times at which deaths occur before t. This is the Nelson estimator of the cumulative hazard, and leads to the Altshuler or Fleming–Harrington estimator of the survivor curve

$$\tilde{S}(t) = \exp -\widehat{H}(t) \tag{12.4}$$

The two estimators are related by the approximation $\exp x \approx 1 - x$ for small x, so they will be nearly equal for large risk sets. The S functions follow Fleming and Harrington in breaking ties in (12.3), so if there were 3 deaths when the risk set contained 12 people, $3/12$ is replaced by $1/12 + 1/11 + 1/10$.

Similar arguments to those used to derive the two estimators lead to the standard error formula for the Kaplan–Meier estimator

$$\text{var}\left(\widehat{S}(t)\right) = \widehat{S}(t)^2 \sum \frac{d_j}{r(t_j)[r(t_j) - d_j]} \tag{12.5}$$

often called Greenwood's formula after its version for life tables, and

$$\text{var}\left(\widehat{H}(t)\right) = \sum \frac{d_j}{r(t_j)[r(t_j) - d_j]} \tag{12.6}$$

We leave it to the reader to check that Greenwood's formula reduces to (12.2) in the absence of ties and censoring. Note that if censoring can occur, both the Kaplan–Meier and Nelson estimators are biased; the bias results from the inability to give a sensible estimate when the risk set is empty.

Tsiatis (1981) suggested the denominator $r(t_j)^2$ rather than $r(t_j)[r(t_j) - d_j]$ on asymptotic grounds. Both Fleming & Harrington (1991) and Andersen *et al.* (1993) give a rigorous derivation of these formulae (and corrected versions for mixed distributions), as well as calculations of bias and limit theorems that justify asymptotic normality. Klein (1991) discussed the bias and small-sample behaviour of the variance estimators; his conclusions for $\widehat{H}(t)$ are that the bias is negligible and the Tsiatis form of the standard error is accurate (for practical use) provided the expected size of the risk set at t is at least five. For the Kaplan–Meier estimator Greenwood's formula is preferred, and is accurate enough (but biased downwards) again provided the expected size of the risk set is at least five.

We can use these formulae to indicate confidence intervals based on asymptotic normality, but we must decide on what scale to compute them. By default the function `survfit` computes confidence intervals on the log survivor (or cumulative hazard) scale, but linear and complementary log-log scales are also available (via the `conf.type` argument). These choices give

$$\widehat{S}(t) \exp\left[\pm k_\alpha \text{ s.e.}(\widehat{H}(t))\right]$$

$$\widehat{S}(t) \left[1 \pm k_\alpha \text{ s.e.}(\widehat{H}(t))\right]$$

$$\exp\left\{\widehat{H}(t) \exp\left[\pm k_\alpha \frac{\text{s.e.}(\widehat{H}(t))}{\widehat{H}(t)}\right]\right\}$$

the last having the advantage of taking values in $(0, 1)$. Bie, Borgan & Liestøl (1987) and Borgan & Liestøl (1990) considered these and an arc-sine transformation; their results indicate that the complementary log-log interval is quite satisfactory for sample sizes as small as 25.

We do not distinguish clearly between log-survivor curves and cumulative hazards, which differ only by sign, yet the natural estimator of the first is the Kaplan–Meier estimator on log scale, and for the second it is the Nelson estimator. This is particularly true for confidence intervals, which we would expect to transform just by a change of sign. Fortunately, practical differences only emerge for very small risk sets, and are then swamped by the very large variability of the estimators.

The function `survfit` also handles censored data, and uses the Kaplan–Meier estimator by default. We try it on the `gehan` data:

```
> attach(gehan)
> Surv(time, cens)
 [1]  1  10  22   7   3  32+ 12  23   8  22  17   6   2  16
[15] 11  34+  8  32+ 12  25+  2  11+  5  20+  4  19+ 15   6
[29]  8  17+ 23  35+  5   6  11  13   4   9+  1   6+  8  10+
> plot(log(time) ~ pair)
# product-limit estimators with Greenwood's formula for errors:
> gehan.surv <- survfit(Surv(time, cens) ~ treat, data = gehan,
      conf.type = "log-log")
> summary(gehan.surv)
    ....
> plot(gehan.surv, conf.int=T, lty=3:2, log=T,
      xlab="time of remission (weeks)", ylab="survival")
> lines(gehan.surv, lty=3:2, lwd=2, cex=2)
> legend(25, 0.1 , c("control","6-MP"), lty=2:3, lwd=2)
```

which calculates and plots (as shown in Figure 12.1) the product-limit estimators for the two groups, giving standard errors calculated using Greenwood's formula. (Confidence intervals are plotted automatically if there is only one group.) Other options are available, including `error="tsiatis"` and

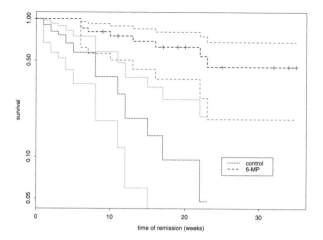

Figure 12.1: Survivor curves (on log scale) for the two groups of the gehan data. The crosses (on the 6-MP curve) represent censoring times. The thicker lines are the estimates, the thinner lines pointwise 95% confidence intervals.

type="fleming-harrington" (which can be abbreviated to the first character). Note that the plot method has a log argument that plots $\widehat{S}(t)$ on log scale, effectively showing the negative cumulative hazard.

Testing survivor curves

We can test for differences between the groups in the gehan example by

```
> survdiff(Surv(time, cens) ~ treat, data=gehan)
                N Observed Expected (O-E)^2/E (O-E)^2/V
treat=6-MP 21       9     19.3      5.46      16.8
treat=control 21   21     10.7      9.77      16.8

 Chisq= 16.8  on 1 degrees of freedom, p= 4.17e-05
```

This is one of a family of tests with parameter ρ defined by Fleming & Harrington (1981) and Harrington & Fleming (1982). The default $\rho = 0$ corresponds to the *log-rank test*. Suppose t_j are the observed death times. If we condition on the risk set and the number of deaths D_j at time t_j the mean of the number of deaths D_{jk} in group k is clearly $E_{jk} = D_k r_k(t_j)/r(t_j)$ under the null hypothesis (where $r_k(t_j)$ is the number from group k at risk at time j). The statistic used is $(O_k - E_k) = \sum_j \widehat{S}(t_j-)^{\rho} [D_{jk} - E_{jk}]$,[2] and from this we compute a statistic $(O - E)^T V^{-1}(O - E)$ with an approximately chi-squared distribution. There are a number of different approximations to the variance matrix V, the one used being the weighted sum over death times of the variance matrices of $D_{jk} - E_{jk}$ computed from the hypergeometric distribution. The sum of $(O_k - E_k)^2/E_k$

---

[2] $\widehat{S}(t-)$ is the Kaplan–Meier estimate of survival just prior to t, ignoring the grouping.

provides a conservative approximation to the chi-squared statistic. The final column is $(O - E)^2$ divided by the diagonal of V; the final line gives the overall statistic computed from the full quadratic form.

The argument `rho=1` of `survdiff` corresponds approximately to the Peto–Peto modification (Peto & Peto, 1972) of the Wilcoxon test, and is more sensitive to early differences in the survivor curves.

A warning: tests of differences between groups are often used inappropriately. The `gehan` dataset has no other covariates, but where there are covariates the differences between the groups may reflect or be masked by differences in the covariates. Thus for the `leuk` dataset

```
> survdiff(Surv(time) ~ ag, data=leuk)
              N Observed Expected (O-E)^2/E (O-E)^2/V
ag=absent  16       16      9.3      4.83      8.45
ag=present 17       17     23.7      1.90      8.45

 Chisq= 8.4  on 1 degrees of freedom, p= 0.00365
```

is inappropriate as there are differences in distribution of `wbc` between the two groups. A model is needed to adjust for the covariates.

12.2   Parametric models

Parametric models for survival data have fallen out of fashion with the advent of less parametric approaches such as the Cox proportional hazard models considered in the next section, but they remain a very useful tool, particularly in exploratory work (as usually they can be fitted very much faster than the Cox models).

The simplest parametric model is the exponential distribution with hazard $\lambda_i > 0$. The natural way to relate this to a covariate vector x for the case (including a constant if required) and to satisfy the positivity constraint is to take

$$\log \lambda_i = \beta^T x_i, \qquad \lambda_i = e^{\beta^T x_i}$$

For the Weibull distribution the hazard function is

$$h(t) = \lambda^\alpha \alpha t^{\alpha-1} = \alpha t^{\alpha-1} \exp(\alpha \beta^T x) \qquad (12.7)$$

if we again make λ an exponential function of the covariates, and so we have the first appearance of the *proportional hazards* model

$$h(t) = h_0(t) \exp \beta^T x \qquad (12.8)$$

which we consider again later. This identification suggests re-parametrizing the Weibull by replacing λ^α by λ, but as this just rescales the coefficients we can move easily from one parametrization to the other.

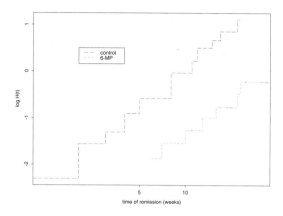

Figure 12.2: A log-log plot of cumulative hazard for the gehan dataset.

The Weibull is also a member of the class of *accelerated life* models, which have survival time T such that $T \exp \beta^T x$ has a fixed distribution; that is, time is speeded up by the factor $\exp \beta^T x$ for an individual with covariate x. This corresponds to replacing t in the survivor function and hazard by $t \exp \beta^T x$, and for models such as the exponential, Weibull and log-logistic with parametric dependence on λt, this corresponds to taking $\lambda = \exp \beta^T x$. For all accelerated-life models we will have

$$\log T = \log T_0 - \beta^T x \tag{12.9}$$

for a random variable T_0 whose distribution does not depend on x, so these are naturally considered as regression models.

For the Weibull the cumulative hazard is linear on a log-log plot, which provides a useful diagnostic aid. For example, for the gehan data

```
> plot(gehan.surv, lty=3:4, col=2:3, fun="cloglog",
    xlab="time of remission (weeks)", ylab="log H(t)")
> legend(2, 0.5, c("control","6-MP"), lty=4:3, col=3:2)
```

(the fun argument needs survival5) we see excellent agreement with the proportional hazards hypothesis and with a Weibull baseline (Figure 12.2).

The function survreg[3] fits parametric survival models of the form

$$\ell(T) \sim \beta^T x + \sigma \epsilon \tag{12.10}$$

where $\ell()$ is usually a log transformation. The distribution argument specifies the distribution of ϵ and $\ell()$, and σ is known as the *scale*. In survival5 the distribution can be weibull (the default) exponential, rayleigh, lognormal or loglogistic, all with a log transformation, or extreme, logistic, gaussian or t with an identity transformation.

---

[3] survReg in S-PLUS 2000.

The default for `distribution` corresponds to the model

$$\log T \sim \boldsymbol{\beta}^T \boldsymbol{x} + \sigma \log E$$

for a standard exponential E whereas our Weibull parametrization corresponds to

$$\log T \sim -\log \lambda + \frac{1}{\alpha} \log E$$

Thus `survreg` uses a log-linear Weibull model for $-\log \lambda$ and the scale factor σ estimates $1/\alpha$. The exponential distribution comes from fixing $\sigma = \alpha = 1$.

We consider exponential analyses, followed by Weibull and log-logistic regression analyses.[4]

```
> options(contrasts=c("contr.treatment", "contr.poly"))
> survreg(Surv(time) ~ ag*log(wbc), leuk, dist="exponential")
  ....
Coefficients:
 (Intercept)     ag log(wbc) ag:log(wbc)
      4.3433  4.135 -0.15402    -0.32781
  ....
Loglik(model)= -145.7   Loglik(intercept only)= -155.5
       Chisq= 19.58 on 3 degrees of freedom, p= 0.00021
> summary(survreg(Surv(time) ~ ag + log(wbc), leuk, dist="exp"))
              Value Std. Error     z        p
 (Intercept)  5.815      1.263  4.60 4.15e-06
          ag  1.018      0.364  2.80 5.14e-03
    log(wbc) -0.304      0.124 -2.45 1.44e-02

> summary(survreg(Surv(time) ~ ag + log(wbc), leuk)) # Weibull
  ....
              Value Std. Error     z        p
 (Intercept)  5.8524     1.323  4.425 9.66e-06
          ag  1.0206     0.378  2.699 6.95e-03
    log(wbc) -0.3103     0.131 -2.363 1.81e-02
  Log(scale)  0.0399     0.139  0.287 7.74e-01

Scale= 1.04

Weibull distribution
Loglik(model)= -146.5   Loglik(intercept only)= -153.6
       Chisq= 14.18 on 2 degrees of freedom, p= 0.00084
  ....
> summary(survreg(Surv(time) ~ ag + log(wbc), leuk,
                  dist="loglogistic"))
              Value Std. Error     z        p
 (Intercept)  8.027      1.701  4.72 2.37e-06
          ag  1.155      0.431  2.68 7.30e-03
```

---

[4] Earlier releases of `survreg` gave log-likelihoods for $\log T$ that differ from these by an additive constant depending only on the data.

```
       log(wbc) -0.609    0.176 -3.47 5.21e-04
      Log(scale) -0.374    0.145 -2.58 9.74e-03

Scale= 0.688

Log logistic distribution
Loglik(model)= -146.6   Loglik(intercept only)= -155.4
        Chisq= 17.58 on 2 degrees of freedom, p= 0.00015
```

The Weibull analysis shows no support for non-exponential shape. For later reference, in the proportional hazards parametrization (12.8) the estimate of the coefficients is $\widehat{\boldsymbol{\beta}} = -(5.85, 1.02, -0.310)^T/1.04 = (-5.63, -0.981, 0.298)^T$. The log-logistic distribution, which is an accelerated life model but not a proportional hazards model (in our parametrization), gives a considerably more significant coefficient for log(wbc). Its usual scale parameter τ (as defined on page 367) is estimated as $1/0.688 \approx 1.45$.

We can test for a difference in groups within the Weibull model by the Wald test (the 'z' value for ag) or we can perform a likelihood ratio test by the anova method.

```
> anova(survreg(Surv(time) ~ log(wbc), leuk),
     survreg(Surv(time) ~ ag + log(wbc), leuk))
  ....
          Terms Resid. Df  -2*LL Test Df Deviance Pr(Chi)
1      log(wbc)      28 299.35
2 ag + log(wbc)      27 293.00  +ag  1   6.3572 0.01169
```

The likelihood ratio test statistic is somewhat less significant than the result given by survdiff.

An extension is to allow different scale parameters σ for each group, by adding a strata argument to the formula. For example

```
> summary(survreg(Surv(time) ~ strata(ag) + log(wbc), data=leuk))
  ....
            Value Std. Error     z       p
(Intercept) 7.499    1.475  5.085 3.68e-07
   log(wbc) -0.422    0.149 -2.834 4.59e-03
  ag=absent 0.152    0.221  0.688 4.92e-01
 ag=present 0.142    0.216  0.658 5.11e-01

Scale:
 ag=absent ag=present
      1.16       1.15

Weibull distribution
Loglik(model)= -149.7   Loglik(intercept only)= -153.2
  ....
```

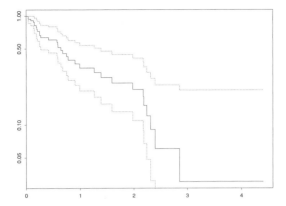

Figure 12.3: A log plot of $S(t)$ (equivalently, a linear plot of $-H(t)$) for the `leuk` dataset with pointwise confidence intervals.

If the accelerated-life model holds, $T \exp(-\beta^T x)$ has the same distribution for all subjects, being standard Weibull, log-logistic and so on. Thus we can get some insight into what the common distribution should be by studying the distribution of $(T_i \exp(-\widehat{\beta}^T x_i))$. Another way to look at this is that the residuals from the regression are $\log T_i - \widehat{\beta}^T x_i$ which we have transformed back to the scale of time. For the `leuk` data we could use, for example,

```
leuk.wei <- survreg(Surv(time) ~ ag + log(wbc), leuk)
ntimes <- leuk$time * exp(-leuk.wei$linear.predictors)
plot(survfit(Surv(ntimes)), log=T)
```

The result (Figure 12.3) is plausibly linear, confirming the suitability of an exponential model. If we wished to test for a general Weibull distribution, we should plot $\log(-\log \widehat{S}(t))$ against $\log t$. (This is provided by the `fun="cloglog"` argument to `plot.survfit` in `survival5`.)

Moving on to the `gehan` dataset, which includes right censoring, we find

```
> survreg(Surv(time, cens) ~ factor(pair) + treat, gehan,
      dist="exp")
    ....
Loglik(model)= -101.6   Loglik(intercept only)= -116.8
        Chisq= 30.27 on 21 degrees of freedom, p= 0.087
> summary(survreg(Surv(time, cens) ~ treat, gehan, dist="exp"))
            Value Std. Error      z        p
(Intercept)  3.69      0.333  11.06 2.00e-28
      treat -1.53      0.398  -3.83 1.27e-04

Scale fixed at 1

Exponential distribution
Loglik(model)= -108.5   Loglik(intercept only)= -116.8
        Chisq= 16.49 on 1 degrees of freedom, p= 4.9e-05
```

```
> summary(survreg(Surv(time, cens) ~ treat, gehan))
              Value Std. Error       z       p
(Intercept)  3.516      0.252   13.96 2.61e-44
      treat -1.267      0.311   -4.08 4.51e-05
 Log(scale) -0.312      0.147   -2.12 3.43e-02

Scale= 0.732

Weibull distribution
Loglik(model)= -106.6   Loglik(intercept only)= -116.4
        Chisq= 19.65 on 1 degrees of freedom, p= 9.3e-06
```

There is no evidence of close matching of pairs. The difference in log hazard between treatments is $-(-1.267)/0.732 = 1.73$ with a standard error of $0.42 = 0.311/0.732$.

Finally, we consider the motors data, which are analysed by Kalbfleisch & Prentice (1980, §3.8.1). According to Nelson & Hahn (1972), the data were collected to assess survival at $130°C$, for which they found a median of 34 400 hours and a 10 percentile of 17 300 hours.

```
> plot(survfit(Surv(time, cens) ~ factor(temp), motors),
    conf.int=F)
> motor.wei <- survreg(Surv(time, cens) ~ temp, motors)
> summary(motor.wei)
               Value Std. Error      z        p
(Intercept) 16.3185    0.62296   26.2 3.03e-151
       temp -0.0453    0.00319  -14.2  6.74e-46
 Log(scale) -1.0956    0.21480   -5.1  3.38e-07

Scale= 0.334

Weibull distribution
Loglik(model)= -147.4   Loglik(intercept only)= -170
        Chisq= 44.32 on 1 degrees of freedom, p= 2.8e-11
    ....
> unlist(predict(motor.wei, data.frame(temp=130), se.fit=T))
    fit se.fit
  33813 7506.3
```

(In earlier versions the prediction is on log scale and the standard errors are computed by `predict.glm` and will not be appropriate.) The `predict` method by default predicts the mean of the distribution. We can obtain predictions for quantiles by

```
> predict(motor.wei, data.frame(temp=130), type="quantile",
    p=c(0.5, 0.1))
[1] 29914 15935
```

We can also use `predict` to find standard errors, but we prefer to compute confidence intervals on log-time scale by

```
> t1 <-  predict(motor.wei, data.frame(temp=130),
                type = "uquantile", p = 0.5, se = T)
> exp(c(t1$fit - 2*t1$se, t1$fit + 2*t1$se))
[1] 19517 45849
> t1 <-  predict(motor.wei, data.frame(temp=130),
                type = "uquantile", p = 0.1, se = T)
> exp(c(t1$fit - 2*t1$se, t1$fit + 2*t1$se))
[1] 10258 24752
```

Nelson & Hahn worked with $z = 1000/(\text{temp} + 273.2)$. We leave the reader to try this; it gives slightly larger quantiles.

Function `censorReg`

S-PLUS 4.5 and S-PLUS 5.0 introduced further functions for parametric survival analysis from a different author; these have a very substantial overlap with `survreg` but are more general in that they allow *truncation* as well as *censoring*. Either or both of censoring and truncation occur when subjects are only observed for part of the time axis. An observation T_i is right-censored if it is known only that $T_i > U_i$ for a censoring time U_i, and left-censored if it is only known that $T_i \leqslant L_i$. (Both left- and right-censoring can occur in a study, but not for the same individual.) Interval censoring is usually taken to refer to subjects known to have an event in $(L_i, U_i]$, but with the time of the event otherwise unknown. Truncation is similar but subtly different. For left and right truncation, subjects with events before L_i or after U_i are not included in the study, and interval truncation refers to both left and right truncation. (Notice the inconsistency with interval censoring.)

The new analogue to `survreg` is `censorReg`. Confusingly, this uses `"logexponential"` and `"lograyleigh"` for what are known to `survreg` as the `"exponential"` and `"rayleigh"` distributions and are accelerated-life models for those distributions.

Let us consider a simple example using `gehan`. We can fit a Weibull model by

```
> options(contrasts=c("contr.treatment", "contr.poly"))
> summary(censorReg(censor(time, cens) ~ treat, gehan))
    . . . .
Coefficients:
            Est. Std.Err. 95% LCL 95% UCL z-value  p-value
(Intercept) 3.52    0.252    3.02   4.009   13.96 2.61e-44
      treat -1.27    0.311   -1.88  -0.658   -4.08 4.51e-05

Extreme value distribution: Dispersion (scale) = 0.73219
Observations: 42 Total; 12 Censored
-2*Log-Likelihood: 213
```

which agrees with our results on page 377.

The potential advantages of `censorReg` come from its wider range of options. As noted previously, it allows truncation, by specifying a call to `censor` with a `truncation` argument. Distributions can be fitted with a *threshold*, that is, a parameter $\gamma > 0$ such that the failure-time model is fitted to $T - \gamma$ (and hence no failures can occur before time γ).

There is a `plot` method for `censorReg` that produces up to seven figures, but at the time of writing would usually fail on correctly-specified fits.

A `strata` argument in a `censorReg` model has a completely different meaning: it fits separate models at each level of the stratifying factor, unlike `survreg` which has common regression coefficients across strata.

12.3   Cox proportional hazards model

Cox (1972) introduced a less parametric approach known as proportional hazards. There is a baseline hazard function $h_0(t)$ that is modified multiplicatively by covariates (including group indicators), so the hazard function for any individual case is

$$h(t) = h_0(t) \exp \beta^T x$$

and the interest is mainly in the proportional factors rather than the baseline hazard. Note that the cumulative hazards will also be proportional, so we can examine the hypothesis by plotting survivor curves for sub-groups on log scale. Later we allow the covariates to depend on time.

The parameter vector β is estimated by maximizing a partial likelihood. Suppose one death occurred at time t_j. Then conditional on this event the probability that case i died is

$$\frac{h_0(t) \exp \beta^T x_i}{\sum_l I(T_l \geqslant t) h_0(t) \exp \beta^T x_l} = \frac{\exp \beta^T x_i}{\sum_l I(T_l \geqslant t) \exp \beta^T x_l} \tag{12.11}$$

which does not depend on the baseline hazard. The partial likelihood for β is the product of such terms over all observed deaths, and usually contains most of the information about β (the remainder being in the observed times of death). However, we need a further condition on the censoring (Fleming & Harrington, 1991, pp. 138–9) that it is independent and *uninformative* for this to be so; the latter means that the likelihood for censored observations in $[t, t + \Delta t)$ does not depend on β.

The correct treatment of ties causes conceptual difficulties as they are an event of probability zero for continuous distributions. Formally (12.11) may be corrected to include all possible combinations of deaths. As this increases the computational load, it is common to employ the Breslow approximation[5] in which each death is always considered to precede all other events at that time. Let

---

[5] First proposed by Peto (1972).

$\tau_i = I(T_i \geq t)\exp \boldsymbol{\beta}^T \boldsymbol{x}_i$, and suppose there are d deaths out of m possible events at time t. Breslow's approximation uses the term

$$\prod_{i=1}^{d} \frac{\tau_i}{\sum_1^m \tau_j}$$

in the partial likelihood at time t. Other options are Efron's approximation

$$\prod_{i=1}^{d} \frac{\tau_i}{\sum_1^m \tau_j - \frac{i}{d}\sum_1^d \tau_j}$$

and the 'exact' partial likelihood

$$\prod_{i=1}^{d} \tau_i \Big/ \sum \prod_{k=1}^{d} \tau_{j_k}$$

where the sum is over subsets of $1, \ldots, m$ of size d. One of these terms is selected by the `method` argument of the function `coxph`, with default `efron`.

The baseline cumulative hazard $H_0(t)$ is estimated by rescaling the contributions to the number at risk by $\exp \widehat{\boldsymbol{\beta}}^T \boldsymbol{x}$ in (12.3). Thus in that formula $r(t) = \sum I(T_i \geq t)\exp \widehat{\boldsymbol{\beta}}^T \boldsymbol{x}_i$.

The Cox model is easily extended to allow different baseline hazard functions for different groups, and this is automatically done if they are declared as `strata`. (The model formula may also include an `offset` term.) For our leukaemia example we have:

```
> attach(leuk)
> leuk.cox <- coxph(Surv(time) ~ ag + log(wbc), leuk)
> summary(leuk.cox)
    ....
          coef exp(coef) se(coef)     z      p
      ag -1.069    0.343    0.429 -2.49 0.0130
log(wbc)  0.368    1.444    0.136  2.70 0.0069

         exp(coef) exp(-coef) lower .95 upper .95
      ag     0.343      2.913     0.148     0.796
log(wbc)     1.444      0.692     1.106     1.886

Rsquare= 0.377   (max possible= 0.994 )
Likelihood ratio test= 15.6  on 2 df,   p=0.000401
Wald test            = 15.1  on 2 df,   p=0.000537
Score (logrank) test = 16.5  on 2 df,   p=0.000263

> update(leuk.cox, ~ . -ag)
    ....
Likelihood ratio test=9.19  on 1 df, p=0.00243   n= 33
```

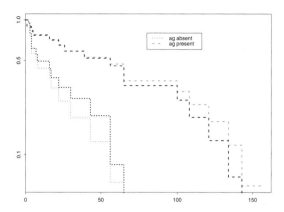

Figure 12.4: Log-survivor curves for the `leuk` dataset. The thick lines are from a Cox model with two strata, the thin lines Kaplan–Meier estimates that ignore the blood counts.

```
> leuk.coxs <- coxph(Surv(time) ~ strata(ag) + log(wbc), leuk)
> leuk.coxs
    ....
        coef exp(coef) se(coef)    z     p
log(wbc) 0.391     1.48    0.143 2.74 0.0062
    ....
Likelihood ratio test=7.78  on 1 df, p=0.00529  n= 33

> leuk.coxs1 <- coxph(Surv(time) ~ strata(ag) + log(wbc) +
    ag:log(wbc), leuk)
> leuk.coxs1
    ....
            coef exp(coef) se(coef)     z    p
  log(wbc) 0.183     1.20    0.188 0.978 0.33
ag:log(wbc) 0.456     1.58    0.285 1.598 0.11
    ....
> plot(survfit(Surv(time) ~ ag), lty=2:3, log=T)
> lines(survfit(leuk.coxs), lty=2:3, lwd=3)
> legend(80, 0.8, c("ag absent", "ag present"), lty=2:3)
```

The 'likelihood ratio test' is actually based on (log) partial likelihoods, not the full likelihood, but has similar asymptotic properties. The tests show that there is a significant effect of `wbc` on survival, but also that there is a significant difference between the two `ag` groups (although as Figure 12.4 shows, this is less than before adjustment for the effect of `wbc`).

Note how `survfit` can take the result of a fit of the proportional hazard model. In the first fit the hazards in the two groups differ only by a factor whereas later they are allowed to have separate baseline hazards (which look very close to proportional). There is marginal evidence for a difference in slope within the two strata. Note how straight the log-survivor functions are in Figure 12.4, confirming the good fit of the exponential model for these data. The Kaplan–Meier survivor curves refer to the populations; those from the `coxph` fit refer to a patient in the stratum with an average `log(wbc)` for the whole dataset. This example shows why it is inappropriate just to test (using `survdiff`) the difference between the two groups: part of the difference is attributable to the lower `wbc` in the `ag` absent group.

The test statistics refer to the whole set of covariates. The likelihood ratio test statistic is the change in deviance on fitting the covariates over just the baseline hazard (by strata); the score test is the expansion at the baseline, and so does not need the parameters to be estimated (although this has been done). The R^2 measure quoted by `summary.coxph` is taken from Nagelkerke (1991).

The general proportional hazards model gives estimated (non-intercept) coefficients $\hat{\beta} = (-1.07, 0.37)^T$, compared to the Weibull fit of $(-0.98, 0.30)^T$ (on page 376). The log-logistic had coefficients $(-1.16, 0.61)^T$ which under the approximations of Solomon (1984) would be scaled by $\tau/2$ to give $(-0.79, 0.42)^T$ for a Cox proportional-hazards fit if the log-logistic regression model (an accelerated-life model) were the true model.

We next consider the Gehan data. We saw before that the pairing has a negligible effect for the exponential model. Here the effect is a little larger, with $P \approx 8\%$. The Gehan data have a large number of (mainly pairwise) ties.

```
> gehan.cox <- coxph(Surv(time, cens) ~ treat, gehan)
> summary(gehan.cox)
    . . . .
      coef exp(coef) se(coef)    z       p
treat 1.57      4.82    0.412 3.81 0.00014

      exp(coef) exp(-coef) lower .95 upper .95
treat      4.82      0.208      2.15      10.8
    . . . .
Likelihood ratio test= 16.4  on 1 df,   p=5.26e-05
    . . . .
> coxph(Surv(time, cens) ~ treat, gehan, method="exact")
    . . . .
      coef exp(coef) se(coef)    z       p
treat 1.63      5.09    0.433 3.76 0.00017

Likelihood ratio test=16.2  on 1 df, p=5.54e-05  n= 42

# The next fit is slow
> coxph(Surv(time, cens) ~ treat + factor(pair), gehan,
    method="exact")
    . . . .
Likelihood ratio test=45.5  on 21 df, p=0.00148  n= 42
    . . . .
> 1 - pchisq(45.5 - 16.2, 20)
[1] 0.082018
```

Finally we consider the `motors` data. The exact fit for the motors dataset is much the slowest, as it has large groups of ties.

```
> motor.cox <- coxph(Surv(time, cens) ~ temp, motors)
> motor.cox
    ....
      coef exp(coef) se(coef)    z       p
temp 0.0918      1.1   0.0274 3.36 0.00079
    ....
> coxph(Surv(time, cens) ~ temp, motors, method="breslow")
    ....
      coef exp(coef) se(coef)    z      p
temp 0.0905     1.09   0.0274 3.3 0.00098
    ....
> coxph(Surv(time, cens) ~ temp, motors, method="exact")
    ....
      coef exp(coef) se(coef)    z      p
temp 0.0947      1.1   0.0274 3.45 0.00056
    ....
> plot( survfit(motor.cox, newdata=data.frame(temp=200),
    conf.type="log-log") )
> summary( survfit(motor.cox, newdata=data.frame(temp=130)) )
time n.risk n.event survival  std.err lower 95% CI upper 95% CI
 408     40       4    1.000 0.000254        0.999            1
 504     36       3    1.000 0.000499        0.999            1
1344     28       2    0.999 0.001910        0.995            1
1440     26       1    0.998 0.002698        0.993            1
1764     20       1    0.996 0.005327        0.986            1
2772     19       1    0.994 0.007922        0.978            1
3444     18       1    0.991 0.010676        0.971            1
3542     17       1    0.988 0.013670        0.962            1
3780     16       1    0.985 0.016980        0.952            1
4860     15       1    0.981 0.020697        0.941            1
5196     14       1    0.977 0.024947        0.929            1
```

The function survfit has a special method for coxph objects that plots the mean and confidence interval of the survivor curve for an average individual (with average values of the covariates). As we see, this can be overridden by giving new data, as shown in Figure 12.5. The non-parametric method is unable to extrapolate to 130°C as none of the test examples survived long enough to estimate the baseline hazard beyond the last failure at 5 196 hours.

Residuals

The concept of a residual is a difficult one for binary data, especially as here the event may not be observed because of censoring. A straightforward possibility is to take

$$r_i = \delta_i - \widehat{H}(t_i)$$

which is known as the *martingale residual* after a derivation from the mathematical theory given by Fleming & Harrington (1991, §4.5). They show that it is

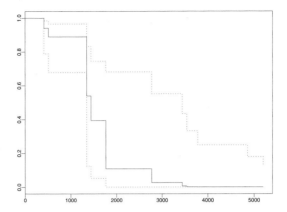

Figure 12.5: The survivor curve for a motor at $200°$C estimated from a Cox proportional hazards model (solid line) with pointwise 95% confidence intervals (dotted lines).

appropriate for checking the functional form of the proportional hazards model, for if

$$h(t) = h_0(t)\phi(x^*)\exp\boldsymbol{\beta}^T\boldsymbol{x}$$

for an (unknown) function of a covariate x^* then

$$E[R \mid X^*] \approx [\phi(X^*) - \overline{\phi}] \sum \delta_i/n$$

and this can be estimated by smoothing a plot of the martingale residuals versus x^*, for example, using `lowess` or the function `scatter.smooth` based on loess. (The term $\overline{\phi}$ is a complexly weighted mean.) The covariate x^* can be one not included in the model, or one of the terms to check for non-linear effects.

The martingale residuals are the default output of `residuals` on a `coxph` fit.

The martingale residuals can have a very skewed distribution, as their maximum value is 1, but they can be arbitrarily negative. The *deviance residuals* are a transformation

$$\text{sign}\,(r_i)\sqrt{2[-r_i - \delta_i\log(\delta_i - r_i)]}$$

which reduces the skewness, and for a parametric survival model when squared and summed give (approximately) the deviance. Deviance residuals are best used in plots that will indicate cases not fitted well by the model.

The *Schoenfeld residuals* (Schoenfeld, 1982) are defined at death times as $\boldsymbol{x}_i - \overline{\boldsymbol{x}}(t_i)$ where $\overline{\boldsymbol{x}}(s)$ is the mean weighted by $\exp\widehat{\boldsymbol{\beta}}^T\boldsymbol{x}$ of the $\boldsymbol{x}$ over only the cases still in the risk set at time s. These residuals form a matrix with one row for each case that died and a column for each covariate. The scaled Schoenfeld residuals (`type="scaledsch"`) are the I^{-1} matrix multiplying the Schoenfeld residuals, where I is the (partial) information matrix at the fitted parameters in the Cox model.

The *score residuals* are the terms of efficient score for the partial likelihood, this being a sum over cases of

$$L_i = \left[\boldsymbol{x}_i - \overline{\boldsymbol{x}}(t_i)\right]\delta_i - \int_0^{t_i} \left[\boldsymbol{x}_i(s) - \overline{\boldsymbol{x}}(s)\right]\hat{h}(s)\,\mathrm{d}s$$

Thus the score residuals form an $n \times p$ matrix. They can be used to examine leverage of individual cases by computing (approximately) the change in $\widehat{\boldsymbol{\beta}}$ if the observation were dropped; `type="dfbeta"` gives this whereas `type="dfbetas"` scales by the standard errors for the components of $\widehat{\boldsymbol{\beta}}$.

Tests of proportionality of hazards

Once a type of departure from the base model is discovered or suspected, the proportional hazards formulation is usually flexible enough to allow an extended model to be formulated and the significance of the departure tested within the extended model. Nevertheless, some approximations can be useful, and are provided by the function `cox.zph` for departures of the type

$$\boldsymbol{\beta}(t) = \boldsymbol{\beta} + \boldsymbol{\theta}g(t)$$

for some postulated smooth function g. Grambsch & Therneau (1994) show that the scaled Schoenfeld residuals for case i have, approximately, mean $g(t_i)\boldsymbol{\theta}$ and a computable variance matrix.

The function `cox.zph` has both `print` and `plot` methods. The printed output gives an estimate of the correlation between $g(t_i)$ and the scaled Schoenfeld residuals and a chi-squared test of $\boldsymbol{\theta} = 0$ for each covariate, and an overall chi-squared test. The plot method gives a plot for each covariate, of the scaled Schoenfeld residuals against $g(t)$ with a spline smooth and pointwise confidence bands for the smooth. (Figure 12.8 on page 390 is an example.)

The function g has to be specified. The default in `cox.zph` is $1 - \widehat{S}(t)$ for the Kaplan–Meier estimator, with options for the ranks of the death times, $g \equiv 1$ and $g = \log$ as well as a user-specified function. (The x-axis of the plots is labelled by the death times, not $\{g(t_i)\}$.)

12.4 Further examples

VA lung cancer data

S-PLUS supplies[6] the dataset `cancer.vet` on a Veteran's Administration lung cancer trial used by Kalbfleisch & Prentice (1980), but as it has no on-line help, it is not obvious what it is! It is a matrix of 137 cases with right-censored survival time and the covariates

treatment   standard or test

---

[6] It is missing in S-PLUS 5.0 and so is supplied in library MASS for that version.

celltype	one of four cell types
Karnofsky score	of performance on scale 0–100, with high values for relatively well patients
diagnosis	time since diagnosis in months at entry to trial
age	in years
therapy	logical for prior therapy

As there are several covariates, we use the Cox model to establish baseline hazards.

```
> VA.temp <- as.data.frame(cancer.vet)
> dimnames(VA.temp)[[2]] <- c("treat", "cell", "stime",
    "status", "Karn", "diag.time","age","therapy")
> attach(VA.temp)
> VA <- data.frame(stime, status, treat=factor(treat), age,
    Karn, diag.time, cell=factor(cell), prior=factor(therapy))
> detach()
> VA.cox <- coxph(Surv(stime, status) ~ treat + age  + Karn +
    diag.time + cell + prior, VA)
> VA.cox
              coef exp(coef) se(coef)        z       p
  treat   2.95e-01     1.343  0.20755  1.41945 1.6e-01
    age  -8.71e-03     0.991  0.00930 -0.93612 3.5e-01
   Karn  -3.28e-02     0.968  0.00551 -5.95801 2.6e-09
diag.time 8.18e-05     1.000  0.00914  0.00895 9.9e-01
  cell2   8.62e-01     2.367  0.27528  3.12970 1.7e-03
  cell3   1.20e+00     3.307  0.30092  3.97474 7.0e-05
  cell4   4.01e-01     1.494  0.28269  1.41955 1.6e-01
  prior   7.16e-02     1.074  0.23231  0.30817 7.6e-01

Likelihood ratio test=62.1  on 8 df, p=1.8e-10  n= 137

> VA.coxs <- coxph(Surv(stime, status) ~ treat + age + Karn +
    diag.time + strata(cell) + prior, VA)
> VA.coxs
              coef exp(coef) se(coef)       z       p
  treat   0.28590     1.331  0.21001  1.361 1.7e-01
    age  -0.01182     0.988  0.00985 -1.201 2.3e-01
   Karn  -0.03826     0.962  0.00593 -6.450 1.1e-10
diag.time -0.00344     0.997  0.00907 -0.379 7.0e-01
  prior   0.16907     1.184  0.23567  0.717 4.7e-01

Likelihood ratio test=44.3  on 5 df, p=2.04e-08  n= 137

> plot(survfit(VA.coxs), log=T, lty=1:4, col=2:5)
> legend(locator(1), c("squamous", "small", "adeno", "large"),
    lty=1:4, col=2:5)
> plot(survfit(VA.coxs), fun="cloglog", lty=1:4, col=2:5)
> cKarn <- factor(cut(VA$Karn, 5))
> VA.cox1 <- coxph(Surv(stime, status) ~ strata(cKarn) + cell, VA)
```

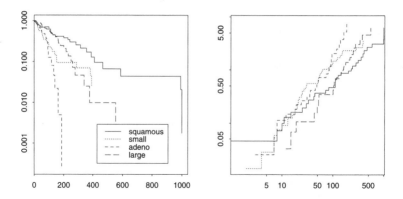

Figure 12.6: Cumulative hazard functions for the cell types in the VA lung cancer trial. The left-hand plot is labelled by survival probability on log scale. The right-hand plot is on log-log scale.

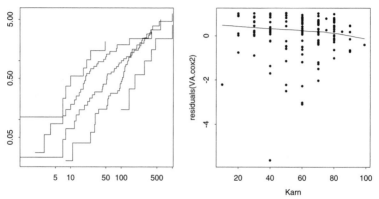

Figure 12.7: Diagnostic plots for the Karnofsky score in the VA lung cancer trial: left: Log-log cumulative hazard plot for five groups; right: Martingale residuals versus Karnofsky score, with a smoothed fit.

```
> plot(survfit(VA.cox1), fun="cloglog")
> VA.cox2 <- coxph(Surv(stime, status) ~ Karn + strata(cell), VA)
> scatter.smooth(VA$Karn, residuals(VA.cox2))
> detach()
```

Figures 12.6 and 12.7 show some support for proportional hazards among the cell types (except perhaps squamous), and suggest a Weibull or even exponential distribution.

```
> VA.wei <- survreg(Surv(stime, status) ~ treat + age + Karn +
      diag.time + cell + prior, VA)
> summary(VA.wei, cor=F)
    . . . .
                Value Std. Error        z        p
(Intercept)  3.262014      0.66253   4.9236  8.50e-07
```

```
       treat -0.228523    0.18684 -1.2231 2.21e-01
         age  0.006099    0.00855  0.7131 4.76e-01
        Karn  0.030068    0.00483  6.2281 4.72e-10
   diag.time -0.000469    0.00836 -0.0561 9.55e-01
       cell2 -0.826185    0.24631 -3.3542 7.96e-04
       cell3 -1.132725    0.25760 -4.3973 1.10e-05
       cell4 -0.397681    0.25475 -1.5611 1.19e-01
       prior -0.043898    0.21228 -0.2068 8.36e-01
  Log(scale) -0.074599    0.06631 -1.1250 2.61e-01

Scale= 0.928

Weibull distribution
Loglik(model)= -715.6   Loglik(intercept only)= -748.1
        Chisq= 65.08 on 8 degrees of freedom, p= 4.7e-11

> VA.exp <- survreg(Surv(stime, status) ~ Karn + cell, VA,
    dist="exp")
> summary(VA.exp, cor=F)
               Value Std. Error     z        p
(Intercept)   3.4222    0.35463  9.65 4.92e-22
       Karn   0.0297    0.00486  6.11 9.97e-10
      cell2  -0.7102    0.24061 -2.95 3.16e-03
      cell3  -1.0933    0.26863 -4.07 4.70e-05
      cell4  -0.3113    0.26635 -1.17 2.43e-01

Exponential distribution
Loglik(model)= -717   Loglik(intercept only)= -751.2
        Chisq= 68.5 on 4 degrees of freedom, p= 4.7e-14
```

Note that `scale` does not differ significantly from one, so an exponential distribution is an appropriate summary.

```
> cox.zph(VA.coxs)
             rho  chisq       p
     treat -0.0607  0.545 0.46024
       age  0.1734  4.634 0.03134
      Karn  0.2568  9.146 0.00249
  diag.time  0.1542  2.891 0.08909
     prior -0.1574  3.476 0.06226
    GLOBAL      NA 13.488 0.01921
> par(mfrow=c(3,2)); plot(cox.zph(VA.coxs))
```

Closer investigation does show some suggestion of time-varying coefficients in the Cox model. The plot is Figure 12.8. Note that some coefficients which are not significant in the basic model show evidence of varying with time. This suggests that the model with just `Karn` and `cell` may be too simple, and that we need to consider interactions. We automate the search of interactions using `stepAIC`, which has methods for both `coxph` and `survreg` fits. With hindsight, we centre the data.

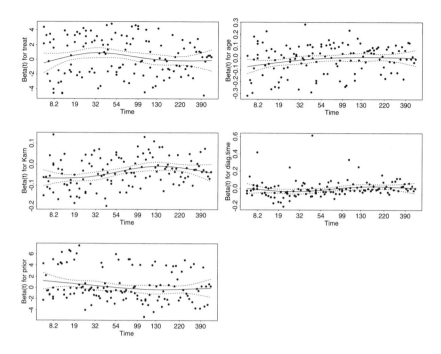

Figure 12.8: Diagnostics plots from `cox.zph` of the constancy of the coefficients in the proportional hazards model `VA.coxs`. Each plot is of a component of the Schoenfeld residual against a non-linear scale of time. A spline smoother is shown, together with ± 2 standard deviations.

```
> VA$Karnc <- VA$Karn - 50
> VA.coxc <- update(VA.cox, ~ . - Karn + Karnc)
> VA.cox2 <- stepAIC(VA.coxc, ~ .^2)
> VA.cox2$anova
Initial Model:
Surv(stime, status) ~ treat + age + diag.time + cell + prior
                      + Karnc

Final Model:
Surv(stime, status) ~ treat + diag.time + cell + prior + Karnc
      + prior:Karnc + diag.time:cell + treat:prior + treat:Karnc
```

	Step	Df	Deviance	Resid. Df	Resid. Dev	AIC
1				129	948.79	964.79
2	+prior:Karnc	1	9.013	128	939.78	957.78
3	+diag.time:cell	3	11.272	125	928.51	952.51
4	-age	1	0.415	126	928.92	950.92
5	+treat:prior	1	2.303	125	926.62	950.62
6	+treat:Karnc	1	2.904	124	923.72	949.72

(The 'deviances' here are minus twice log partial likelihoods.) Applying `stepAIC` to `VA.wei` leads to the same sequence of steps. As the variables `diag.time` and `Karn` are not factors, this will be easier to interpret this fit using nesting:

```
> VA.cox3 <- update(VA.cox2, ~ treat/Karnc + prior*Karnc
    + treat:prior + cell/diag.time)
> VA.cox3
                      coef exp(coef) se(coef)      z       p
         treat     0.8065     2.240  0.27081  2.978 2.9e-03
         prior     0.9191     2.507  0.31568  2.912 3.6e-03
         Karnc    -0.0107     0.989  0.00949 -1.129 2.6e-01
         cell2     1.7068     5.511  0.37233  4.584 4.6e-06
         cell3     1.5633     4.775  0.44205  3.536 4.1e-04
         cell4     0.7476     2.112  0.48136  1.553 1.2e-01
Karnc %in% treat  -0.0187     0.981  0.01101 -1.695 9.0e-02
   prior:Karnc    -0.0481     0.953  0.01281 -3.752 1.8e-04
   treat:prior    -0.7264     0.484  0.41833 -1.736 8.3e-02
 cell1diag.time    0.0532     1.055  0.01595  3.333 8.6e-04
 cell2diag.time   -0.0245     0.976  0.01293 -1.896 5.8e-02
 cell3diag.time    0.0161     1.016  0.04137  0.388 7.0e-01
 cell4diag.time    0.0150     1.015  0.04033  0.373 7.1e-01
```

Thus the hazard increases with time since diagnosis in squamous cells, only, and the effect of the Karnofsky score is only pronounced in the group with prior therapy. We tried replacing `diag.time` with a polynomial, with negligible benefit. Using `cox.zph` shows a very significant change with time in the coefficients of the `treat*Karn` interaction.

```
> cox.zph(VA.cox3)
                      rho     chisq         p
         treat    0.18012   6.10371  0.013490
         prior    0.07197   0.76091  0.383044
         Karnc    0.27220  14.46103  0.000143
         cell2    0.09053   1.31766  0.251013
         cell3    0.06247   0.54793  0.459164
         cell4    0.00528   0.00343  0.953318
Karnc %in% treat -0.20606   7.80427  0.005212
   prior:Karnc   -0.04017   0.26806  0.604637
   treat:prior   -0.13061   2.33270  0.126682
 cell1diag.time   0.11067   1.62464  0.202446
 cell2diag.time  -0.01680   0.04414  0.833596
 cell3diag.time   0.09713   1.10082  0.294086
 cell4diag.time   0.16912   3.16738  0.075123
        GLOBAL         NA  25.52734  0.019661

> par(mfrow=c(2,2))
> plot(cox.zph(VA.cox3), var=c(1,3,7))
```

Stanford heart transplants

This set of data is analysed by Kalbfleisch & Prentice (1980, §5.5.3). (The data given in Kalbfleisch & Prentice are rounded, but the full data are supplied as data frame heart.) It is on survival from early heart transplant operations at Stanford. The new feature is that patients may change treatment during the study, moving from the control group to the treatment group at transplantation, so some of the covariates such as waiting time for a transplant are time-dependent (in the simplest possible way). Patients who received a transplant are treated as two cases, before and after the operation, so cases in the transplant group are in general both right-censored and left-truncated. This is handled by Surv by supplying entry and exit times. For example, patient 4 has the rows

```
start  stop event      age         year     surgery transplant
 0.0   36.0   0  -7.73716632  0.49007529      0          0
36.0   39.0   1  -7.73716632  0.49007529      0          1
```

which show that he waited 36 days for a transplant and then died after 3 days. The proportional hazards model is fitted from this set of cases, but some summaries need to take account of the splitting of patients.

The covariates are age (in years minus 48), year (after 1 October 1967) and an indicator for previous surgery. Rather than use the six models considered by Kalbfleisch & Prentice, we do our own model selection.

```
> coxph(Surv(start, stop, event) ~ transplant*
        (age + surgery + year), heart)
    ....
Likelihood ratio test=18.9  on 7 df, p=0.00852  n= 172
> coxph(Surv(start, stop, event) ~ transplant*(age+year) +
        surgery, heart)
    ....
Likelihood ratio test=18.4  on 6 df, p=0.0053  n= 172
> stan <- coxph(Surv(start, stop, event) ~ transplant*year +
        age + surgery, heart)
> stan
    ....
                    coef exp(coef) se(coef)     z      p
    transplant  -0.6213     0.537   0.5311  -1.17  0.240
          year  -0.2526     0.777   0.1049  -2.41  0.016
           age   0.0299     1.030   0.0137   2.18  0.029
       surgery  -0.6641     0.515   0.3681  -1.80  0.071
transplant:year   0.1974     1.218   0.1395   1.42  0.160

Likelihood ratio test=17.1  on 5 df, p=0.00424  n= 172

> stan1 <- coxph(Surv(start, stop, event) ~ strata(transplant)+
        year + year:transplant + age + surgery, heart)
> plot(survfit(stan1), conf.int=T, log=T, lty=c(1,3), col=2:3)
> legend(locator(1), c("before", "after"), lty=c(1,3), col=2:3)

> attach(heart)
> plot(year[transplant==0], residuals(stan1, collapse=id),
        xlab = "year", ylab="martingale residual")
```

```
> lines(lowess(year[transplant==0],
      residuals(stan1, collapse=id)))
> sresid <- resid(stan1, type="dfbeta", collapse=id)
> detach()
> -100 * sresid %*% diag(1/stan1$coef)
```

This analysis suggests that survival rates over the study improved *prior* to transplantation, which Kalbfleisch & Prentice suggest could be due to changes in recruitment. The diagnostic plots of Figure 12.9 show nothing amiss. The `collapse` argument is needed as those patients who received transplants are treated as two cases, and we need the residual per patient.

Now consider predicting the survival of future patient aged 50 on 1 October 1971 with prior surgery, transplanted after six months.

```
# Survivor curve for the "average" subject
> summary(survfit(stan))
#  follow-up for two years
> stan2 <- data.frame(start=c(0,183), stop=c(183,2*365),
      event=c(0,0), year=c(4,4), age=c(50,50)-48, surgery=c(1,1),
      transplant=c(0,1))
> summary(survfit(stan, stan2, individual=T,
      conf.type="log-log"))
```

time	n.risk	n.event	survival	std.err	lower 95% CI	upper 95% CI
....						
165	43	1	0.654	0.11509	0.384	0.828
186	41	1	0.643	0.11602	0.374	0.820
188	40	1	0.632	0.11697	0.364	0.812
207	39	1	0.621	0.11790	0.353	0.804
219	38	1	0.610	0.11885	0.343	0.796
263	37	1	0.599	0.11978	0.332	0.788
285	35	2	0.575	0.11524	0.325	0.762
308	33	1	0.564	0.11618	0.314	0.753
334	32	1	0.552	0.11712	0.302	0.744
340	31	1	0.540	0.11799	0.291	0.735
343	29	1	0.527	0.11883	0.279	0.725
584	21	1	0.511	0.12018	0.263	0.713
675	17	1	0.492	0.12171	0.245	0.699

The argument `individual=T` is needed to avoid averaging the two cases (which are the same individual). (Earlier versions give incorrect results for the survival probabilities and subsequent columns.)

Australian AIDS survival

The data on the survival of AIDS patients within Australia are of unusually high quality within that field, and jointly with Dr Patty Solomon we have studied survival up to 1992.[7] There are a large number of difficulties in defining survival

---

[7] We are grateful to the Australian National Centre in HIV Epidemiology and Clinical Research for making these data available to us.

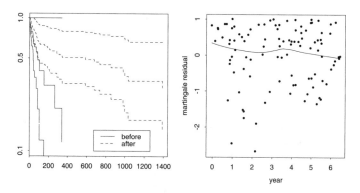

Figure 12.9: Plots for the Stanford heart transplant study: left: log survivor curves for the two groups; right: martingale residuals against calendar time.

from AIDS (acquired immunodeficiency syndrome), in part because as a syndrome its diagnosis is not clear-cut and has almost certainly changed with time. (To avoid any possible confusion, we are studying survival from AIDS and not the HIV infection which is generally accepted as the cause of AIDS.)

The major covariates available were the reported transmission category, and the state or territory within Australia. The AIDS epidemic had started in New South Wales and then spread, so the states have different profiles of cases in calendar time. A factor that was expected to be important in survival is the widespread availability of zidovudine (AZT) to AIDS patients from mid-1987 which has enhanced survival, and the use of zidovudine for HIV-infected patients from mid-1990, which it was thought might delay the onset of AIDS without necessarily postponing death further.

The transmission categories were:

hs	male homosexual or bisexual contact
hsid	as **hs** and also intravenous drug user
id	female or heterosexual male intravenous drug user
het	heterosexual contact
haem	haemophilia or coagulation disorder
blood	receipt of blood, blood components or tissue
mother	mother with or at risk of HIV infection
other	other or unknown

The data file gave data on all patients whose AIDS status was diagnosed prior to January 1992, with their status then. Since there is a delay in notification of death, some deaths in late 1991 would not have been reported and we adjusted the endpoint of the study to 1 July 1991. A total of 2 843 patients were included, of whom about 1 770 had died by the end date. The file contained an ID number, the dates of first diagnosis, birth and death (if applicable), as well as the state and the coded transmission category. We combined the states ACT and NSW (as Australian Capital Territory is a small enclave within New South Wales), and to maintain confidentiality the dates have been jittered and the smallest states combined. Only the transformed file `Aids2` is included in our library.

As there are a number of patients who are diagnosed at (strictly, after) death, there are a number of zero survivals. The software used to have problems with these, so all deaths were shifted by 0.9 days to occur after other events the same day. To transform `Aids2` to a form suitable for time-dependent-covariate analysis we used

```
time.depend.covar <- function(data) {
    id <- row.names(data); n <- length(id)
    events <- c(0, 10043, 11139, 12053) # julian days
    crit1 <- matrix(events[1:3], n, 3 ,byrow=T)
    crit2 <- matrix(events[2:4], n, 3, byrow=T)
    diag <- matrix(data$diag,n,3); death <- matrix(data$death,n,3)
    incid <- (diag < crit2) & (death >= crit1); incid <- t(incid)
    indr <- col(incid)[incid]; indc <- row(incid)[incid]
    ind <- cbind(indr, indc); idno <- id[indr]
    state <- data$state[indr]; T.categ <- data$T.categ[indr]
    age <- data$age[indr]; sex <- data$sex[indr]
    late <- indc - 1
    start <- t(pmax(crit1 - diag, 0))[incid]
    stop <- t(pmin(crit2, death + 0.9) - diag)[incid]
    status <- matrix(as.numeric(data$status),n,3)-1 # 0/1
    status[death > crit2] <- 0; status <- status[ind]
    levels(state) <- c("NSW", "Other", "QLD", "VIC")
    levels(T.categ) <- c("hs", "hsid", "id", "het", "haem",
                  "blood", "mother", "other")
    levels(sex) <- c("F", "M")
    data.frame(idno, zid=factor(late), start, stop, status,
               state, T.categ, age, sex)
}
Aids3 <- time.depend.covar(Aids2)
```

The factor `zid` indicates whether the patient is likely to have received zidovudine at all, and if so whether it might have been administered during HIV infection.

Our analysis was based on a proportional hazards model that allowed a proportional change in hazard from 1 July 1987 to 30 June 1990 and another from 1 July 1990; the results show a halving of hazard from 1 July 1987 but a nonsignificant change in 1990.

```
> attach(Aids3)
> aids.cox <- coxph(Surv(start, stop, status)
      ~ zid + state + T.categ + sex + age, data=Aids3)
> summary(aids.cox)
```

	coef	exp(coef)	se(coef)	z	p
zid1	-0.69087	0.501	0.06578	-10.5034	0.0e+00
zid2	-0.78274	0.457	0.07550	-10.3675	0.0e+00
stateOther	-0.07246	0.930	0.08964	-0.8083	4.2e-01

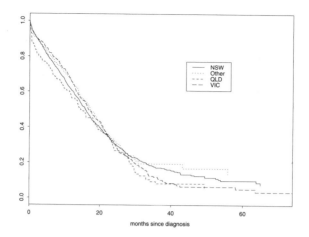

Figure 12.10: Survival of AIDS patients in Australia by state.

```
      stateQLD   0.18315    1.201  0.08752    2.0927 3.6e-02
      stateVIC   0.00464    1.005  0.06134    0.0756 9.4e-01
 T.categhsid  -0.09937    0.905  0.15208   -0.6534 5.1e-01
   T.categid  -0.37979    0.684  0.24613   -1.5431 1.2e-01
  T.categhet  -0.66592    0.514  0.26457   -2.5170 1.2e-02
 T.categhaem   0.38113    1.464  0.18827    2.0243 4.3e-02
T.categblood   0.16856    1.184  0.13763    1.2248 2.2e-01
T.categmother  0.44448    1.560  0.58901    0.7546 4.5e-01
T.categother   0.13156    1.141  0.16380    0.8032 4.2e-01
          sex   0.02421    1.025  0.17557    0.1379 8.9e-01
          age   0.01374    1.014  0.00249    5.5060 3.7e-08
 . . . .
Likelihood ratio test= 185   on 14 df,    p=0
```

The effect of `sex` is nonsignificant, and so dropped in further analyses. There is no detected difference in survival during 1990.

Note that Queensland has a significantly elevated hazard relative to New South Wales (which has over 60% of the cases), and that the intravenous drug users have a longer survival, whereas those infected via blood or blood products have a shorter survival, relative to the first category who form 87% of the cases. We can use stratified Cox models to examine these effects (Figures 12.10 and 12.11).

```
> aids1.cox <- coxph(Surv(start, stop, status)
    ~ zid + strata(state) + T.categ + age, data=Aids3)
> aids1.surv <- survfit(aids1.cox)
> aids1.surv
              n events mean se(mean) median 0.95LCL 0.95UCL
  state=NSW 1780   1116  639     17.6    481     450     509
state=Other  249    142  658     42.2    525     436     618
  state=QLD  226    149  519     33.5    439     357     568
  state=VIC  588    355  610     26.3    508     424     618
```

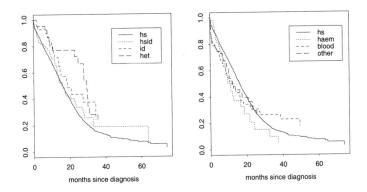

Figure 12.11: Survival of AIDS patients in Australia by transmission category.

```
> plot(aids1.surv, mark.time=F, lty=1:4, col=2:5, xscale=365.25/12,
    xlab="months since diagnosis")
> legend(locator(1), levels(state), lty=1:4, col=2:5)

> aids2.cox <- coxph(Surv(start, stop, status)
    ~ zid + state + strata(T.categ) + age, data=Aids3)
> aids2.surv <- survfit(aids2.cox)
> aids2.surv
```

	n	events	mean	se(mean)	median	0.95LCL	0.95UCL
T.categ=hs	2465	1533	633	15.6	492	473.9	515
T.categ=hsid	72	45	723	86.7	493	396.9	716
T.categ=id	48	19	653	54.3	568	427.9	NA
T.categ=het	40	17	775	57.3	897	753.9	NA
T.categ=haem	46	29	431	53.9	337	209.9	657
T.categ=blood	94	76	583	86.1	358	151.9	666
T.categ=mother	7	3	395	92.6	655	0.9	NA
T.categ=other	70	40	421	40.7	369	24.9	NA

```
> par(mfrow=c(1,2))
> plot(aids2.surv[1:4], mark.time=F, lty=1:4, col=2:5,
    xscale=365.25/12, xlab="months since diagnosis")
> legend(locator(1), levels(T.categ)[1:4], lty=1:4, col=2:5)

> plot(aids2.surv[c(1,5,6,8)], mark.time=F, lty=1:4, col=2:5,
    xscale=365.25/12, xlab="months since diagnosis")
> legend(locator(1), levels(T.categ)[c(1,5,6,8)], lty=1:4, col=2:5)
```

We now consider the possible non-linear dependence of log-hazard on age.
First we consider the martingale residual plot.

```
cases <- diff(c(0,idno)) != 0
aids.res <- residuals(aids.cox, collapse=idno)
scatter.smooth(age[cases], aids.res, xlab = "age",
    ylab="martingale residual")
```

This shows a slight rise in residual with age over 60, but no obvious effect. The next step is to augment a linear term in age by a step function, with breaks chosen from prior experience. The contrast is chosen explicitly to take as base level the 31–40 age group.

```
age2 <- cut(age, c(-1, 15, 30, 40, 50, 60, 100))
c.age <- factor(as.numeric(age2), labels=c("0-15", "16-30",
    "31-40", "41-50", "51-60", "61+"))
table(c.age)
 0-15 16-30 31-40 41-50 51-60 61+
   39  1022  1583   987   269  85
tmp <- diag(6)
dimnames(tmp) <- list(levels(c.age), levels(c.age))
contrasts(c.age) <- tmp[, -3]

summary(coxph(Surv(start, stop, status) ~ zid + state
    + T.categ + age + c.age, data=Aids3))
    . . . .
                 coef exp(coef) se(coef)         z        p
    . . . .
        age  0.00922     1.009  0.00818    1.1266 2.6e-01
  c.age0-15  0.49909     1.647  0.36411    1.3707 1.7e-01
 c.age16-30 -0.01963     0.981  0.09592   -0.2047 8.4e-01
 c.age41-50 -0.00482     0.995  0.09714   -0.0496 9.6e-01
 c.age51-60  0.19814     1.219  0.18199    1.0887 2.8e-01
   c.age61+  0.41369     1.512  0.30821    1.3422 1.8e-01
    . . . .
Likelihood ratio test= 193  on 18 df,   p=0
    . . . .
detach()
```

which is not a significant improvement in fit. Beyond this we could fit a smooth function of age via splines, but to save computational time we defer this to the parametric analysis, which we now consider. From the survivor curves the obvious model is the Weibull. Since this is both a proportional hazards model and an accelerated-life model, we can include the effect of the introduction of zidovudine by assuming a doubling of survival after July 1987. With 'time' computed on this basis we find

```
make.aidsp <- function(){
    cutoff <- 10043
    btime <- pmin(cutoff, Aids2$death) - pmin(cutoff, Aids2$diag)
    atime <- pmax(cutoff, Aids2$death) - pmax(cutoff, Aids2$diag)
    survtime <- btime + 0.5*atime
    status <- as.numeric(Aids2$status)
    data.frame(survtime, status=status - 1, state=Aids2$state,
        T.categ=Aids2$T.categ, age=Aids2$age, sex=Aids2$sex)
}
Aidsp <- make.aidsp()
attach(Aidsp)
```

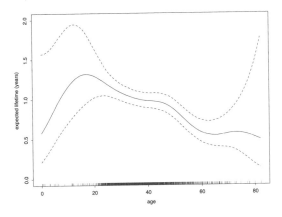

Figure 12.12: Predicted survival versus age of a NSW hs patient (solid line), with point-wise 95% confidence intervals (dashed lines) and a rug of all observed ages.

```
aids.wei <- survreg(Surv(survtime + 0.9, status) ~ state
    + T.categ + sex + age, data=Aidsp)
summary(aids.wei, correlation=F)
    ....
Coefficients:
                  Value Std. Error        z          p
  (Intercept)   6.41825     0.2098  30.5970  1.34e-205
    stateOther  0.09387     0.0931   1.0079   3.13e-01
      stateQLD -0.18213     0.0913  -1.9956   4.60e-02
      stateVIC -0.00750     0.0637  -0.1177   9.06e-01
   T.categhsid  0.09363     0.1582   0.5918   5.54e-01
     T.categid  0.40132     0.2552   1.5727   1.16e-01
    T.categhet  0.67689     0.2744   2.4667   1.36e-02
   T.categhaem -0.34090     0.1956  -1.7429   8.14e-02
  T.categblood -0.17336     0.1429  -1.2131   2.25e-01
 T.categmother -0.40186     0.6123  -0.6563   5.12e-01
  T.categother -0.11279     0.1696  -0.6649   5.06e-01
           sex -0.00426     0.1827  -0.0233   9.81e-01
           age -0.01374     0.0026  -5.2862   1.25e-07
    Log(scale)  0.03969     0.0193   2.0572   3.97e-02

Scale= 1.04
```

Note that we continue to avoid zero survival. This shows good agreement with the parameters for the Cox model. The parameter α (the reciprocal of the scale) is close to one. For practical purposes the exponential is a good fit, and the parameters are little changed.

We also considered parametric non-linear functions of age by using a spline function. We use the P-splines of Eilers & Marx (1996) as this is implemented in both survreg and coxph in survival5: it can be seen as a convenient approximation to smoothing splines. For useful confidence intervals we include the constant term in the predictions, which are for a NSW hs patient. Note that for valid prediction with `pspline` the range of the new data must exactly match that of the old data.

```
> survreg(Surv(survtime+0.9, status) ~ state + T.categ
   + age, data=Aidsp)
   ....
Scale= 1.0405

Weibull distribution
Loglik(model)= -12111   Loglik(intercept only)= -12140

> aids.ps <- survreg(Surv(survtime+0.9,status) ~   state
   + T.categ + pspline(age, df=6), data=Aidsp)
> aids.ps
   ....
                                  coef se(coef)      se2 Chisq   DF
              (Intercept)   4.83189 0.82449  0.60594 34.34 1.00
   ....
pspline(age, df = 6), lin -0.01362 0.00251  0.00251 29.45 1.00
pspline(age, df = 6), non                           9.82 5.04
                                      p
   ....
pspline(age, df = 6), lin 5.8e-08
pspline(age, df = 6), non 8.3e-02
   ....
> zz <- predict(aids.ps, data.frame(
   state=factor(rep("NSW",83), levels=levels(Aidsp$state)),
   T.categ=factor(rep("hs", 83), levels=levels(Aidsp$T.categ)),
   age=0:82), se=T, type="linear")
> plot(0:82, exp(zz$fit)/365.25, type="l", ylim=c(0,2),
   xlab="age", ylab = "expected lifetime (years)")
> lines(0:82, exp(zz$fit+1.96*zz$se.fit)/365.25, lty=3, col=2)
> lines(0:82, exp(zz$fit-1.96*zz$se.fit)/365.25, lty=3, col=2)
> rug(Aidsp$age+runif(length(Aidsp$age), -0.5, 0.5), 0.015)
```

The results (Figure 12.12) suggest that a non-linear in age term is not worthwhile, although there are too few young people to be sure. We predict log-time to get confidence intervals on that scale.

Chapter 13

Time Series Analysis

There are now a large number of books on time series. Our philosophy and notation are close to those of the applied book by Diggle (1990) (from which some of our examples are taken). Brockwell & Davis (1991) and Priestley (1981) provide more theoretical treatments, and Bloomfield (1976) and Priestley are particularly thorough on spectral analysis. Brockwell & Davis (1996) is an excellent low-level introduction to the theory.

Functions for time series have been included in S for some years, and further time-series support was one of the earliest enhancements of S-PLUS. In the current system regularly spaced time series are of class `rts`, and are created by the function `rts`. S-PLUS 5.x introduced a new set of time series classes aimed at event- and calendar-based series. These supersede the older classes `cts` for series of dates and `its` for irregularly spaced series, but like them are only useful for manipulating and plotting such time series: no analysis functions are provided.

Our first running example is `lh`, a series of 48 observations at 10-minute intervals on luteinizing hormone levels for a human female taken from Diggle (1990). This was created by

```
> lh <- rts(scan(n=48))
2.4 2.4 2.4 2.2 2.1 1.5 2.3 2.3 2.5 2.0 1.9 1.7
2.2 1.8 3.2 3.2 2.7 2.2 2.2 1.9 1.9 1.8 2.7 3.0
2.3 2.0 2.0 2.9 2.9 2.7 2.7 2.3 2.6 2.4 1.8 1.7
1.5 1.4 2.1 3.3 3.5 3.5 3.1 2.6 2.1 3.4 3.0 2.9
```

Printing it gives

```
> lh
 1: 2.4 2.4 2.4 2.2 2.1 1.5 2.3 2.3 2.5 2.0 1.9 1.7 2.2 1.8
15: 3.2 3.2 2.7 2.2 2.2 1.9 1.9 1.8 2.7 3.0 2.3 2.0 2.0 2.9
29: 2.9 2.7 2.7 2.3 2.6 2.4 1.8 1.7 1.5 1.4 2.1 3.3 3.5 3.5
43: 3.1 2.6 2.1 3.4 3.0
 start deltat frequency
     1      1         1
```

which shows the attribute vector `tspar` of the class `rts`, which is used for plotting and other computations. The components are the `start`, the label for the first observation, `deltat` (Δt), the increment between observations and

401

frequency, the reciprocal of `deltat`. Note that the final index can be deduced from the attributes and length. Any of `start`, `deltat`, `frequency` and `end` can be specified in the call to `rts`, provided they are specified consistently.

Our second example is a seasonal series. Our dataset `deaths` gives monthly deaths in the UK from a set of common lung diseases for the years 1974 to 1979, from Diggle (1990). This was read into by

```
> deaths <- rts(scan(n=72), start=1974, frequency=12,
               units="months")
3035 2552 2704 2554 2014 1655 1721 1524 1596 2074 2199 2512
2933 2889 2938 2497 1870 1726 1607 1545 1396 1787 2076 2837
2787 3891 3179 2011 1636 1580 1489 1300 1356 1653 2013 2823
3102 2294 2385 2444 1748 1554 1498 1361 1346 1564 1640 2293
2815 3137 2679 1969 1870 1633 1529 1366 1357 1570 1535 2491
3084 2605 2573 2143 1693 1504 1461 1354 1333 1492 1781 1915
> deaths
        Jan  Feb  Mar  Apr  May  Jun  Jul  Aug  Sep  Oct  Nov
1974:  3035 2552 2704 2554 2014 1655 1721 1524 1596 2074 2199
1975:  2933 2889 2938 2497 1870 1726 1607 1545 1396 1787 2076
1976:  2787 3891 3179 2011 1636 1580 1489 1300 1356 1653 2013
1977:  3102 2294 2385 2444 1748 1554 1498 1361 1346 1564 1640
1978:  2815 3137 2679 1969 1870 1633 1529 1366 1357 1570 1535
1979:  3084 2605 2573 2143 1693 1504 1461 1354 1333 1492 1781
       ....
  start   deltat frequency
   1974 0.083333        12
Time units :   months
```

Note how the specification of `units="months"` has triggered a special form of labelling of the print. Quarterly data (with both `frequency=4` and `units="quarters"`) are also treated specially. (Specifying `units` causes serious problems in S-PLUS 5.0.)

There is a series of functions to extract aspects of the time base:

```
> tspar(deaths)
  start   deltat frequency
   1974 0.083333        12
attr(, "units"):
[1] "months"
> start(deaths)
[1] 1974
> end(deaths)
[1] 1979.9
> frequency(deaths)
 frequency
        12
> units(deaths)
[1] "months"

> cycle(deaths)
```

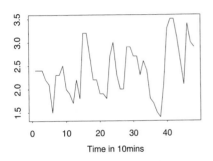
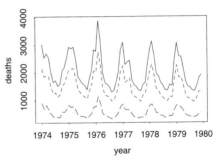

Figure 13.1: Plots by `ts.plot` of `lh` and the three series on deaths by lung diseases. In the right-hand plot the dashed series is for males, the long dashed series for females and the solid line for the total.

```
        Jan Feb Mar Apr May Jun Jul Aug Sep Oct Nov Dec
1974:    1   2   3   4   5   6   7   8   9   10  11  12
1975:    1   2   3   4   5   6   7   8   9   10  11  12
1976:    1   2   3   4   5   6   7   8   9   10  11  12
1977:    1   2   3   4   5   6   7   8   9   10  11  12
1978:    1   2   3   4   5   6   7   8   9   10  11  12
1979:    1   2   3   4   5   6   7   8   9   10  11  12
 start    deltat frequency
  1974 0.083333        12
Time units :   months
```

Time series can be plotted by `plot`, but the functions `ts.plot`, `ts.lines` and `ts.points` are provided for time-series objects. All can plot several related series together. For example, the `deaths` series is the sum of two series `mdeaths` and `fdeaths` for males and females. Figure 13.1 was created by

```
> par(mfrow = c(2,2))
> ts.plot(lh)
> ts.plot(deaths, mdeaths, fdeaths, lty=c(1,3,4), xlab="year",
    ylab="deaths")
```

The functions `ts.union` and `ts.intersect` bind together multiple time series. The time axes are aligned and only observations at times that appear in all the series are retained with `ts.intersect`; with `ts.union` the combined series covers the whole range of the components, possibly as NA values. The result is a matrix with `tspar` attributes set, or a data frame if argument `dframe=T` is set.

The function `window` extracts a sub-series of a single or multiple time series, by specifying `start` and/or `end`.

The function `lag` shifts the time axis of a series back by k positions, default one. Thus `lag(deaths, k=3)` is the series of deaths shifted one quarter into the past. This can cause confusion, as most people think of lags as shifting time and not the series: that is, the current value of a series lagged by one year is last year's, not next year's.

The function `diff` takes the difference between a series and its lagged values, and so returns a series of length $n - k$ with values lost from the beginning (if $k > 0$) or end. (The argument `lag` specifies k and defaults to one. Note that the lag is used in the usual sense here, so `diff(deaths, lag=3)` is equal to `deaths - lag(deaths, k=-3)`!) The function `diff` has an argument `differences` which causes the operation to be iterated. For later use, we denote the dth difference of series X_t by $\nabla^d X_t$, and the dth difference at lag s by $\nabla_s^d X_t$.

The function `aggregate` can be used to change the frequency of the time base. For example, to obtain quarterly sums or annual means of `deaths`:

```
> aggregate(deaths, 4, sum)
          1    2    3    4
1974: 8291 6223 4841 6785
     ....
> aggregate(deaths, 1, mean)
1974: 2178.3 2175.1 2143.2 1935.8 1995.9 1911.5
```

Each of the functions `lag`, `diff` and `aggregate` can also be applied to multiple time series objects formed by `ts.union` or `ts.intersect`.

13.1 Second-order summaries

The theory for time series is based on the assumption of second-order stationarity after removing any trends (which will include seasonal trends). Thus second moments are particularly important in the practical analysis of time series. We assume that the series X_t runs throughout time, but is observed only for $t = 1, \ldots, n$. We use the notations X_t and $X(t)$ interchangeably. The series has a mean μ, often taken to be zero, and the covariance and correlation

$$\gamma_t = \operatorname{cov}(X_{t+\tau}, X_\tau), \qquad \rho_t = \operatorname{corr}(X_{t+\tau}, X_\tau)$$

do not depend on τ. The covariance is estimated for $t > 0$ from the $n - t$ observed pairs $(X_{1+t}, X_1), \ldots, (X_n, X_{n-t})$. If we just take the standard correlation or covariance of these pairs we use different estimates of the mean and variance for each of the subseries $X_{1+t}, \ldots, X_n$ and $X_1, \ldots, X_{n-t}$, whereas under our assumption of second-order stationarity these have the same mean and variance. This suggests the estimators

$$c_t = \frac{1}{n} \sum_{s=\max(1,-t)}^{\min(n-t,n)} [X_{s+t} - \overline{X}][X_s - \overline{X}], \qquad r_t = \frac{c_t}{c_0}$$

Note that we use divisor n even though there are $n - |t|$ terms. This is to ensure that the sequence (c_t) is the covariance sequence of some second-order stationary time series.[1] Note that all of γ, ρ, c, r are symmetric functions ($\gamma_{-t} = \gamma_t$ and so on).

The function `acf` computes and by default plots the sequences (c_t) and (r_t), known as the *autocovariance* and *autocorrelation* functions. The argument `type` controls which is used, and defaults to the correlation.

Our definitions are easily extended to several time series observed over the same interval. Let

$$\gamma_{ij}(t) = \text{cov}\left(X_i(t + \tau), X_j(\tau)\right)$$

$$c_{ij}(t) = \frac{1}{n} \sum_{s=\max(1,-t)}^{\min(n-t,n)} [X_i(s+t) - \overline{X_i}][X_j(s) - \overline{X_j}]$$

which are not symmetric in t for $i \neq j$. These forms are used by `acf` for multiple time series:

```
acf(lh)
acf(lh, type="covariance")
acf(deaths)
acf(ts.union(mdeaths, fdeaths))
```

The `type` may be abbreviated in any unique way, for example `cov`. The output is shown in Figures 13.2 and 13.3. Note that approximate 95% confidence limits are shown for the autocorrelation plots; these are for an independent series for which $\rho_t = I(t = 0)$. As with a time series *a priori* one is expecting autocorrelation, these limits must be viewed with caution. In particular, if any ρ_t is non-zero, all the limits are invalid.

Note that for a series with a non-unit frequency such as `deaths` the lags are expressed in the basic time unit, here years. The function `acf` chooses the number of lags to plot unless this is specified by the argument `lag.max`. Plotting can be suppressed by setting argument `plot=F`. The function returns a list which can be plotted subsequently by `acf.plot`.

The plots of the `deaths` series show the pattern typical of seasonal series, and the autocorrelations do not damp down for large lags. Note how one of the cross-series is only plotted for negative lags. We have $c_{ji}(t) = c_{ij}(-t)$, so the cross terms are needed for all lags, whereas the terms for a single series are symmetric about 0. The labels are confusing: the plot in row 2 column 1 shows c_{12} for negative lags, a reflection of the plot of c_{21} for positive lags.

Spectral analysis

The spectral approach to second-order properties is better able to separate short-term and seasonal effects, and also has a sampling theory which is easier to use for non-independent series.

---

[1] That is, the covariance sequence is positive-definite.

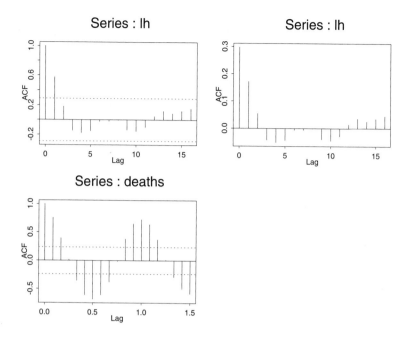

Figure 13.2: acf plots for the series `lh` and `deaths`. The top row shows the autocorrelation (left) and autocovariance (right).

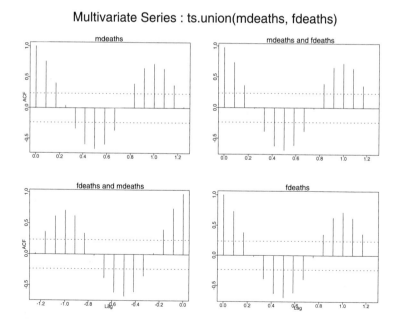

Figure 13.3: Autocorrelation plots for the multiple time series of male and female deaths.

We only give a brief treatment; extensive accounts are given by Bloomfield (1976) and Priestley (1981). Be warned that accounts differ in their choices of where to put the constants in spectral analysis; we have tried to follow S-PLUS as far as possible.

The covariance sequence of a second-order stationary time series can always be expressed as

$$\gamma_t = \frac{1}{2\pi} \int_{-\pi}^{\pi} e^{i\omega t} \, dF(\omega)$$

for the *spectrum* F, a finite measure on $(-\pi, \pi]$. Under mild conditions that exclude purely periodic components of the series, the measure has a density known as the *spectral density* f, so

$$\gamma_t = \frac{1}{2\pi} \int_{-\pi}^{\pi} e^{i\omega t} f(\omega) \, d\omega = \int_{-1/2}^{1/2} e^{2\pi i \omega_f t} f(2\pi\omega_f) \, d\omega_f \tag{13.1}$$

where in the first form the frequency ω is in units of radians/time and in the second form ω_f is in units of cycles/time, and in both cases time is measured in units of Δt. If the time series object has a frequency greater than one and time is measured in the base units, the spectral density will be divided by frequency.

The Fourier integral can be inverted to give

$$f(\omega) = \sum_{-\infty}^{\infty} \gamma_t e^{-i\omega t} = \gamma_0 \left[1 + 2 \sum_{1}^{\infty} \rho_t \cos(\omega t) \right] \tag{13.2}$$

By the symmetry of γ_t, $f(-\omega) = f(\omega)$, and we need only consider f on $(0, \pi)$. Equations (13.1) and (13.2) are the first place the differing constants appear. Bloomfield and Brockwell & Davis omit the factor $1/2\pi$ in (13.1) which therefore appears in (13.2).

The basic tool in estimating the spectral density is the *periodogram*. For a frequency ω we effectively compute the squared correlation between the series and the sine/cosine waves of frequency ω by

$$I(\omega) = \left| \sum_{t=1}^{n} e^{-i\omega t} X_t \right|^2 / n = \frac{1}{n} \left[\left\{ \sum_{t=1}^{n} X_t \sin(\omega t) \right\}^2 + \left\{ \sum_{t=1}^{n} X_t \cos(\omega t) \right\}^2 \right]$$
$$\tag{13.3}$$

Frequency 0 corresponds to the mean, which is normally removed. The frequency π corresponds to a cosine series of alternating ± 1 with no sine series. Bloomfield (but not Brockwell & Davis) has a factor $1/2\pi$ in the definition of the periodogram. S-PLUS appears to divide by the frequency to match its view of the spectral density.

The periodogram is related to the autocovariance function by

$$I(\omega) = \sum_{-\infty}^{\infty} c_t e^{-i\omega t} = c_0 \left[1 + 2 \sum_{1}^{\infty} r_t \cos(\omega t) \right]$$

$$c_t = \frac{1}{2\pi} \int_{-\pi}^{\pi} e^{i\omega t} I(\omega) \, d\omega$$

and so conveys the same information. However, each form makes some of that information easier to interpret.

Asymptotic theory shows that $I(\omega) \sim f(\omega)E$ where E has a standard exponential distribution, except for $\omega = 0$ and $\omega = \pi$. Thus if $I(\omega)$ is plotted on log scale, the variation about the spectral density is the same for all $\omega \in (0, \pi)$ and is given by a Gumbel distribution (as that is the distribution of $\log E$). Furthermore, $I(\omega_1)$ and $I(\omega_2)$ will be asymptotically independent at distinct frequencies. Indeed if ω_k is a *Fourier frequency* of the form $\omega_k = 2\pi k/n$, then the periodogram at two Fourier frequencies will be approximately independent for large n. Thus although the periodogram itself does not provide a consistent estimator of the spectral density, if we assume that the latter is smooth, we can average over adjacent independently distributed periodogram ordinates and obtain a much less variable estimate of $f(\omega)$. A kernel smoother is used of the form

$$\hat{f}(\omega) = \frac{1}{h} \int K\left(\frac{\lambda - \omega}{h}\right) I(\lambda)\, d\lambda$$

$$\approx \frac{2\pi}{nh} \sum_k K\left(\frac{\omega_k - \omega}{h}\right) I(\omega_k) = \sum_k g_k I(\omega_k)$$

for a probability density K. The parameter h controls the degree of smoothing. To see its effect we approximate the mean and variance of $\hat{f}(\omega)$:

$$\mathrm{var}\left(\hat{f}(\omega)\right) \approx \sum_k g_k^2 f(\omega_k)^2 \approx f(\omega)^2 \sum_k g_k^2 \approx \frac{2\pi}{nh} f(\omega)^2 \int K(x)^2\, dx$$

$$E\left(\hat{f}(\omega)\right) \approx \sum_k g_k f(\omega_k) \approx f(\omega) + \frac{f''(\omega)}{2} \sum_k g_k(\omega_k - \omega)^2$$

$$\mathrm{bias}\left(\hat{f}(\omega)\right) \approx \frac{f''(\omega)}{2} h^2 \int x^2 K(x)\, dx$$

so as h increases the variance decreases but the bias increases. We see that the ratio of the variance to the squared mean is approximately $g^2 = \sum_k g_k^2$. If $\hat{f}(\omega)$ had a distribution proportional to χ_ν^2, this ratio would be $2/\nu$, so $2/g^2$ is referred to as the equivalent degrees of freedom. Bloomfield and S-PLUS refer to $\sqrt{2\,\mathrm{bias}\left(\hat{f}(\omega)\right)/f''(\omega)}$ as the *bandwidth*, which is proportional to h.

To understand these quantities, consider a simple moving average over $2m + 1$ Fourier frequencies centred on a Fourier frequency ω. Then the variance is $f(\omega)^2/(2m + 1)$ and the equivalent degrees of freedom are $2(2m + 1)$, as we would expect on averaging $2m + 1$ exponential (or χ_2^2) variates. The bandwidth is approximately

$$\frac{(2m + 1)2\pi}{n} \frac{1}{\sqrt{12}}$$

and the first factor is the width of the window in frequency space. (Since S-PLUS works in cycles rather than radians, the bandwidth is about $(2m + 1)/n\sqrt{12}$ frequency on its scale.) The bandwidth is thus a measure of the size of the smoothing window, but rather smaller than the effective width.

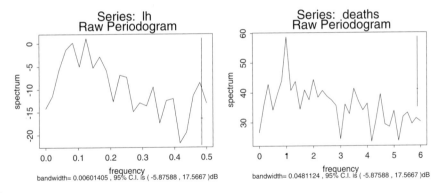

Figure 13.4: Periodogram plots for `lh` and `deaths`.

The workhorse function for spectral analysis is `spectrum`, which with its default options computes and plots the periodogram on log scale. The function `spectrum` calls `spec.pgram` to do most of the work. (Note: `spectrum` by default removes a linear trend from the series before estimating the spectral density.) For our examples we can use:

```
par(mfrow=c(2,2))
spectrum(lh)
spectrum(deaths)
```

with the result shown in Figure 13.4.

Note how elaborately labelled the figures are. The plots are on log scale, in units of *decibels*; that is, the plot is of $10 \log_{10} I(\omega)$. The function `spec.pgram` returns the bandwidth and (equivalent) degrees of freedom as components `bandwidth` and `df`.

The function `spectrum` also produces smoothed plots, using repeated smoothing with modified Daniell smoothers (Bloomfield, 1976), which are moving averages giving half weight to the end values of the span. Trial-and-error is needed to choose the spans (Figures 13.5 and 13.6):

```
par(mfrow=c(2,2))
spectrum(lh)
spectrum(lh, spans=3)
spectrum(lh, spans=c(3,3))
spectrum(lh, spans=c(3,5))

spectrum(deaths)
spectrum(deaths, spans=c(3,3))
spectrum(deaths, spans=c(3,5))
spectrum(deaths, spans=c(5,7))
```

The spans should be odd integers, and it helps to produce a smooth plot if they are different and at least two are used. The width of the centre mark on the 95% confidence interval indicator indicates the bandwidth.

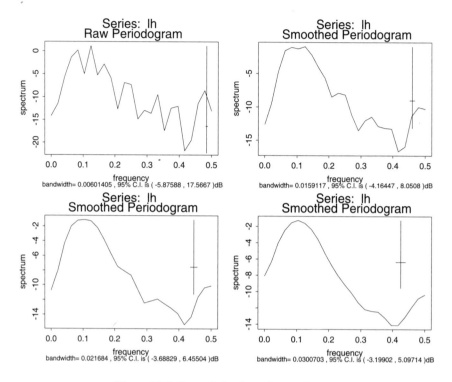

Figure 13.5: Spectral density estimates for lh.

The periodogram has other uses. If there are periodic components in the series the distribution theory given previously does not apply, but there will be peaks in the plotted periodogram. Smoothing will reduce those peaks, but they can be seen quite clearly by plotting the *cumulative periodogram*

$$U(\omega) = \sum_{0<\omega_k\leqslant\omega} I(\omega_k) \Big/ \sum_{1}^{\lfloor n/2 \rfloor} I(\omega_k)$$

against ω. The cumulative periodogram is also very useful as a test of whether a particular spectral density is appropriate, as if we replace $I(\omega)$ by $I(\omega)/f(\omega)$, $U(\omega)$ should be a straight line. Furthermore, asymptotically, the maximum deviation from that straight line has a distribution given by that of the Kolmogorov–Smirnov statistic, with a 95% limit approximately $1.358/[\sqrt{m}+0.11+0.12/\sqrt{m}]$ where $m = \lfloor n/2 \rfloor$ is the number of Fourier frequencies included. This is particularly useful for a residual series with f constant in testing if the series is uncorrelated.

The distribution theory can be made more accurate, and the peaks made sharper, by *tapering* the de-meaned series (Bloomfield, 1976). The magnitude of the first α and last α of the series is tapered down towards zero by a cosine

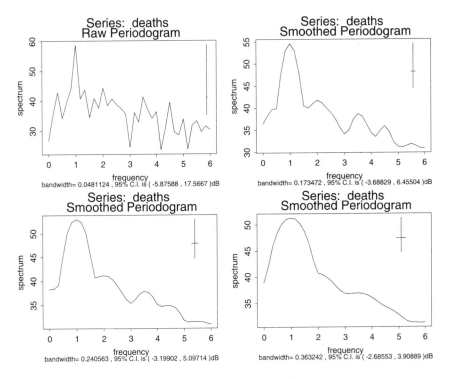

Figure 13.6: Spectral density estimates for deaths.

bell; that is, X_t is replaced by

$$X_t' = \begin{cases} (1 - \cos\frac{\pi(t-0.5)}{\alpha n})X_t & t \leqslant \alpha n \\ X_t & \alpha n < t < (1-\alpha)n \\ (1 - \cos\frac{\pi(n-t+0.5)}{\alpha n})X_t & t \geqslant (1-\alpha)n \end{cases}$$

The proportion α is controlled by the parameter `taper` of `spec.pgram`, and defaults to 10%. It should rarely need to be altered. (The taper function `spec.taper` can be called directly if needed.) Tapering does increase the variance of the periodogram and hence the spectral density estimate, by about 12% for the default taper, but it will decrease the bias near peaks very markedly, if those peaks are not at Fourier frequencies.

We cannot show this directly using `spectrum`, since that plots frequency zero, and as the taper is reduced, the mean of the tapered series and hence $I(0)$ goes to zero. We need to use a taper for the plot used to set the scale:

```
spectrum(deaths)
deaths.spc <- spec.pgram(deaths, taper=0)
lines(deaths.spc$freq, deaths.spc$spec, lty=3)
```

In this example tapering does not help resolve the peaks, but they are at Fourier frequencies. S-PLUS does not supply a function for the cumulative periodogram,

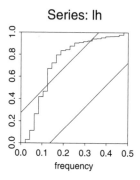
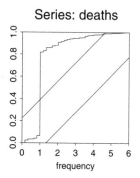

Figure 13.7: Cumulative periodogram plots for `lh` and `deaths`.

but we wrote `cpgram` using the `fft` function to compute a discrete Fourier transform. The results for our examples are shown in Figure 13.7, with 95% confidence bands.

```
cpgram(lh)
cpgram(deaths)
```

13.2 ARIMA models

In the late 1960s Box and Jenkins advocated a methodology for time series based on finite-parameter models for the second-order properties, so this approach is often named after them. Let ϵ_t denote a series of uncorrelated random variables with mean zero and variance σ^2. A moving average process of order q (MA(q)) is defined by

$$X_t = \sum_0^q \beta_j \epsilon_{t-j} \tag{13.4}$$

an autoregressive process of order p (AR(p)) is defined by

$$X_t = \sum_1^p \alpha_i X_{t-i} + \epsilon_t \tag{13.5}$$

and an ARMA(p, q) process is defined by

$$X_t = \sum_1^p \alpha_i X_{t-i} + \sum_0^q \beta_j \epsilon_{t-j} \tag{13.6}$$

(S-PLUS reverses the sign of the MA coefficients for $j > 0$.)

Note that we do not need both σ^2 and β_0 for a MA(q) process, and we take $\beta_0 = 1$. Some authors put the regression terms of (13.5) and (13.6) on the left-hand side and reverse the sign of α_i. Any of these processes can be given mean μ by adding μ to each observation.

An ARIMA(p, d, q) process (where the I stands for integrated) is a process whose dth difference $\nabla^d X$ is an ARMA(p, q) process.

Equation (13.4) will always define a second-order stationary time series, but (13.5) and (13.6) need not. They need the condition that all the (complex) roots of the polynomial

$$\phi_\alpha(z) = 1 - \alpha_1 z - \cdots - \alpha_p z^p$$

lie outside the unit disc. (The function `polyroot` can be used to check this.) However, there are in general 2^q sets of coefficients in (13.4) that give the same second-order properties, and it is conventional to take the set with roots of

$$\phi_\beta(z) = 1 + \beta_1 z + \cdots + \beta_q z^q$$

outside the unit disc. Let B be the backshift or lag operator defined by $BX_t = X_{t-1}$. Then we conventionally write an ARMA process as

$$\phi_\alpha(B)X = \phi_\beta(B)\epsilon \tag{13.7}$$

The function `arima.sim` simulates an ARIMA process. Simple usage is of the form

```
ts.sim <- arima.sim(list(order=c(1,1,0), ar=0.7), n=200)
```

which generates a series whose first differences follow an AR(1) process.

Model identification

A lot of attention has been paid to *identifying* ARMA models, that is choosing plausible values of p and q by looking at the second-order properties. Much of the literature is reviewed by de Gooijer *et al.* (1985). Nowadays it is computationally feasible to fit all plausible models and choose on the basis of their goodness of fit, but some simple diagnostics are still useful. For an MA(q) process we have

$$\gamma_k = \sigma^2 \sum_{i=0}^{q-|k|} \beta_i \beta_{i+|k|}$$

which is zero for $|k| > q$, and this may be discernible from plots of the ACF. For an AR(p) process the population autocovariances are generally all non-zero, but they satisfy the Yule–Walker equations

$$\rho_k = \sum_{1}^{p} \alpha_i \rho_{k-i}, \qquad k > 0 \tag{13.8}$$

This motivates the *partial autocorrelation function*. The partial correlation between X_s and X_{s+t} is the correlation after regression on $X_{s+1}, \ldots, X_{s+t-1}$, and is zero for $t > p$ for an AR(p) process. The PACF can be estimated by solving the Yule–Walker equations (13.8) with $p = t$ and ρ replaced by r, and is given by the `type="partial"` option of `acf`:

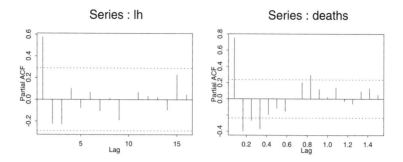

Figure 13.8: Partial autocorrelation plots for the series `lh` and `deaths`.

```
acf(lh, type="partial")
acf(deaths, type="partial")
```

as shown in Figure 13.8. These are short series, so no definitive pattern emerges, but `lh` might be fitted well by an AR(1) or perhaps an AR(3) process.

Model fitting

Selection among ARMA processes can be done by Akaike's information criterion (AIC) which penalizes the deviance by twice the number of parameters; the model with the smallest AIC is chosen. (All likelihoods considered assume a Gaussian distribution for the time series.) Fitting can be done by the functions `ar` or `arima.mle`:

```
> lh.ar1 <- ar(lh, F, 1)
> cpgram(lh.ar1$resid, main="AR(1) fit to lh")
> lh.ar <- ar(lh, order.max=9)
> lh.ar$order
[1] 3
> lh.ar$aic
 [1] 18.30668  0.99567  0.53802  0.00000  1.49036  3.21280
 [7]  4.99323  6.46950  8.46258  8.74120
> cpgram(lh.ar$resid, main="AR(3) fit to lh")
> lh1 <- lh - mean(lh)
> lh.arima1 <- arima.mle(lh1, model=list(order=c(1,0,0)),
      n.cond=3)
> arima.diag(lh.arima1)
> lh.arima3 <- arima.mle(lh1, model=list(order=c(3,0,0)),
      n.cond=3)
> arima.diag(lh.arima3)
> lh.arima11 <- arima.mle(lh1, model=list(order=c(1,0,1)),
      n.cond=3)
> arima.diag(lh.arima11)
```

This first fits an AR(1) process and obtains, after removing the mean,

$$X_t = 0.576X_{t-1} + \epsilon_t$$

AR(1) fit to lh

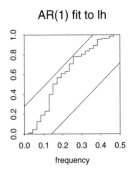

AR(3) fit to lh

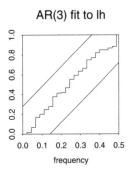

Figure 13.9: Cumulative periodogram plots for residuals of AR models fitted to `lh`.

with $\sigma^2 = 0.208$. It then uses AIC to choose the order among AR processes, selects $p = 3$ and fits

$$X_t = 0.653X_{t-1} - 0.064X_{t-2} - 0.227X_{t-3} + \epsilon_t$$

with $\sigma^2 = 0.196$ and AIC is reduced by 0.996. (For `ar` the component `aic` is the excess over the best fitting model, and it starts from $p = 0$.) The function `ar` by default fits the model by solving the Yule–Walker equations (13.8) with ρ replaced by r. An alternative is to use `method="burg"`.

The function `arima.mle` fits by maximum likelihood and does not include a mean. Confusingly, the `loglik` component returned by `arima.mle` is a measure of the deviance (minus twice the log-likelihood). Equally confusingly, the likelihood considered is not the full likelihood but a likelihood conditional on a set of starting values for the AR and difference terms, so the AIC given cannot be compared between models unless the component `n.cond` is the same.[2] The conditional likelihood conditions on $p + d$ starting values for a non-seasonal series, or `n.cond` if this is larger. The fitted models are

$$X_t = 0.586(0.121)X_{t-1} + \epsilon_t$$

with $\sigma^2 = 0.211$ and AIC $= 59.61$,

$$X_t = 0.658(0.145)X_{t-1} - 0.066(0.174)X_{t-2} - 0.234(0.145)X_{t-3} + \epsilon_t$$

with $\sigma^2 = 0.190$ and AIC $= 59.09$ and

$$X_t = 0.463(0.218)X_{t-1} + \epsilon_t + 0.200(0.241)\epsilon_{t-1}$$

with $\sigma^2 = 0.205$ and AIC $= 60.41$, which shows the MA term is not worthwhile.

The diagnostic plots are shown in Figures 13.9 and 13.10. The cumulative periodograms of the residuals show that the AR(1) process has not removed all

---

[2] This is both unnecessary and unfortunate, as the full likelihood could be computed quite easily; see Brockwell & Davis (1991, §8.7).

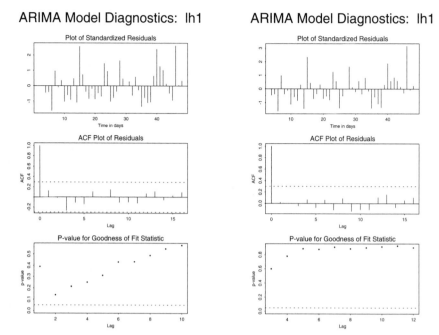

Figure 13.10: Diagnostic plots for ARIMA models fitted to `lh`: (left) AR(1) and (right) AR(3).

the correlation. The bottom panel of the `arima.diag` plots the P-value for the Box & Pierce (1970) *portmanteau test*

$$Q_K = n \sum_1^K c_k^2 \qquad (13.9)$$

applied to the residuals (and not the Ljung–Box test as a comment in the function suggests). Here the maximum K is set by the parameter `gof.lag` (which defaults to 10) plus $p+q$. Note that although an AR(3) model fits better according to AIC, this is not clear-cut from the diagnostic plots. Although the AIC is smaller, a formal test of the difference in deviances, 4.52, using a χ_2^2 distribution is not significant.

The function `arima.diag` can produce (standardized) residuals from a fit by `arima.mle` by setting `plot=F`, `acf.resid=F`, `gof.lag=0` and `resid=T` or `std.resid=T`.

The function `arima.mle` can also include differencing and so fit an ARIMA model (the middle integer in `order` is d).

Forecasting

Forecasting is relatively straightforward using the function `arima.forecast`:

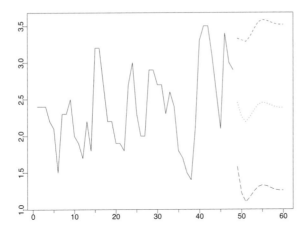

Figure 13.11: Forecasts for 12 periods (of 10mins) ahead for the series lh. The dashed curves are approximate pointwise 95% confidence intervals.

```
lh.fore <- arima.forecast(lh1, n=12, model=lh.arima3$model)
lh.fore$mean <- lh.fore$mean + mean(lh)
ts.plot(lh, lh.fore$mean, lh.fore$mean+2*lh.fore$std.err,
    lh.fore$mean-2*lh.fore$std.err)
```

(see Figure 13.11) but the standard errors do not include the effect of estimating the mean and the parameters of the ARIMA model.

Spectral densities via AR processes

The spectral density of an ARMA process has a simple form: it is given by

$$f(\omega) = \sigma^2 \left| \frac{1 + \sum_s \beta_s e^{-is\omega}}{1 - \sum_t \alpha_t e^{-it\omega}} \right|^2 \tag{13.10}$$

and so we can estimate the spectral density by substituting parameter estimates in (13.10). It is most usual to fit high-order AR models, both because they can be fitted rapidly, and since they can produce peaks in the spectral density estimate by small values of $|1 - \sum \alpha_t e^{-it\omega}|$ (which correspond to nearly non-stationary fitted models since there must be roots of $\phi_\alpha(z)$ near $e^{-i\omega}$).

This procedure is implemented by `spectrum` with `method="ar"`, which calls `spec.ar`. Although popular because it often produces visually pleasing spectral density estimates, it is not recommended (for example, Thomson, 1990).

Regression terms

The `arima` family of functions can also handle regressions with ARIMA residual processes, that is, models of the form

$$X_t = \sum \gamma_i Z_t^{(i)} + \eta_t, \qquad \phi_\alpha(B)\Delta^d \eta_t = \phi_\beta(B)\epsilon \tag{13.11}$$

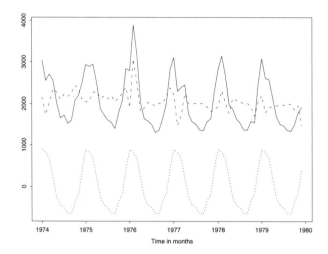

Figure 13.12: stl decomposition for the deaths series (solid line). The dotted series is the seasonal component, the dashed series the remainder.

for one or more external time series $z^{(i)}$. This can be specified for simulation to arima.sim, the parameters γ estimated by arima.mle, and forecasts computed by arima.forecast (provided forecasts of the external series are available). Again, the variability of the parameter estimates $\widehat{\gamma}$ is not taken into account in the computed prediction standard errors.

13.3 Seasonality

For a seasonal series there are two possible approaches. One is to decompose the series, usually into a trend, a seasonal component and a residual, and to apply non-seasonal methods to the residual component. The other is to model all the aspects simultaneously.

Decompositions

Two decomposition algorithms are available. The function sabl dates from 1982 and is being superseded by the function stl. Both are fairly complex, and the on-line documentation should be consulted for full details and references (principally Cleveland *et al.*, 1990).

The function stl can extract a strictly periodic component plus a remainder:

```
deaths.stl <- stl(deaths, "periodic")
ts.plot(deaths, deaths.stl$sea, deaths.stl$rem)
```

as shown in Figure 13.12.

The function monthplot plots the seasonal component of a series decomposed by sabl or stl.

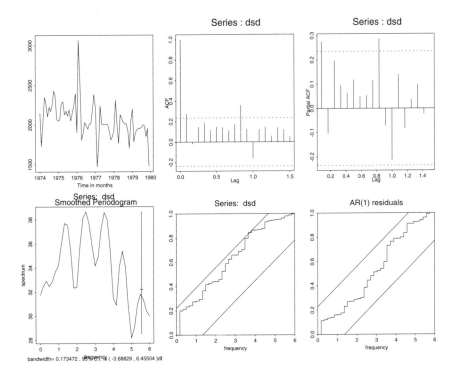

Figure 13.13: Diagnostics for an AR(1) fit to the remainder of an `stl` decomposition of the `deaths` series.

We now return to complete the analysis of the `deaths` series by analysing the non-seasonal component. The results are shown in Figure 13.13.

```
> dsd <- deaths.stl$rem
> ts.plot(dsd)
> acf(dsd)
> acf(dsd, type="partial")
> spectrum(dsd, span=c(3,3))
> cpgram(dsd)
> dsd.ar <- ar(dsd)
> dsd.ar$order
[1] 1
> dsd.ar$aic
  [1] 3.64856  0.00000  1.22644  0.40857  1.75586  3.46936
    ....
> dsd.ar$ar
[1,] 0.27469
> cpgram(dsd.ar$resid, main="AR(1) residuals")
> dsd.rar <- ar.gm(dsd)
> dsd.rar$ar
    ....
  [1] 0.41493
```

The large jump in the cumulative periodogram at the lowest (non-zero) Fourier frequency is caused by the downward trend in the series. The spectrum has dips at the integers since we have removed the seasonal component and hence all of the frequency and its multiples. (The dip at 1 is obscured by the peak at the Fourier frequency to its left; see the cumulative periodogram.) The plot of the remainder series shows exceptional values for February–March 1976 and 1977. As there are only six cycles, the seasonal pattern is difficult to establish at all precisely.

The robust AR-fitting function ar.gm helps to overcome the effect of outliers such as February 1976, and produces a considerably higher AR coefficient. The values for February 1976 and 1977 are heavily down-weighted. Unfortunately its output does not mesh well with the other time-series functions.

Seasonal ARIMA models

The function diff allows differences at lags greater than one, so for a monthly series the difference at lag 12 is the difference from this time last year. Let s denote the period, often 12. We can then consider ARIMA models for the sub-series sampled s apart, for example, for all Januaries. This corresponds to replacing B by B^s in the definition (13.7). Thus an ARIMA $(P, D, Q)_s$ process is a seasonal version of an ARIMA process. However, we may include both seasonal and non-seasonal terms, obtaining a process of the form

$$\Phi_{AR}(B)\Phi_{SAR}(B^s)Y = \Phi_{MA}(B)\Phi_{SMA}(B^s)\epsilon, \qquad Y = (I-B)^d(I-B^s)^D X$$

If we expand this, we see that it is an ARMA $(p + sP, q + sQ)$ model for Y_t, but parametrized in a special way with large numbers of zero coefficients. It can still be fitted as an ARMA process, and arima.mle can handle models specified in this form, by specifying extra terms in the argument model with the period set. (Examples follow.)

To identify a suitable model we first look at the seasonally differenced series. Figure 13.14 suggests that this may be over-differencing, but that the non-seasonal term should be an AR(2).

```
> deaths.diff <- diff(deaths, 12)
> acf(deaths.diff, 30)
> acf(deaths.diff, 30, type="partial")
> ar(deaths.diff)
$order:
[1] 12
     . . . .
$aic:
 [1]  7.8143  9.5471  5.4082  7.3929  8.5839 10.1979 12.1388
 [8] 14.0201 15.7926 17.2504  8.9905 10.9557  0.0000  1.6472
[15]  2.6845  4.4097  6.4047  8.3152
# this suggests the seasonal effect is still present.
> deaths.arima1 <- arima.mle(deaths, model=list(
      list(order=c(2,0,0)), list(order=c(0,1,0), period=12)) )
```

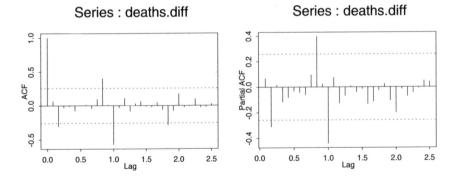

Figure 13.14: Autocorrelation and partial autocorrelation plots for the seasonally differenced `deaths` series. The negative values at lag 12 suggest over-differencing.

```
> deaths.arima1$aic
[1] 847.41
> deaths.arima1$model[[1]]$ar  # the non-seasonal part
[1]  0.12301 -0.30522
> sqrt(diag(deaths.arima1$var.coef))
[1] 0.12504 0.12504
> arima.diag(deaths.arima1, gof.lag=24)
# suggests need a seasonal AR term
> deaths1 <- deaths - mean(deaths)
> deaths.arima2 <- arima.mle(deaths1, model=list(
      list(order=c(2,0,0)),  list(order=c(1,0,0), period=12)) )
> deaths.arima2$aic
[1] 845.38
> deaths.arima2$model[[1]]$ar # non-seasonal part
[1]  0.21601 -0.25356
> deaths.arima2$model[[2]]$ar # seasonal part
[1] 0.82943
> sqrt(diag(deaths.arima2$var.coef))
[1] 0.12702 0.12702 0.07335
> arima.diag(deaths.arima2, gof.lag=24)
> cpgram(arima.diag(deaths.arima2, plot=F, resid=T)$resid)
> deaths.arima3 <- arima.mle(deaths, model=list(
      list(order=c(2,0,0)), list(order=c(1,1,0), period=12)) )
> deaths.arima3$aic  # not comparable to those above
[1] 638.21
> deaths.arima3$model[[1]]$ar
[1]  0.41212 -0.26938
> deaths.arima3$model[[2]]$ar
[1] -0.7269
> sqrt(diag(deaths.arima3$var.coef))
[1] 0.14199 0.14199 0.10125
> arima.diag(deaths.arima3, gof.lag=24)
> arima.mle(deaths1, model=list(list(order=c(2,0,0)),
      list(order=c(1,0,0), period=12)), n.cond=26 )$aic
```

```
[1] 664.14
> deaths.arima4 <- arima.mle(deaths1, model=list(
    list(order=c(2,0,0)), list(order=c(2,0,0), period=12)) )
> deaths.arima4$aic
[1] 634.07
> deaths.arima4$model[[1]]$ar
[1]   0.47821 -0.27873
> deaths.arima4$model[[2]]$ar
[1] 0.17346 0.63474
> sqrt(diag(deaths.arima4$var.coef))
[1] 0.14160 0.14160 0.11393 0.11393
```

The AR-fitting suggests a model of order 12 (of up to 16) which indicates that seasonal effects are still present. The diagnostics from the ARIMA($(2,0,0) \times (0,1,0)_{12}$) model suggest problems at lag 12. Dropping the differencing in favour of a seasonal AR term gives an AIC that favours the second model. However, the diagnostics suggest that there is still seasonal structure in the residuals, so we include differencing and a seasonal AR term. This gives a seasonal model of the form

$$(I + 0.727B^s)(I - B^s)X = (I - 0.273B^s - 0.727B^{2s})X = \epsilon$$

which suggests a seasonal AR(2) term might be more appropriate.

The best fit found is an ARIMA($(2,0,0) \times (2,0,0)_{12}$) model, but this does condition on the first 26 observations, including the exceptional value for February 1976 (see Figure 13.1). Fitting the ARIMA($(2,0,0) \times (1,0,0)_{12}$) model to the same series shows that the seasonal AR term does help the fit considerably.

Trading days

Economic and financial series can be affected by the number of trading days in the month or quarter in question. The function `arima.td` computes a multiple time series with seven series giving the number of days in the months and the differences between the number of Saturdays, Sundays, Mondays, Tuesdays, Wednesdays and Thursdays, and the number of Fridays in the month or quarter. This can be used as a regression term in an ARIMA model. It also has other uses. For the `deaths` series we might consider dividing by the number of days in the month to find daily death rates; we already have enough problems with February!

13.4   Nottingham temperature data

We now consider a substantial example. The data are mean monthly air temperatures ($^\circ$F) at Nottingham Castle for the months January 1920–December 1939, from 'Meteorology of Nottingham', in *City Engineer and Surveyor*. They also occur in Anderson (1976). We use the years 1920–1936 to forecast the years 1937–1939 and compare with the recorded temperatures. The data are series `nottem` in our library.

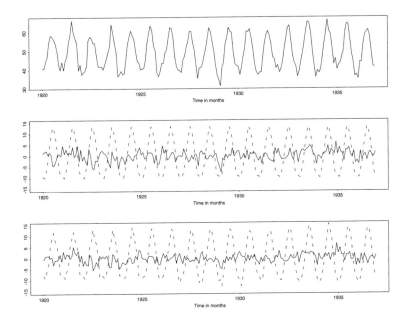

Figure 13.15: Plots of the first 17 years of the `nottem` dataset. Top is the data, middle the `stl` decomposition with a seasonal periodic component and bottom the `stl` decomposition with a 'local' seasonal component.

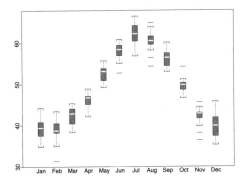

Figure 13.16: Monthly boxplots of the first 17 years of the `nottem` dataset.

```
nott <- window(nottem, end=c(1936,12))
ts.plot(nott)
nott.stl <- stl(nott, "period")
ts.plot(nott.stl$rem-49, nott.stl$sea,
    ylim = c(-15, 15), lty=c(1,3))
nott.stl <- stl(nott, 5)
ts.plot(nott.stl$rem-49, nott.stl$sea,
    ylim = c(-15, 15), lty=c(1,3))
boxplot(split(nott, cycle(nott)), names=month.abb)
```

Figures 13.15 and 13.16 show clearly that February 1929 is an outlier. It *is* correct—it was an exceptionally cold month in England. The `stl` plots show that the seasonal pattern is fairly stable over time. Since the value for February 1929 will distort the fitting process, we altered it to a low value for February of 35°. We first model the remainder series:

```
> nott[110] <- 35
> nott.stl <- stl(nott, "period")
> nott1 <- nott.stl$rem - mean(nott.stl$rem)
> acf(nott1)
> acf(nott1,, "partial")
> cpgram(nott1)
> ar(nott1)$aic
 [1] 13.67432  0.00000  0.11133  2.07849  3.40381  5.40125
> plot(0:23, ar(nott1)$aic, xlab="order", ylab="AIC",
     main="AIC for AR(p)")
> nott1.ar1 <- arima.mle(nott1, model=list(order=c(1,0,0)))
> nott1.ar1$model$ar:
[1] 0.27255
> sqrt(nott1.ar1$var.coef)
ar(1) 0.067529
> nott1.fore <- arima.forecast(nott1, n=36,
     model=nott1.ar1$model)
> nott1.fore$mean <- nott1.fore$mean + mean(nott.stl$rem) +
                     as.vector(nott.stl$sea[1:36])
> ts.plot(window(nottem, 1937), nott1.fore$mean,
     nott1.fore$mean+2*nott1.fore$std.err,
     nott1.fore$mean-2*nott1.fore$std.err, lty=c(3,1,2,2))
> title("via Seasonal Decomposition")
```

(see Figures 13.17 and 13.18) all of which suggest an AR(1) model. (Remember that a seasonal term has been removed, so we expect negative correlation at lag 12.) The confidence intervals in Figure 13.18 for this method ignore the variability of the seasonal terms. We can easily make a rough adjustment. Each seasonal term is approximately the mean of 17 approximately independent observations (since 0.2725516^{12} is negligible). Those observations have variance about 5.05 about the seasonal term, so the seasonal term has standard error about $\sqrt{5.05/17} = 0.55$, compared to the 2.25 for the forecast. The effect of estimating the seasonal terms is in this case negligible. (Note that the forecast errors are correlated with errors in the seasonal terms.)

We now move to the Box–Jenkins methodology of using differencing:

```
> acf(diff(nott,12), 30)
> acf(diff(nott,12), 30, "partial")
> cpgram(diff(nott,12))
> nott.arima1 <- arima.mle(nott,
     model=list(list(order=c(1,0,0)), list(order=c(2,1,0),
     period=12)))
> nott.arima1
```

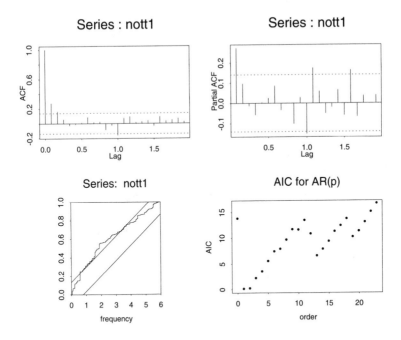

Figure 13.17: Summaries for the remainder series of the `nottem` dataset.

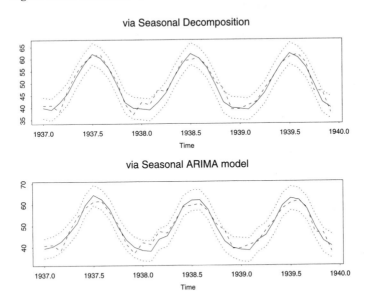

Figure 13.18: Forecasts (solid), true values (dashed) and approximate 95% confidence intervals for the `nottem` series. The upper plot is via a seasonal decomposition, the lower plot via a seasonal ARIMA model.

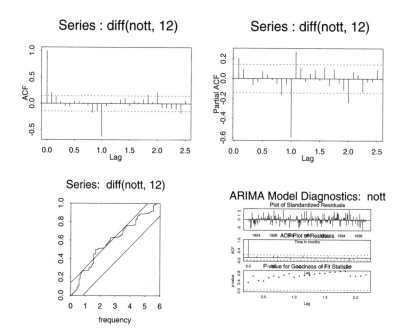

Figure 13.19: Seasonal ARIMA modelling of the `nottem` series. The top row shows the ACF and partial ACF of the yearly differences, which suggest an AR model with seasonal and non-seasonal terms. The bottom row shows the cumulative periodogram of the differenced series and the output of `arima.diag` for the fitted model.

```
$model[[1]]$ar:
[1] 0.32425
$model[[2]]$ar:
[1] -0.87576 -0.31311
> sqrt(diag(nott.arima1$var.coef))
[1] 0.073201 0.073491 0.073491
> arima.diag(nott.arima1, gof.lag=24)
> nott.fore <- arima.forecast(nott, n=36,
    model=nott.arima1$model)
> ts.plot(window(nottem, 1937), nott.fore$mean,
    nott.fore$mean+2*nott.fore$std.err,
    nott.fore$mean-2*nott.fore$std.err, lty=c(3,1,2,2))
> title("via Seasonal ARIMA model")
```

(see Figures 13.18 and 13.19) which produces slightly wider confidence intervals, but happens to fit the true values slightly better. The fitted model is

$$(I + 0.875B^s + 0.313B^{2s})(I - B^s)(I - 0.324B)X = \epsilon$$

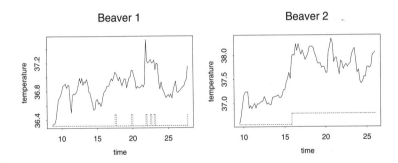

Figure 13.20: Plots of temperature (solid) and activity (dashed) for two beavers. The time is shown in hours since midnight of the first day of observation.

13.5   Regression with autocorrelated errors

We touched briefly on the use of regression terms with the functions `arima.mle` and `arima.forecast` on page 417. In this section we consider a number of ways to use S-PLUS facilities to study regression with autocorrelated errors. They are most pertinent when the regression rather than time-series prediction is of primary interest.

Our main example is taken from Reynolds (1994). She describes a small part of a study of the long-term temperature dynamics of beaver (*Castor canadensis*) in north-central Wisconsin. Body temperature was measured by telemetry every 10 minutes for four females, but data from one period of less than a day for each of two animals is used there (and here). Columns indicate the day (December 12–13, 1990 and November 3–4, 1990 for the two examples), time (hhmm on a 24-hour clock), temperature ($°C$) and a binary index of activity (0 = animal inside retreat; 1 = animal outside retreat).

Figure 13.20 shows the two series. The first series has a missing observation (at 22:20), and this and the pattern of activity suggests that it is easier to start with beaver 2.

```
beav1 <- beav1; beav2 <- beav2
attach(beav1)
beav1$hours <- 24*(day-346) + trunc(time/100) + (time%%100)/60
detach(); attach(beav2)
beav2$hours <- 24*(day-307) + trunc(time/100) + (time%%100)/60
detach()
par(mfrow=c(2,2))
plot(beav1$hours, beav1$temp, type="l", xlab="time",
    ylab="temperature", main="Beaver 1")
usr <- par("usr"); usr[3:4] <- c(-0.2, 8); par(usr=usr)
lines(beav1$hours, beav1$activ, type="s", lty=2)
plot(beav2$hours, beav2$temp, type="l", xlab="time",
    ylab="temperature", main="Beaver 2")
usr <- par("usr"); usr[3:4] <- c(-0.2, 8); par(usr=usr)
lines(beav2$hours, beav2$activ, type="s", lty=2)
```

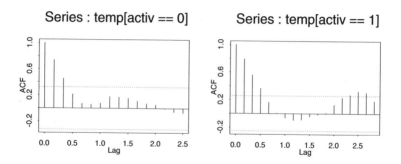

Figure 13.21: ACF of the beaver 2 temperature series before and after activity begins.

Beaver 2

Looking at the series before and after activity begins suggests a moderate amount of autocorrelation, confirmed by the plots in Figure 13.21.

```
attach(beav2)
temp <- rts(temp, start=8+2/3, frequency=6, units="hours")
activ <- rts(activ, start=8+2/3, frequency=6, units="hours")
acf(temp[activ==0]); acf(temp[activ==1]) # also look at PACFs
ar(temp[activ==0]); ar(temp[activ==1])
```

Fitting an $AR(p)$ model to each part of the series selects $AR(1)$ models with coefficients 0.74 and 0.79, so a common $AR(1)$ model for the residual series looks plausible.

We begin by fitting by `arima.mle` a common $AR(1)$ model to find the baseline deviance of -122.86. Adding a term for activity gives a deviance of -140.78, and coefficients of 37.284 (°C) for the mean while inactive and 0.584 for the change in mean due to activity. The AR coefficient is estimated as $0.8255(0.056)$. We might consider that we need a smoother transition in temperature at a change in activity, but as there is only one transition we cannot learn much.

```
arima.mle(temp, xreg=rep(1, length(temp)), model=list(ar=0.75))
arima.mle(temp, xreg=cbind(1, activ), model=list(ar=0.75))
```

We can test for diurnal variation by adding a sine-wave term. This reduces the deviance to -142.55, giving an increase in AIC for two further parameters.

```
dreg <- cbind(sin(2*pi*hours/24), cos(2*pi*hours/24))
arima.mle(temp, xreg=cbind(1, activ,dreg), model=list(ar=0.75))
```

How can we find standard errors for the regression parameter estimates? Unfortunately, `arima.mle` does not give them. We could use a plot of the log-likelihood to find a confidence region. Or we could use first principles and compute $(X^T\widehat{\Sigma}^{-1}X)^{-1}$, assuming that the covariance structure $\widehat{\Sigma}$ of the fitted AR

model is the true structure. However, the most promising idea is to use *pre-whitening*. Our model is

$$y = X\beta + \eta, \qquad (I - \alpha B)\eta = \epsilon,$$

and so

$$(I - \alpha B)y = (I - \alpha B)X\beta + \epsilon$$

which we can fit by least squares, using $(I - \widehat{\alpha}B)X$ as the regressors. This is sometimes known as the Cochrane–Orcutt scheme (Cochrane & Orcutt, 1949), and gives standard errors of 0.096 and 0.098. The mean when active is estimated as 37.868(0.076). These standard errors ignore the estimation of the AR model but are often reliable enough.

```
alpha <- 0.8255
stemp <- temp - alpha*lag(temp, -1)
X <- cbind(1, activ); sX <- X[-1, ] - alpha*X[-100, ]
beav2.ls <- lm(stemp ~ -1 + sX)
beav2.sls <- summary(beav2.ls)
Coefficients:
          Value Std. Error t value Pr(>|t|)
    sX   37.284    0.096    389.071   0.000
 sXactiv  0.584    0.098      5.934   0.000
sqrt(t(c(1,1)) %*% beav2.sls$cov %*% c(1,1)) * beav2.sls$sigma
          [,1]
[1,] 0.076229
plot(hours[-1], residuals(beav2.ls))
detach(); rm(temp, activ)
```

Looking at the residuals from this regression compares each observation with the prediction from the current activity and the immediate past observation. No particular pattern emerges.

Using gls

We can use the function gls from library nlme version 3[3] to fit this model and obtain standard errors.

```
> library(nlme3, first=T)  # may be needed
> beav2.gls <- gls(temp ~ activ, data=beav2,
                corr=corAR1(), method="ML")
> summary(beav2.gls)
    ....
Correlation Structure: AR(1)
 Parameter estimate(s):
     Phi
 0.87318

Coefficients:
```

---
[3] See page 471; with earlier versions we can use lme: see the on-line scripts

```
             Value Std.Error t-value p-value
(Intercept) 37.19      0.11   328.75       0
      activ  0.61      0.11     5.65       0
```

There are some end effects due to the sharp initial rise in temperature:

```
> summary(update(beav2.gls, subset=6:100))
   ....
Correlation Structure: AR(1)
 Parameter estimate(s):
     Phi
 0.83803
Fixed effects: temp ~ activ
             Value Std.Error DF t-value p-value
(Intercept) 37.25       0.1 93  386.68       0
      activ  0.60       0.1 93    6.07       0
```

and REML estimates of the standard errors are somewhat larger.

Beaver 1

Applying the same ideas to beaver 2, we can select an initial covariance model based on the observations before the first activity at 17:30. The autocorrelations again suggest an $AR(1)$ model, whose coefficient is fitted as 0.82. We included as regressors activity now and 10, 20 and 30 minutes ago. This gives a mean when inactive of $36.859(0.032)$.

```
attach(beav1)
temp <- rts(c(temp[1:82], NA, temp[83:114]), start=9.5,
            frequency=6, units="hours")
activ <- rts(c(activ[1:82], NA, activ[83:114]), start=9.5,
            frequency=6, units="hours")
acf(temp[1:53]) # and also type="partial"
ar(temp[1:53])

act <- c(rep(0, 10), activ)
X <- cbind(1, act=act[11:125], act1 = act[10:124],
           act2 = act[9:123], act3 = act[8:122])
arima.mle(temp, xreg=X, model=list(ar=0.82))
$model$ar:
[1] 0.8025

alpha <- 0.80
stemp <- temp - alpha*lag(temp, -1)
sX <- X[-1, ] - alpha * X[-115,]
beav1.ls <- lm(stemp ~ -1 + sX, na.action=na.omit)
summary(beav1.ls, cor=F)
Coefficients:
        Value Std. Error t value Pr(>|t|)
  sX   36.856     0.039   939.833    0.000
```

```
  sXact    0.254   0.039      6.464   0.000
  sXact1   0.171   0.051      3.352   0.001
  sXact2   0.162   0.051      3.148   0.002
  sXact3   0.105   0.043      2.448   0.016
  detach(); rm(temp, activ)
```

All the terms are significant, and adding earlier activity does not improve the model. (Note that we need to be careful with missing values here: which rows contain missing values depend on which terms are included in the model.)

Our analysis shows that there is a difference in temperature between activity and inactivity, and that temperature may build up gradually with activity (which seems physiologically reasonable). A *caveat* is that the apparent outlier at 21:50, at the start of a period of activity, contributes considerably to this conclusion.

13.6 Exercises

13.1.   Our dataset `accdeaths` gives monthly accidental deaths in the USA 1973–8, from Brockwell & Davis (1991). Find a suitable ARIMA model, and predict the deaths for the first six months of 1979.

13.2.   Dataset `austres` is a quarterly series of the number of Australian residents from March 1971 to March 1994. It comes from Brockwell & Davis (1996) who analyse the percentage quarterly changes. Explore suitable models in S-PLUS.

13.3.   Repeat Exercise 9.1 as a time series problem.

13.4.   Use the information gained in the analysis of `beav1` in Section 13.5 to refine the analysis for `beav2`.

13.5.   Consider the problem of estimating the effect of seat belt legislation on road accident casualties in the UK considered by Harvey & Durbin (1986). The data (from Harvey, 1989) are in the series `drivers`.

Chapter 14

Spatial Statistics

Spatial statistics is a recent and graphical subject that is ideally suited to implementation in S; S itself includes one spatial interpolation method, `akima`, and `loess` which can be used for two-dimensional smoothing, but the specialist methods of spatial statistics have been added and are given in our library `spatial`. The main references for spatial statistics are Ripley (1981, 1988), Diggle (1983), Upton & Fingleton (1985) and Cressie (1991). Not surprisingly, our notation is closest to that of Ripley (1981).

The S-PLUS module[1] S+SPATIALSTATS provides more comprehensive (and more polished) facilities for spatial statistics than those provided in our library `spatial`. Details of how to work through our examples in that module may be found in the on-line complements to this book. (See page 467 for where to obtain these.)

14.1   Spatial interpolation and smoothing

We provide three examples of datasets for spatial interpolation. The dataset `topo` contains 52 measurements of topographic height (in feet) within a square of side 310 feet (labelled in 50 feet units). The datasets `shkap` and `npr1` are measurements on oil fields in the (then) USSR and in the USA. Both contain permeability measurements (a measure of the ease of oil flow in the rock) and `npr1` also has porosity (the volumetric proportion of the rock which is pore space).

Suppose we are given n observations $Z(x_i)$ and we wish to map the process $Z(x)$ within a region D. (The sample points x_i are usually, but not always, within D.) Although our treatment is quite general, our S code assumes D to be a two-dimensional region, which covers the majority of examples. There are however applications to the terrestrial sphere and in three dimensions in mineral and oil applications.

---

[1] S-PLUS modules are additional-cost products; contact your S-PLUS distributor for details.

Trend surfaces

One of the earliest methods was fitting *trend surfaces*, polynomial regression surfaces of the form

$$f((x,y)) = \sum_{r+s \leqslant p} a_{rs}x^r y^s \qquad (14.1)$$

where the parameter p is the order of the surface. There are $P = (p+1)(p+2)/2$ coefficients. Originally (14.1) was fitted by least squares, and could for example be fitted using `lm` with `poly` which will give polynomials in one or more variables. There will however be difficulties in prediction, and `predict.gam` must be used to ensure that the correct orthogonal polynomials are generated. This is rather inefficient in applications such as ours in which the number of points at which prediction is needed may far exceed n. Our function `surf.ls` implicitly rescales x and y to $[-1, 1]$, which ensures that the first few polynomials are far from collinear. We show some low-order trend surfaces for the `topo` dataset in Figure 14.1, generated by:

```
library(spatial, first=T)   # avoid name clashes
par(mfrow=c(2,2), pty="s")
topo.ls <- surf.ls(2, topo)
trsurf <- trmat(topo.ls, 0, 6.5, 0, 6.5, 30)
contour(trsurf, levels=seq(600,1000,25), xlab="", ylab="")
points(topo)
title("Degree=2")
topo.ls <- surf.ls(3, topo)
    ....
topo.ls <- surf.ls(4, topo)
    ....
topo.ls <- surf.ls(6, topo)
    ....
```

Figure 14.1 shows trend surfaces for the `topo` dataset. The highest degree, 6, has 28 coefficients fitted from 52 points. The higher-order surfaces begin to show the difficulties of fitting by polynomials in two or more dimensions, when inevitably extrapolation is needed at the edges.

There are several other ways to show trend surfaces in S-PLUS. Figure 14.2 shows a greyscale plot from `levelplot` and a perspective plot from `wireframe`. They were generated by

```
topo.ls <- surf.ls(4, topo)
trsurf <- trmat(topo.ls, 0, 6.5, 0, 6.5, 30)
trsurf[c("x", "y")] <- expand.grid(x=trsurf$x, y=trsurf$y)
plt1 <- levelplot(z ~ x * y, trsurf, aspect=1,
          at = seq(650, 1000, 10),   xlab = "", ylab = "")
plt2 <- wireframe(z ~ x * y, trsurf, aspect=c(1, 0.5),
          screen = list(z = -30, x = -60))
print(plt1, position = c(0, 0, 0.5, 1), more=T)
print(plt2, position = c(0.45, 0, 1, 1))
```

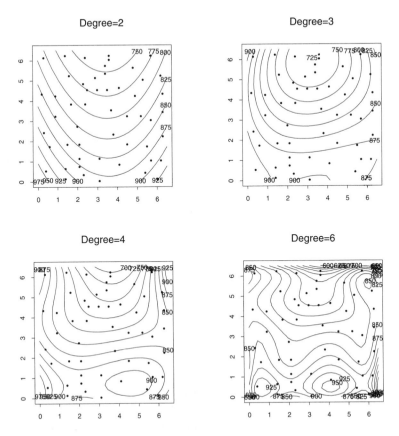

Figure 14.1: Trend surfaces for the topo dataset, of degrees 2, 3, 4 and 6.

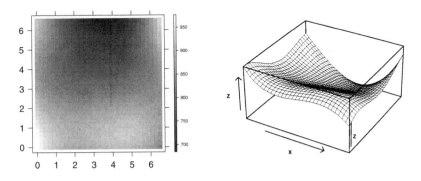

Figure 14.2: The quartic trend surfaces for the topo dataset.

Users of S-PLUS 4.x can use the rotatable 3D-plots in the GUI graphics.

One difficulty with fitting trend surfaces is that in most applications the observations are not regularly spaced, and sometimes they are most dense where the surface is high (for example, in mineral prospecting). This makes it important to take the spatial correlation of the errors into consideration. We thus suppose that

$$Z(\boldsymbol{x}) = \boldsymbol{f}(\boldsymbol{x})^T \boldsymbol{\beta} + \epsilon(\boldsymbol{x})$$

for a parametrized trend term such as (14.1) and a zero-mean spatial stochastic process $\epsilon(\boldsymbol{x})$ of errors. We assume that $\epsilon(\boldsymbol{x})$ possesses second moments, and has covariance matrix

$$C(\boldsymbol{x}, \boldsymbol{y}) = \text{cov}\left(\epsilon(\boldsymbol{x}), \epsilon(\boldsymbol{y})\right)$$

(this assumption is relaxed slightly later). Then the natural way to estimate β is by *generalized least squares*, that is, to minimize

$$[Z(\boldsymbol{x}_i) - \boldsymbol{f}(\boldsymbol{x}_i)^T \boldsymbol{\beta}]^T [C(\boldsymbol{x}_i, \boldsymbol{x}_j)]^{-1} [Z(\boldsymbol{x}_i) - \boldsymbol{f}(\boldsymbol{x}_i)^T \boldsymbol{\beta}]$$

We need some simplified notation. Let $\boldsymbol{Z} = F\boldsymbol{\beta} + \boldsymbol{\epsilon}$ where

$$F = \begin{bmatrix} \boldsymbol{f}(\boldsymbol{x}_1)^T \\ \vdots \\ \boldsymbol{f}(\boldsymbol{x}_n)^T \end{bmatrix}, \qquad \boldsymbol{Z} = \begin{bmatrix} Z(\boldsymbol{x}_1) \\ \vdots \\ Z(\boldsymbol{x}_n) \end{bmatrix}, \qquad \boldsymbol{\epsilon} = \begin{bmatrix} \epsilon(\boldsymbol{x}_1) \\ \vdots \\ \epsilon(\boldsymbol{x}_n) \end{bmatrix}$$

and let $K = [C(\boldsymbol{x}_i, \boldsymbol{x}_j)]$. We assume that K is of full rank. Then the problem is to minimize

$$[\boldsymbol{Z} - F\boldsymbol{\beta}]^T K^{-1} [\boldsymbol{Z} - F\boldsymbol{\beta}] \tag{14.2}$$

The Choleski decomposition (Golub & Van Loan, 1989; Nash, 1990) finds a lower-triangular matrix L such that $K = LL^T$. (The S function `chol` is unusual in working with $U = L^T$.) Then minimizing (14.2) is equivalent to

$$\min_{\boldsymbol{\beta}} \| L^{-1} [\boldsymbol{Z} - F\boldsymbol{\beta}] \|^2$$

which reduces the problem to one of ordinary least squares. To solve this we use the QR decomposition (Golub & Van Loan, 1989) of $L^{-1}F$ as

$$QL^{-1}F = \begin{bmatrix} R \\ 0 \end{bmatrix}$$

for an orthogonal matrix Q and upper-triangular $P \times P$ matrix R. Write

$$QL^{-1}\boldsymbol{Z} = \begin{bmatrix} \boldsymbol{Y}_1 \\ \boldsymbol{Y}_2 \end{bmatrix}$$

as the upper P and lower $n - P$ rows. Then $\widehat{\boldsymbol{\beta}}$ solves

$$R\widehat{\boldsymbol{\beta}} = \boldsymbol{Y}_1$$

which is easy to compute as R is triangular.

Trend surfaces for the `topo` data fitted by generalized least squares are shown later (Figure 14.5), where we discuss the choice of the covariance function C.

Local trend surfaces

We have commented on the difficulties of using polynomials as global surfaces. There are two ways to make their effect local. The first is to fit a polynomial surface for each predicted point, using only the nearby data points. The function `loess` is of this class, and provides a wide range of options. By default it fits a quadratic surface by weighted least squares, the weights ensuring that 'local' data points are most influential. We only give details for the span parameter α less than one. Let $q = \lfloor \alpha n \rfloor$, and let δ denote the Euclidean distance to the qth nearest point to x. Then the weights are

$$w_i = \left[1 - \left(\frac{d(x, x_i)}{\delta} \right)^3 \right]_+^3$$

for the observation at x_i. ($[\quad]_+$ denotes the positive part.) Full details of `loess` are given by Cleveland, Grosse & Shyu (1992). For our example we have:

```
par(mfcol=c(2,2), pty="s")
topo.loess <- loess(z ~ x * y, topo, degree=2, span = 0.25,
    normalize = F)
topo.mar <- list(x=seq(0, 6.5, 0.1), y=seq(0, 6.5, 0.1))
topo.lo <- predict(topo.loess, expand.grid(topo.mar), se = T)
contour(topo.mar$x, topo.mar$y, topo.lo$fit,
    levels = seq(700,1000,25), xlab="fit", ylab="")
points(topo)
contour(topo.mar$x,topo.mar$y,topo.lo$se.fit,
    levels = seq(5, 25, 5), xlab="standard error", ylab="")
points(topo)
topo.loess <- loess(z ~ x * y, topo, degree=1, span = 0.25,
    normalize = F)
    ....
```

We turn normalization off to use Euclidean distance on unscaled variables. Note that the predictions from `loess` are confined to the range of the data in each of the x and y directions even though we requested them to cover the square: this is a side-effect of the algorithms used. The standard-error calculations are slow[2]; `loess` is much faster without them.

Although `loess` allows a wide range of smoothing via its parameter span, it is designed for exploratory work and has no way to choose the smoothness except to 'look good'.

The Dirichlet tessellation of a set of points is the set of *tiles*, each of which is associated with a data point, and is the set of points nearer to that data point than any other. There is an associated triangulation, the Delaunay triangulation, in which data points are connected by an edge of the triangulation if and only if their Dirichlet tiles share an edge. (Algorithms and examples are given in Ripley (1981, §4.3). There is S and FORTRAN code in library `delaunay` available from

[2] Especially on S-PLUS 5.x; replace 0.1 by 0.25 to speed up the calculations.

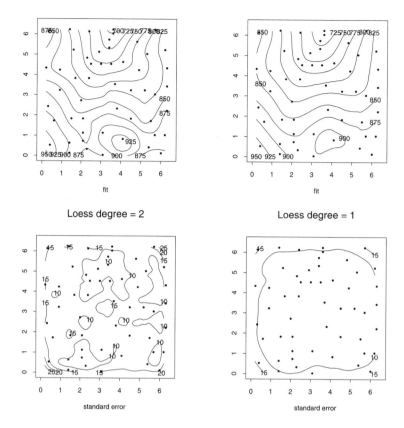

Figure 14.3: loess surfaces and prediction standard errors for the topo dataset.

statlib.) Akima's (1978) fitting method fits a fifth-order trend surface within
each triangle of the Delaunay triangulation; details are given in Ripley (1981,
§4.3). The S implementation is the function interp; Akima's example is in
datasets akima.x, akima.y and akima.z. The method is forced to interpolate
the data, and has no flexibility at all to choose the smoothness of the surface. The
arguments ncp and extrap control details of the method: see the on-line help
for details. For Figure 14.4 we used

```
par(mfrow=c(1,2), pty="s")
contour(interp(topo$x, topo$y, topo$z),xlab="interp default",
        ylab="", levels = seq(600,1000,25))
points(topo)
topo.mar <- list(x = seq(0, 6.5, 0.1), y=seq(0, 6.5, 0.1))
contour(interp(topo$x, topo$y, topo$z, topo.mar$x, topo.mar$y,
    ncp=4, extrap=T), xlab="interp", ylab="",
    levels = seq(600,1000,25))
points(topo)
```

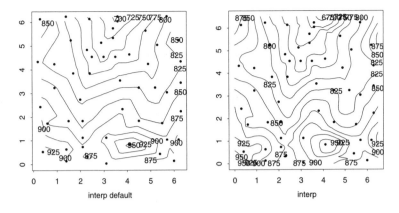

Figure 14.4: `interp` surfaces for the `topo` dataset.

14.2   Kriging

Kriging is the name of a technique developed by Matheron in the early 1960s for mining applications which has been independently discovered many times. Journel & Huijbregts (1978) give a comprehensive guide to its application in the mining industry. In its full form, *universal kriging*, it amounts to fitting a process of the form

$$Z(x) = f(x)^T \beta + \epsilon(x)$$

by generalized least squares, predicting the value at x of both terms and taking their sum. Thus it differs from trend-surface prediction which predicts $\epsilon(x)$ by zero. In what is most commonly termed *kriging*, the trend surface is of degree zero, that is, a constant.

Our derivation of the predictions is given by Ripley (1981, pp. 48–50). Let $k(x) = [C(x, x_i)]$. The computational steps are as follows.

1. Form $K = [C(x_i, y_i)]$, with Cholesky decomposition L.

2. Form F and Z.

3. Minimize $\|L^{-1}Z - L^{-1}F\beta\|^2$, reducing $L^{-1}F$ to R.

4. Form $W = Z - F\widehat{\beta}$, and y such that $L(L^T y) = W$.

5. Predict $Z(x)$ by $\widehat{Z}(x) = y^T k(x) + f(x)^T \widehat{\beta}$, with error variance given by $C(x, x) - \|e\|^2 + \|g\|^2$ where

$$Le = k(x), \qquad R^T g = f(x) - (L^{-1}F)^T e.$$

This recipe involves only linear algebra and so can be implemented in S, but our C version is about 10 times faster. For the `topo` data we have (Figure 14.5):

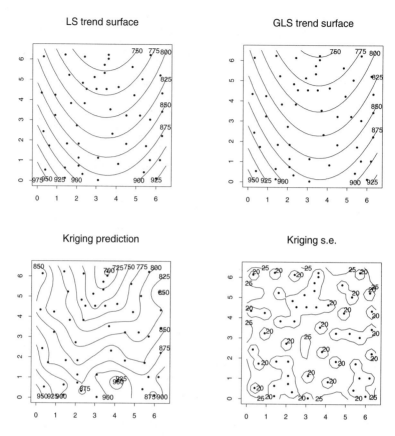

Figure 14.5: Trend surfaces by least squares and generalized least squares, and a kriged surface and standard error of prediction, for the `topo` dataset.

```
topo.ls <- surf.ls(2, topo)
trsurf <- trmat(topo.ls, 0, 6.5, 0, 6.5, 30)
contour(trsurf, levels=seq(600, 1000, 25), xlab="", ylab="")
points(topo)
    ....
topo.gls <- surf.gls(2, expcov, topo, d=0.7)
trsurf <- con2tr(trmat(topo.ls, 0, 6.5, 0, 6.5, 30))
    ....
prsurf <- con2tr(prmat(topo.gls, 0, 6.5, 0, 6.5, 50))
    ....
sesurf <- con2tr(semat(topo.gls, 0, 6.5, 0, 6.5, 30))
    ....
```

Covariance estimation

To use either generalized least squares or kriging we have to know the covariance function C. We assume that

$$C(\boldsymbol{x}, \boldsymbol{y}) = c(d(\boldsymbol{x}, \boldsymbol{y})) \tag{14.3}$$

where $d()$ is Euclidean distance. (An extension known as *geometric anisotropy* can be incorporated by rescaling the variables, as we did for the Mahalanobis distance in Chapter 11.) We can compute a *correlogram* by dividing the distance into a number of bins and finding the covariance between pairs whose distance falls into that bin, then dividing by the overall variance.

Choosing the covariance is very much an iterative process, as we need the covariance of the residuals, and the fitting of the trend surface by generalized least squares depends on the assumed form of the covariance function. Furthermore, as we have residuals their covariance function is a biased estimator of c. In practice it is important to get the form right for small distances, for which the bias is least.

Although $c(0)$ must be one, there is no reason why $c(0+)$ should not be less than one. This is known in the kriging literature as a *nugget effect* since it could arise from a very short-range component of the process $Z(\boldsymbol{x})$. A more general explanation is measurement error. In any case, if there is a nugget effect, the predicted surface will have spikes at the data points, and so effectively will not interpolate but smooth.

The kriging literature tends to work with the *variogram* rather than the covariance function. More properly termed the semi-variogram, this is defined by

$$V(\boldsymbol{x}, \boldsymbol{y}) = \frac{1}{2} E[Z(\boldsymbol{x}) - Z(\boldsymbol{y})]^2$$

and is related to C by

$$V(\boldsymbol{x}, \boldsymbol{y}) = \frac{1}{2}[C(\boldsymbol{x}, \boldsymbol{x}) + C(\boldsymbol{y}, \boldsymbol{y})] - C(\boldsymbol{x}, \boldsymbol{y}) = c(0) - c(d(\boldsymbol{x}, \boldsymbol{y}))$$

under our assumption (14.3). However, since different variance estimates will be used in different bins, the empirical versions will not be so exactly related. Much heat and little light emerges from discussions of their comparison.

There are a number of standard forms of covariance functions that are commonly used. A nugget effect can be added to each. The exponential covariance has

$$c(r) = \sigma^2 \exp -r/d$$

the so-called Gaussian covariance is

$$c(r) = \sigma^2 \exp -(r/d)^2$$

and the spherical covariance is in two dimensions

$$c(r) = \sigma^2 \left[1 - \frac{2}{\pi} \left(\frac{r}{d} \sqrt{1 - \frac{r^2}{d^2}} + \sin^{-1} \frac{r}{d} \right) \right]$$

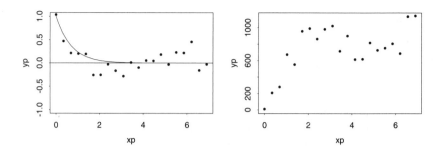

Figure 14.6: Correlogram (left) and variogram (right) for the residuals of `topo` dataset from a least-squares quadratic trend surface.

and in three dimensions (but also valid as a covariance function in two)

$$c(r) = \sigma^2 \left[1 - \frac{3r}{2d} + \frac{r^3}{2d^3} \right]$$

for $r \leqslant d$ and zero for $r > d$. Note that this is genuinely local, since points at a greater distance than d from x are given zero weight at step 5 (although they do affect the trend surface).

We promised to relax the assumption of second-order stationarity slightly. As we only need to predict residuals, we only need a covariance to exist in the space of linear combinations $\sum a_i Z(x_i)$ that are orthogonal to the trend surface. For degree 0, this corresponds to combinations with sum zero. It is possible that the variogram is finite, without the covariance existing, and there are extensions to more general trend surfaces given by Matheron (1973) and reproduced by Cressie (1991, §5.4). In particular, we can always add a constant to c without affecting the predictions (except perhaps numerically). Thus if the variogram v is specified, we work with covariance function $c = \text{const} - v$ for a suitably large constant. The main advantage is in allowing us to use certain functional forms that do not correspond to covariances, such as

$$v(d) = d^\alpha, 0 \leqslant \alpha < 2 \quad \text{or} \quad d^3 - \alpha d$$

The variogram $d^2 \log d$ corresponds to a thin-plate spline in $\mathbb{R}^2$ (see Wahba, 1990 and the review in Cressie, 1991, §3.4.5).

Our functions `correlogram` and `variogram` allow the empirical correlogram and variogram to be plotted and functions `expcov`, `gaucov` and `sphercov` compute the exponential, Gaussian and spherical covariance functions (the latter in two and three dimensions) and can be used as arguments to `surf.gls`. For our running example we have

```
topo.kr <- surf.ls(2, topo)
correlogram(topo.kr, 25)
d <- seq(0, 7, 0.1)
lines(d, expcov(d, 0.7))
variogram(topo.kr, 25)
```

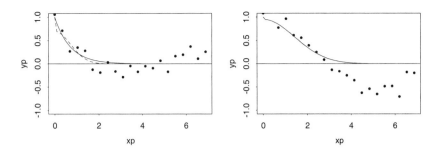

Figure 14.7: Correlograms for the `topo` dataset: (left) residuals from quadratic trend surface showing exponential covariance (solid) and Gaussian covariance (dashed); (right) raw data with fitted Gaussian covariance function.

See Figure 14.6. We then consider fits by generalized least squares.

```
topo.kr <- surf.gls(2, expcov, topo, d=0.7)
correlogram(topo.kr, 25)
lines(d, expcov(d, 0.7))
lines(d, gaucov(d, 1.0, 0.3), lty=3) # try nugget effect

topo.kr <- surf.gls(2, gaucov, topo, d=1, alph=0.3)
prsurf <- prmat(topo.kr, 0, 6.5, 0, 6.5, 50)
contour(prsurf, levels=seq(600, 1000, 25), xlab="fit", ylab="")
points(topo)
sesurf <- semat(topo.kr, 0, 6.5, 0, 6.5, 25)
contour(sesurf, levels=c(15, 20, 25),
        xlab="standard error", ylab="")
points(topo)

topo.kr <- surf.ls(0, topo)
correlogram(topo.kr, 25)
lines(d, gaucov(d, 2, 0.05))

topo.kr <- surf.gls(0, gaucov, topo, d=2, alph=0.05, nx=10000)
prsurf <- prmat(topo.kr, 0, 6.5, 0, 6.5, 50)
contour(prsurf, levels=seq(600, 1000, 25), xlab="fit", ylab="")
points(topo)
sesurf <- semat(topo.kr, 0, 6.5, 0, 6.5, 25)
contour(sesurf, levels=c(15, 20, 25),
        xlab="standard error", ylab="")
points(topo)
```

We first fit a quadratic surface by least squares, then try one plausible covariance function (Figure 14.7). Re-fitting by generalized least squares suggests this function and another with a nugget effect, and we predict the surface from both. The first was shown in Figure 14.5, the second in Figure 14.8. We also consider not using a trend surface but a longer-range covariance function, also shown in Figure 14.8. (The small nugget effect is to ensure numerical stability as without it

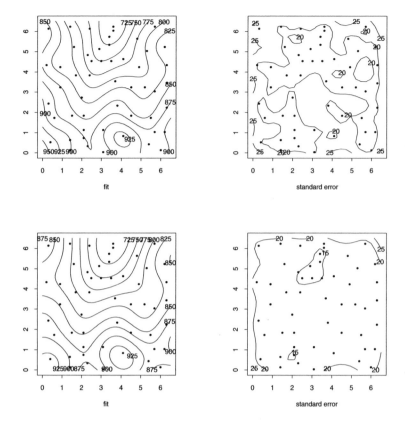

Figure 14.8: Two more kriged surfaces and standard errors of prediction for the topo dataset. The top row uses a quadratic trend surface and a nugget effect. The bottom row is without a trend surface.

the matrix K is very ill-conditioned; the correlations at short distances are very near one. We increased nx for a more accurate lookup table of covariances.)

14.3   Point process analysis

A spatial point pattern is a collection of n points within a region $D \subset \mathbb{R}^2$. The number of points is thought of as random, and the points are considered to be generated by a stationary isotropic point process in $\mathbb{R}^2$. (This means that there is no preferred origin or orientation of the pattern.) For such patterns probably the most useful summaries of the process are the first and second moments of the counts $N(A)$ of the numbers of points within a set $A \subset D$. The first moment can be specified by a single number, the *intensity* λ giving the expected number of points per unit area, obviously estimated by n/a where a denotes the area of D.

The second moment can be specified by Ripley's K function. For example, $\lambda K(t)$ is the expected number of points within distance t of a point of the pattern. The benchmark of complete randomness is the Poisson process, for which $K(t) = \pi t^2$, the area of the search region for the points. Values larger than this indicate clustering on that distance scale, and smaller values indicate regularity. This suggests working with $L(t) = \sqrt{K(t)/\pi}$, which will be linear for a Poisson process.

We only have a single pattern from which to estimate K or L. The definition in the previous paragraph suggests an estimator of $\lambda K(t)$; average over all points of the pattern the number seen within distance t of that point. This would be valid but for the fact that some of the points will be outside D and so invisible. There are a number of edge-corrections available, but that of Ripley (1976) is both simple to compute and rather efficient. This considers a circle centred on the point x and passing through another point y. If the circle lies entirely within D, the point is counted once. If a proportion $p(x, y)$ of the circle lies within D, the point is counted as $1/p$ points. (We may want to put a limit on small p, to reduce the variance at the expense of some bias.) This gives an estimator $\lambda \widehat{K}(t)$ which is unbiased for t up to the circumradius of D (so that it is possible to observe two points $2t$ apart). Since we do not know λ, we estimate it by $\hat{\lambda} = n/a$. Finally

$$\widehat{K}(t) = \frac{a}{n^2} \sum_{x \in D, d(y,x) \leqslant t} \frac{1}{p(x, y)}$$

and obviously we estimate $L(t)$ by $\sqrt{\widehat{K}(t)/\pi}$. We find that on square-root scale the variance of the estimator varies little with t.

Our first example is the Swedish pines data of Ripley (1981, §8.6). This records 72 trees within a 10-metre square. Figure 14.9 shows that $\widehat{L}$ is not straight, and comparison with simulations from a binomial process (a Poisson process conditioned on $N(D) = n$, so n independently uniformly distributed points within D) shows that the lack of straightness is significant. The upper two panels of Figure 14.9 were produced by the following code.

```
library(spatial)
pines <- ppinit("pines.dat")
par(mfrow=c(2,2), pty="s")
plot(pines, xlim=c(0,10), ylim=c(0,10), xlab="", ylab="",
    xaxs="i", yaxs="i")
plot(Kfn(pines,5), type="s", xlab="distance", ylab="L(t)")
lims <- Kenvl(5, 100, Psim(72))
lines(lims$x, lims$l, lty=2)
lines(lims$x, lims$u, lty=2)
```

The function `ppinit` reads the data from the file and also the coordinates of a rectangular domain D. The latter can be reset, or set up for simulations, by the function `ppregion`. (It *must* be set for each session.) The function `Kfn` returns an estimate of $L(t)$ and other useful information for plotting, for distances up to its second argument `fs` (for full-scale).

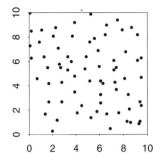

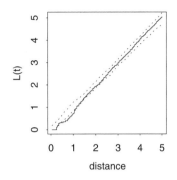

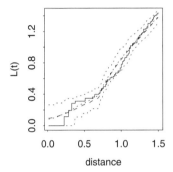

Figure 14.9: The Swedish pines dataset from Ripley (1981), with two plots of $L(t)$. That at the upper right shows the envelope of 100 binomial simulations, that at the lower left the average and the envelope (dotted) of 100 simulations of a Strauss process with $c = 0.2$ and $R = 0.7$. Also shown (dashed) is the average for $c = 0.15$. All units are in metres.

The functions `Kaver` and `Kenvl` return the average and, for `Kenvl`, also the extremes of K-functions (on L scale) for a series of simulations. The function `Psim(n)` simulates the binomial process on n points within the domain D which has already been set.

Alternative processes

We need to consider alternative point processes to the Poisson. One of the most useful for regular point patterns is the so-called Strauss process, which is simulated by `Strauss(n, c, r)`. This has a density of n points proportional to

$$c^{\text{number of } R\text{-close pairs}}$$

and so has $K(t) < \pi t^2$ for $t \leqslant R$ (and up to about $2R$). For $c = 0$ we have a 'hard-core' process that never generates pairs closer than R and so can be envisaged as laying down the centres of non-overlapping discs of diameter $r = R$.

Figure 14.9 also shows the average and envelope of the *L*-plots for a Strauss process fitted to the pines data by Ripley (1981). There the parameters were chosen by trial-and-error based on a knowledge of how the *L*-plot changed with (c, R). Ripley (1988) considers the estimation of *c* for known *R* by the pseudo-likelihood. This is done by our function `pplik` and returns an estimate of about $c = 0.15$ ('about' since it uses numerical integration). As Figure 14.9 shows, the difference between $c = 0.2$ and $c = 0.15$ is small. We used the following code.

```
ppregion(pines)
plot(Kfn(pines,1.5), type="s", xlab="distance", ylab="L(t)")
lims <- Kenvl(1.5, 100, Strauss(72, 0.2, 0.7))
lines(lims$x, lims$a, lty=2)
lines(lims$x, lims$l, lty=2)
lines(lims$x, lims$u, lty=2)
pplik(pines, 0.7)
lines(Kaver(1.5, 100, Strauss(72, 0.15, 0.7)), lty=3)
```

which took about 5 seconds.

The theory is given by Ripley (1988, p. 67). For a point $\xi \in D$ let $t(\xi)$ denote the number of points of the pattern within distance *t* of ξ. Then the pseudo-likelihood estimator solves

$$\frac{\int_D t(\xi) c^{t(\xi)} \, d\xi}{\int_D c^t(\xi) \, d\xi} = \frac{\#(\, R\text{-close pairs}\,)}{n} = \frac{n\widehat{K}(R)}{a}$$

and the left-hand side is an increasing function of *c*. The function `pplik` uses the S-PLUS function `uniroot` to find a solution in the range $(0, 1]$.

Other processes for which simulation functions are provided are the binomial process (`Psim(n)`) and Matérn's sequential spatial inhibition process (`SSI(n, r)`) which sequentially lays down centres of discs of radius *r* that do not overlap existing discs.

14.4   Exercises

14.1.   Repeat Exercise 9.1 as a spatial statistics problem.

Appendix A

Getting Started

A.1 Using S-PLUS under UNIX

S-PLUS versions 3.x and 5.x are based on different 'engines'. Only mix the two with great care: in particular it is almost essential to use separate directories for work done under each.

We use $ to denote the UNIX shell prompt, and assume that the commands to invoke S-PLUS 3.x and S-PLUS 5.x are the defaults, Splus and Splus5.

Getting started

S-PLUS makes use of the file system and for each project we strongly recommend that you have a separate working directory to hold the files for that project.

The suggested procedure for the first occasion on which you use S-PLUS is:

1. Create a separate directory, say, SwS, for this project, which we suppose is 'Statistics with S-PLUS', and make it your working directory.

   ```
   $ mkdir SwS
   $ cd SwS
   ```

   Copy any data files you need to use with S-PLUS to this directory.

2. 3.x Create a subdirectory of SwS called .Data by

   ```
   $ mkdir .Data
   ```

   5.x Within the project directory run

   ```
   $ Splus5 CHAPTER
   ```

   which will create the subdirectories that S-PLUS uses under a directory .Data.

3. Start the S-PLUS system with one of

   ```
   $ Splus    (3.x)
   $ Splus5   (5.x)
   ```

449

4. At this point S commands may be issued (see later). The prompt is > unless the command is incomplete, when it is +. To use our software library issue

```
> library(MASS, first=T)
```

5. To quit the S program the command is

```
>  q()
```

The behaviour if you do not initialize the directory differs by version.

3.x S-PLUS will use the .Data subdirectory of your home directory, or create such a directory for you (with a warning).

5.0 The following options are tried.

(i) If the user's home directory is itself an S-PLUS chapter, it is used.

(ii) A 'temporary' chapter is created in the user's home directory for the current S-PLUS session only. This will have a name of the form Schapterxxxxx and is *not* deleted at the end of the session.

5.1 The following options are tried.

(i) If the directory ~/MySwork is an S-PLUS chapter, it is used.

(ii) If the directory ~/MySwork does not exist, it is created as an S-PLUS chapter and used.

(iii) A 'temporary' chapter is created in the user's home directory for the current S-PLUS session only. This will have a name of the form Schapterxxxxx.

To keep projects separate, we strongly recommend that you *do* create a working directory.

For subsequent sessions the procedure is simpler: make SwS the working directory and start the program as before:

```
$ cd SwS
$ Splus(5)
```

issue S commands, terminating with the command

```
>  q()
```

On the other hand, to start a new project start at step 1.

Getting to the operating system

Sometimes it is necessary to get to the operating system. On a multi-window system we suggest you use another window! If that is not possible, C-shell users can suspend the S-PLUS process. However, it is also possible to issue commands to the operating system by starting the line with !, for example

```
> !date
Thu Feb 17 22:23:03 GMT 1994
```

Note: this starts a new copy of the shell or command interpreter, which inherits its environment from that used to run S-PLUS.

Getting help

There are two ways to access the help system. One is from the command line as described on page 5. Function `help` will put up a pager in the terminal window running S-PLUS to view the help file. If you prefer, a separate help window (which can be left up) can be obtained by

> `help(var, window=T)`

Under 3.x the help page can be printed by

> `help(var, offline=T)`

provided the system manager has set this facility up when the system was installed.

With S-PLUS 3.x we can use `help.start()` to bring up a window listing the help topics available. Items may be selected interactively from a series of menus, and the selection process causes other windows to appear with the help information. These may be scanned at the screen and then either dismissed or sent to a printer. This help system is shut down with `help.off()` and *not* by quitting the help window. If `help` or ? is used when this help system is running, the requests are sent to its window.

Command line editing

Working in a windowing system such as X-windows usually allows you to modify and re-submit previous commands by a cut-and-paste method using the mouse. This is not available when working with a simple terminal, and even with a windowing system some users prefer a keyboard-based recall and edit mechanism.

UNIX versions of S-PLUS have a command line editor that is made available by invoking S-PLUS with the −e flag:

$ `Splus -e`

There are two conventions available for the line editor, either emacs or vi style, set according to the shell environment variable S_CLEDITOR. In a `csh` shell, to get the emacs conventions use

$ `setenv S_CLEDITOR emacs`

and for the vi conventions, use vi instead of emacs. (The default is vi, so setting S_CLEDITOR is not usually necessary in that case.) For a list of the commands use

`?Command.edit`

For users of GNU emacs and Xemacs there is the independently developed ESS package available from

`http://franz.stat.wisc.edu/pub/ESS/`

which provides a comprehensive working environment for S programming. In particular it provides an editing environment tailored to S files with syntax highlighting.

A.2   Using S-PLUS under Windows

Versions 3.x and 4.x (including 2000) of S-PLUS for Windows have quite different user interfaces. We only describe their use under Windows 9x and NT4.0.

There are versions of 4.x called 'Standard Edition' that lack the command-line interface and do not have the capabilities needed for use with this book.

Getting started

1. Create a new folder, say, SWS, for this project, then (S-PLUS 3.x) create a new folder _Data within that folder.

2. Copy any data files you need to use with S-PLUS to the folder SWS.

3. From the Start menu, select Settings, Taskbar, the Start Menu Programs page and click on the Advanced button.

4. Open the S-PLUS folder under Programs.

5. Create a duplicate copy of the S-PLUS for Windows icon, for example, using Copy from the Edit menu. Change the name of this icon to reflect the project for which it will be used.

6. Right-click on the new icon, and select Properties from the pop-up menu.

7. On the page labelled Shortcut,
   (S-PLUS 3.x)   type the complete path to your directory as the field
   Start in, or
   (S-PLUS 4.x)   add at the end of the Target field S_PROJ= followed by the complete path to your directory. If that path contains spaces, enclose it in double quotes, as in

   ```
   S_PROJ="c:\my work\S-Plus project"
   ```

   If you have other files for the project, set the Start in field to their folder. If this is the same place as _Data, you can set S_PROJ=. on the target rather than repeating the path there.

8. Select the project's S-PLUS icon from the Start menu tree.

9. (S-PLUS 4.x)   You will be asked if _Data and _Prefs directories should be created. Click on OK. When the program has initialized, click on the Commands Window button with icon ▨ on the upper toolbar. If you always want a Commands Window at startup, select this from the menus via Options | General Settings... | Startup. (This setting is saved for the project on exiting the program.)

At this point S commands may be issued. The prompt is > unless the command is incomplete, when it is +. The usual Windows command-line recall and editing are available. To use our software library issue

```
library(MASS, first=T)
```

You can exit the program from the File menu, and entering the command q()
in the Commands window will also exit the program.

To keep projects separate, we strongly recommend that you *do* create a working directory.

For subsequent sessions the procedure is much simpler; just launch S-PLUS
by double-clicking on the project's icon. On the other hand, to start a new project
start at step 1.

Getting help

The Windows version of S-PLUS has a standard Windows help system accessible
from the Help item on the main menu, and this can give a tutorial to its use. It is
also possible to use the commands help or ?, for example, by

```
> help(var)
> ?var
```

For inbuilt functions these send the request to the Windows help system. For
many user-written functions the help file is displayed in a pager (default Notepad)
window.

Under S-PLUS 3.x, for some libraries (including ours) it is necessary to specify the library, for example,

```
> help(lda, library="MASS")
```

which uses a Windows help file specific to the library. To make help and ?
behave in the same way for functions in libraries as for inbuilt functions use

```
> library(helpfix, first=T)
```

S-PLUS 4.x has its 'Guides' on-line in PDF format if they were installed from
the CD-ROM: look under the Help menu.

A.3   Customizing your S-PLUS environment

The S environment can be customized in many ways, down to replacing system functions by your own versions. For multi-user systems, all that is normally
desirable is to use the options command to change the defaults of some variables, and if it is appropriate to change these for every session of a project, to use
.First to set them (see the next subsection).

The function options accesses or changes the dataset .Options, which
can also be manipulated directly. Its exact contents will differ between operating
systems and S-PLUS releases, but one example is

```
> unlist(options())
     echo prompt continue width length        keep check digits
  "FALSE" "> "    "+ "      "80"  "48"    "function" "0"    "7"
        memory object.size audit.size          error    show
  "2147483647" "5000000"    "500000"    "dump.calls" "TRUE"
  compact scrap free warn editor expressions reference
  "100000" "500" "1"   "0"   "vi"    "256"          "1"
  contrasts.factor contrasts.ordered   ts.eps  pager
  "contr.helmert"    "contr.poly"        "1e-05" "less"
```

(The function unlist has converted all entries to character strings.) Other options (such as gui) are by default unset. Calling options with no argument or a character vector argument returns a list of the current settings of all options, or those specified. Calling it with one or more name=value pairs resets the values of component name, or sets it if it was unset. For example,

```
> options("width")
$width:
[1] 80
> options("width"=65)
> options(c("length", "width"))
$length:
[1] 48
$width:
[1] 65
```

The meanings of the more commonly used options are given in Table A.1. In S-PLUS 5.x changes made by calling options within a function are local to the current top-level expression unless the argument TEMPORARY=F is set.

There is a similar command, ps.options, to customize the actions of the postscript graphics driver in the UNIX versions of S-PLUS.

Session startup and finishing functions

If the .Data subdirectory of the working directory contains a function .First this function is executed silently at the start of any S session. This allows some automatic customization of the session which may be particular to that working directory. A typical .First function might include commands such as

```
.First <- function()
{
  options(prompt = "> ", continue = "+   ", digits = 5,
     length = 99999, gui = "motif", editor="vi")
  ps.options(paper = "a4", font = 3, pointsize = 10,
     horizontal = F)
  library(MASS, first=T)
}
```

If it were known that the project would always use the same windowing system it might be appropriate to open a graphics window and also a help window with statements such as:

Table A.1: Commonly used options to customize the S environment.

`width`	The page width, in characters. Not always respected.
`length`	The page length, in lines. Used to split up listings of large objects, repeating column headings. Set to a very large value to suppress this.
`digits`	Number of significant digits to use in printing. Set this to 17 for full precision.
`echo`	Logical for whether expressions are echoed before being evaluated. Useful when reading commands from a file.
`prompt`	The primary prompt.
`continue`	The command continuation prompt.
`editor`	The default text editor for `ed` and `fix`.
`error`	Function called to handle errors.
`warn`	The level of strictness in handling warnings. The default, 0, collects them; 1 reports them immediately and 2 makes any warning an error condition.
`memory`	The maximum memory (in bytes) that can be allocated.
`object.size`	The maximum size (in bytes) of any object.
`conflicts.ok`	(S-PLUS 5.x) warn about masked functions when a database is attached.
`indentation`	(S-PLUS 5.x) indentation in code listings, for example, `indentation=" "`.

```
motif("-geometry 600x500-0+0")
help.start()
```

If there is a function `.Last` it is executed when the session is terminated. A typical `.Last` function on a UNIX system where disk space was always in short supply might be:

```
.Last <- function()
{
    unix("rm -f ps.out.*.ps") # remove unwanted PostScript files
    cat("Adios.\n")
}
```

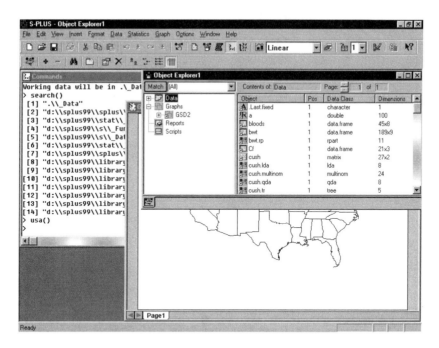

Figure B.1: A snapshot of the interface of S-PLUS 2000 showing three subwindows, from front to back an object explorer, a graphsheet and a commands window. What is displayed as the lower toolbar depends on which subwindow is on top.

```
Script1 - program                                                    _ □ ×
 3    1

lda
lda.default <-
function(x, grouping, prior = proportions, tol = 0.0001, method = c(
    "moment", "mle", "mve", "t"), CV = F, nu = 5, ...)
{
    which.is.max <- function(x)
    {
        d <- (1:length(x))[x == max(x)]
        if(length(d) > 1)
            d <- sample(d, 1)
        d
    }
    if(is.null(dim(x)))
        stop("x is not a matrix")
    n <- nrow(x)
    p <- ncol(x)
    if(n != length(grouping))
        stop("nrow(x) and length(grouping) are different")
    g <- as.factor(grouping)
    lev <- levels(g)
> lda
function(x, ...)
{
    if(is.null(class(x)))
        class(x) <- data.class(x)
    UseMethod("lda", x, ...)
```

Figure B.2: A script subwindow. The definition of `lda` appears in the output pane: it appeared on pressing return at the first line. The definition of `lda.default` was inserted by highlighting the name (as shown), right-clicking and selecting Expand Inplace.

Appendix B

The GUI in Version 4.x

S-PLUS 4.x provides a graphical user interface to the S-PLUS engine, and many new actions. The GUI is highly configurable, but its default state in S-PLUS 2000[1] is shown in Figure B.1. The top toolbar is constant, but the second toolbar and the menu items depend on the type of subwindow that has focus.

The object explorer and the commands window were selected by the two buttons on the top toolbar that are depressed. (What is opened when S-PLUS is launched is set during installation and can be set for each project from the Options menu.) To find out what the buttons mean, hover the mouse pointer over them and read the description in the bottom bar of the main S-PLUS window.

Explanations of how to use a GUI are lengthy and appear in the S-PLUS guides and in the help system. We highlight a few points that we find make working with the GUI easier.

B.1 Subwindows

Instances of these types of subwindow can be launched from a button on the main toolbar or from the New or Open button. Some of the less obvious buttons are shown in Figure B.3.

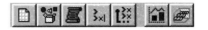

Figure B.3: Some buttons from the main toolbar in S-PLUS 4.5. From left to right these give a data window, object browser (known as an object explorer in S-PLUS 2000), history window, commands windows, commands history windows and 2D and 3D palettes.

The commands window

Commands typed in the commands window are executed immediately. Previous commands can be recalled by using the up and down arrow keys, and edited

---

[1] S-PLUS 4.0 and 4.5 look similar, with raised buttons as shown in Figure B.3.

before submission (by pressing the return key). The commands history button (immediately to the right of the command window button) brings up a dialog with a list of the last few commands, which can be selected and re-submitted. When the commands window has focus, the second toolbar has just one button (with icon representing a pair of axes and a linear plot) that selects editable graphics. This is not recommended for routine use, as it may make the graphics very slow, and plots can be made editable later (see under 'object browsers and explorers'). It is also possible to launch a graphsheet with or without editable graphics from the command line by

```
graphsheet(object.mode="object-oriented")
graphsheet(object.mode="fast")
```

Script windows

You can use a script window rather than a commands window to enter S commands, and this may be most convenient for S programming. A new script window can be opened from the New file button or menu item, and presents a two-part subwindow as shown in Figure B.2. S commands can be typed into the top window and edited there. Pressing return submits the current line. Groups of commands can be selected (in the usual ways in Windows), and submitted by pressing the function key F10 or by the leftmost button on the second line (marked to represent a 'play' key). If text output is produced, this will appear in the bottom part of the subwindow. This output pane is cleared at each submission.

The input part of a script window is associated with a file, conventionally with extension .ssc and double-clicking on .ssc files in Windows Explorer ought to open them in a script window in S-PLUS, launching a new S-PLUS if none is running.

A script window can be launched from the command line to edit a function by the new function Edit (note the difference from edit).

It is the help features that mark a scripts window as different from a commands window. Select a function name by double-clicking on it. Then help on that function is available by pressing the function key F1, and the right-click menu has items Show Dialog... and Expand Inplace to pop up a dialog box for the arguments of the function and to paste in the function body. Versions 4.5 and 2000 have a variety of convenient shortcuts for programmers: auto-indent, auto-insertion of right brace (}) and highlighting of matching left parentheses, brackets, braces and quotes (to)] } " and ').

Scripts can be saved as text files with extension .ssc; use the Save file menu item or button when the script window has focus. They can then be loaded into S-PLUS by opening them in Explorer.

More than one script window can be open at once. To avoid cluttering the screen script windows can be hidden (and unhidden) from the Windows file menu. The Hide item hides the window which has focus, whereas the Unhide... provides a list of windows from which to select.

The ESS package mentioned on page 451 can also be used with NTemacs on Windows, and provides a programming editor that some may prefer (including both of us).

Report windows

A useful alternative to the output pane of a script window is a *report window* which records all the output directed to it and can be saved to a file. The contents of the report window can be edited, so mistakes can be removed and the contents annotated before saving.

Where output is sent is selected by the dialog box brought up by the Text Output Routing... item on the Options menu. This allows separate selections for normal output and warnings/errors; the most useful options are Default and Report.

Object browsers and explorers

There can be one or more object browsers (S-PLUS 4.x) or explorers (S-PLUS 2000) on screen. They provide a two-panel view (see Figure B.1) which will be familiar from many Windows programs. Object views can be expanded down to component level. The right-click menu is context-sensitive: for example, for a linear model fit (of class lm) it has Summary, Plot, Predict and Coefficients items. Double-clicking on a data frame or vector will open it for editing or viewing in a spreadsheet-like *data window*.

If a graphsheet is selected and expanded it will first show its pages (if there are more than one) and then the plotted objects. If the object is labelled CompositeObject then the right-click menu will include the item Convert to Objects which will convert that plot to editable form.

Object browsers and explorers are highly customizable, both in the amount of detail in the right pane and in the databases and classes of objects to be shown. Right-clicking on the background of the left and right panes or on folders or using the Format menu will lead to dialog boxes to customize the format.

The ordering of items in the right pane can be puzzling: click on the heading of a column to sort on that column (as in Windows Explorer).

Data windows

A data window provides a spreadsheet-like view (Figure B.4) of a data frame (or vector or matrix). The scrollbars scroll the table, but the headings remain visible. A region can be selected by dragging (and extended by shift-clicking); including the headings in the selection includes the whole row or column as appropriate.

Entries can be edited and rows and columns inserted or deleted in the usual spreadsheet styles. Toolbar buttons are provided for most of these operations, and for sorting by the selected column. Double-clicking in the top row of a column brings up a format dialog for that column: double-clicking in the top left cell brings up a format dialog for the window that allows the type font and size to be altered.

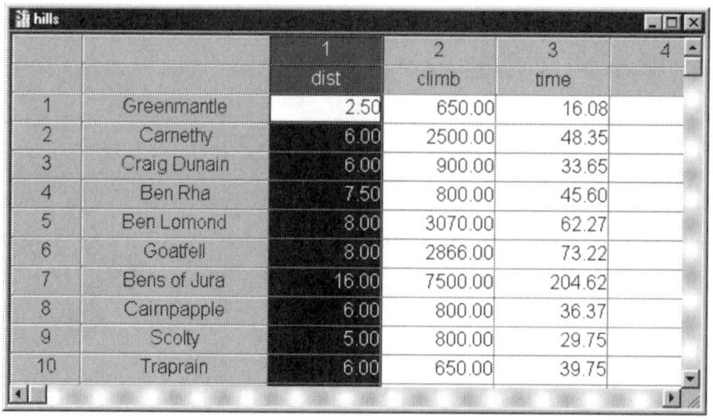

Figure B.4: A data window view of the first few rows of the `hills` dataset with the `dist` column selected.

B.2   Graphics in the GUI

S-PLUS 4.x has a completely separate style of graphics based on menus and toolbars with simple command-line equivalents. The style of the interface is designed for intuitive exploration by experienced Windows users; most of the options are set from dialog boxes brought up by selecting and double-clicking or right-clicking elements of the plot. There are separate dialog boxes for different plot elements: at least the background, plotted objects, axes and any annotations.

This graphical system works in a new graphical device called a *Graph Sheet*. A graphsheet device can be launched from the command window by the function `graphsheet()`, but it will normally be opened from a plot palette button or from the new document icon on the toolbar. Graphsheets can be used as graphics devices with both base and Trellis command-line graphics, and provide limited editing facilities, for example to edit the text of labels and the colour and width of lines, *if* the 'Object-oriented Graphs' button [⊠] has been selected or if the graphsheet was started with argument `object.mode=T`. It is possible to convert graphs to the editable form at a later date using the object browser / explorer (see page 459). Note that command-line graphics will not normally use[2] a graphsheet already opened from the GUI, but will open a new one. Normally the GUI graphics will start a new graphsheet for each plot, whereas the command-line graphics will re-use the existing one.

To produce a hardcopy of a graphsheet to a printer the normal Windows printing facilities can be used: it is also possible to export the graph(s) to a file via the Export Graph... item on the File menu. Finally, graphsheets can be saved (as S-PLUS graph files with extension `.sgr`) and re-imported for editing or additions. Their structure can be browsed in the graphsheets view of the object browser / explorer.

---

[2] An existing graphsheet can be taken over by using its name as the `Name` argument to `graphsheet`.

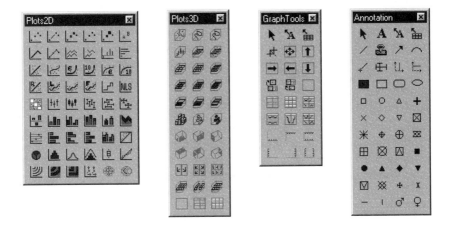

Figure B.5: Four palettes for graphs from S-PLUS 2000. The two plot palettes are launched from buttons on the main toolbar; the Graph Tools and Annotation palettes have buttons on the `graphsheet` toolbar. Some of the items on the Graph Tools palette appear on the Plots 2D palette in versions 4.0 and 4.5.

2D plots

The 2D plot palette (Figure B.5) on the toolbar contains many buttons for different plot types. The basic types of plots are scatter and line plots (of types `"l"`, `"s"` and `"S"`), smoothed plots and various fitted lines. There are also barplots, histograms and pie charts.

Most of these plots work on two columns of data.[3] As they are not specified as arguments to a function, they need to be specified in some other way, and there are many possibilities. The most convenient way will often be to double-click a data frame in the object browser, select the desired columns (in the order x, y, z if needed, ...) and either drag-and-drop onto the appropriate button in the 2D plot palette or just click on the button. The variables may be specified or changed using either of the dialog boxes brought up by double-clicking the plotted points or the background.

All plots can be conditioned in the Trellis style but with a little less flexibility. The default is to use four panels in a 2×2 layout. The conditioning variable(s) can be set in the Multipanel tab of the plot background dialog box or by drag-and-drop (as described in the User's Guide). Alternatively, additional columns can be selected and the Conditioning button selected on the toolbar before generating the plot.

Control over the conditioning is available from the Multipanel tab. This allows the number and layout of panels to be changed, the partitioning of continuous variables to be set (with overlap as in shingles if desired) and strips to be plotted or not. (If they are, they can be selected and their properties edited.)

---

[3] If only one column is selected this is used for y and the index is used for x.

3D plots

The distinction between 2D and 3D plots is by number of axes not data; contour, level and filled contour plots are on the 2D plot palette even though they require *three* columns to be selected. Like Trellis plots, the contour and surface plots require a z coordinate evaluated at a rectangular grid of x and y coordinates, but unlike Trellis they will interpolate irregular data to an automatically chosen grid (controlled from the Gridding tab of the plot dialog box).

The contour and level plots are analogues of `contourplot` and `levelplot`; the filled contour plots are contour plots with the regions between contours filled in different colours. The details of the contours, levels and colours can be altered from the dialog box brought up by double-clicking or right-clicking inside the plot region.

The analogue of the Trellis function `cloud` is a 3D scatterplot from the 3D plot palette. A cloud of points can be represented as points, connected by lines (with or without highlighting the points) or as a 'dropped line scatter' plot where each point is represented by a line segment from (x, y, z) to $(0, 0, z)$. There are also 3D bar charts (sometimes known as Manhattan diagrams) for data on a rectangular grid.

Surfaces can be represented in many ways:

(a) as a wireframe surface, plotted at the grid spacing or at every other grid point;

(b) as a wireframe interpolated to a finer grid (default half the spacing) by splines;

(c) as filled versions of (a) and (b), in which the surface is shown in a solid colour (selected from the Fills tab of the dialog box selected from the surface);

(d) as a draped surface, with levels represented by 8, 16 or 32 colours (many representations intermediate between (c) and (d) can be selected from the Lines and Fills tabs);

(e) as contours or filled contours, in which contour levels are plotted at the appropriate height as horizontal sections of the surface.

To change the representation of a surface, select the surface by clicking on it, then the appropriate button in the 3D plot palette (Figure B.5).

All of these 3D plots can be rotated. First select the plot area by clicking within the plot region delimited by the axes, but not on the surface. Four circles and a triangle will appear. Any of these can be dragged to rotate the plot: the circles give horizontal rotation (about the vertical axis) and the triangle rotates the vertical axis. It may be a good idea to change to a simple view of the surface (such as a coarse wireframe grid) if rotation proves to be slow.

The software also allows multiple (2, 4 or 6) views of the plot from different (equispaced) angles, and these can be rotated simultaneously by rotating one of the panels. Multiple views are selected by buttons on the 3D plot palette. As these are a form of conditioning, a single view is selected again by clicking on the 'no conditioning' button in the palette.

Plots can also be conditioned on the x, y or z variables and shown as a series of 'exploded' views.

All the plots from the 3D palette can be conditioned on additional variables, selecting 2×2 or 2×3 layouts from a palette button, with fine-tuning from the background dialog box.

The shape of the enclosing cuboid (the aspect parameter in 3D Trellis) can be set from the 3D Workbox tab of the plot's dialog box.

Editing plots

Many of the properties of a graph can be altered from dialog boxes. To select a part of the graph (such as an axis or fitted line or label) (left-)click on it. Clicking on any of the data points in a 2D plot will select both the points and the fitted curve. Then either double-clicking or right-clicking will bring up or a tabbed dialog box or a shortcut menu to the tabs from which the properties of that part can be selected.

The plot region or the whole graph can be selected. The dialog box has at least four tabs, Plot Summary (including the data frame used), Position/Size (which includes aspect ratio, with an option for 'proportional units'), Fill/Border (colours, patterns, ...) and Multipanel (for conditioning). The 3D plots add the 3D Workbox tab. When the whole graph is selected it can be resized by dragging the handles, or moved by dragging a point outside the plot region, and similarly for the plot region.

Once an axis label or title is selected, clicking on the text brings up an 'in-place' edit box for replacement text. Double-clicking on the surrounding box enables properties such as font and colour to be altered: these can also be altered from the toolbar when the text is selected.

Legends and titles can be added from the Insert menu; showing a legend can also be toggled from a toolbar button.

There is an *annotation* palette (Figure B.5) which has tools to label points (as in identify) and to add text, a date stamp or various symbols to a graph. This palette is selected from its toolbar button or from the Toolbars item on the Views menu. It provides a simple drawing package with which to enhance graphs.

Multiple graphs

A graphsheet can display more than one graph. To add a graph to an existing graphsheet, ensure that no elements are selected and create the new graph, holding down Shift while the plot button is clicked. (As we saw in the Introduction, doing this while a graph is selected adds to that graph.) It may be necessary to use the Arrange Graphs item on the Format menu to produce a usable layout (as the default might be to overlay the graphs). The graphs can be re-ordered by selecting them in the order required (use shift-click) and then using Arrange Graphs.

Graphsheets can make multiple pages of graphs. The circumstances under which they do so is set by the Auto pages item in the Options | Graph Options ... dialog box, which can be overridden for each graphsheet from the Options tab of its right-click background menu. The default is to create separate pages if sent several frames (plots starting on a new page) during the execution of a single expression, for example, when using `plot(lm.object, ask=T)`. This can be changed to a new graph page for every plot, or to always use the same graph page. To set this up from the command line we can use, for example

```
graphsheet(Name="GStest")
guiModify("GraphSheet", Name="GStest",
        AutoPageMode="Every Graph")
```

In S-PLUS 4.5 and later we can use `guiGetGSName()` for the name of the current graphsheet.

Setting GUI properties from the command-line

As the last example shows, it is possible to change almost all of the settings in the GUI by calls from the S language. The simplest (and in many cases the only) way to find the corresponding S command is to invoke the operation from the GUI and then open a history window (using the toolbar button to the left of that for the command-line window labelled by a scroll). The appropriate command(s) will be recorded in the history window and can be used from an S script with minimal changes (for example, giving the appropriate name for the graphsheet).

To find the existing settings of a graphsheet (say), use

```
nm <- guiGetArgumentNames("GraphSheet")
pr <- guiGetPropertyValue("GraphSheet", Name="GStest")
names(nm) <- pr
print(pr)
```

and this listing gives the corresponding arguments to be used with `guiModify` to alter the settings. However, it does not seem to be possible to find the set of allowable values for the settings except by trying them from a dialog box or menu and examining the history. (From S-PLUS 4.5 the help pages for `guiCreate("LinePlot")` and so on give more details than earlier versions.)

Using command-line graphics on a graphsheet

The traditional command-line graphics operations are mapped to objects in the object-oriented graphics model. In the default 'fast' mode the graphics calls in each top-level S expression are mapped to a single composite object. This is usually what is required, but can have some unexpected side-effects. One is that all the graphics calls from within a function are combined into a single object, and *nothing is displayed* until the function call finishes. Similarly, running a series of simulations in a `for` loop and plotting the results after each one will under 4.x not display any of the results until the whole series of simulations has completed. This is in contrast to the behaviour under S-PLUS 3.3 for Windows and for the UNIX versions, under which the graphics are displayed immediately.

The workaround is to ensure that the graphics calls are split into multiple objects; this can only be done in 4.0 release 2 or later using an undocumented effect of the function `guiLocator`. If this is called with a zero or negative argument it pauses (`guiLocator(-n)` pauses for n seconds), and then plots the pending graphics calls (which will appear in an object-browser view of the graphsheet as a single object). Two side-effects of the call to `guiLocator` are to move the focus to the graphsheet which is thus brought to the front and to start a new expression for the purposes of the Page creation option (the `AutoPageMode` property) of the graphsheet.

B.3 Statistical analysis *via* the GUI

We can do something similar to the analysis in Chapter 1 of the `hills` and `michelson` datasets using the GUI. Unfortunately the details of what to do are quite different in different versions. First attach library `MASS` via the commands window or a script window or in S-PLUS 4.5 or 2000, from the File menu. Then we need to display the data frames from `MASS`.

4.0 and 4.5 Open an object browser, create a new page (from the right-click menu), filter on the location of the `MASS` library, select all classes and click on OK. Select `data.frame` in the left pane and click on the `Object` header in the right pane to sort the items alphabetically.

2000 Open an object explorer, and insert a folder (from the right-click menu). Name the folder `MASS`, select Data as the Data Objects and use the Advanced tab to filter on the location of the `MASS` library. Click on OK.

Next select the `hills` data frame in the browser / explorer. Click on the 2D Plots button on the top toolbar, and select the button for a Scatter Matrix. This gives a plot similar to Figure 1.2, but without the square aspect ratio. To change that, select the plot region (click outside the scatterplot matrix), right-click and select the Position/Size... item. In the dialog box enter equal dimensions for width and height of the graph size and click on OK.

To try out the brush-and-spin plot, select Brush and Spin from the Graph menu, select all the variables in the dialog box (you can accept the default, ALL) and click on OK.

To approximate Figure 1.4, double click on `hills` to open a data window, and select the columns `dist` and `time` in that order. (Hold down Ctrl while clicking on `time`.) Then open the 2D Plots palette and select the Linear Fit button. Then to label points, open the annotations palette (second from right on the graphsheet toolbar), and select the Label Points button (third on the top row). Click near a point on the plot for a label, but shift-click for this label to be permanent so that further points can be labelled. To add the `ltsreg` line, we need to add another *plot*, which we can do by first selecting the existing plot and then shift-clicking on the appropriate button, this time for a robust fit. Finally we need to select the robust line and from the dialog box brought up by the lines shortcut on the right-click menu select a dashed line. (In S-PLUS 2000 we also need to set the plot symbols to open circles for the second plot.)

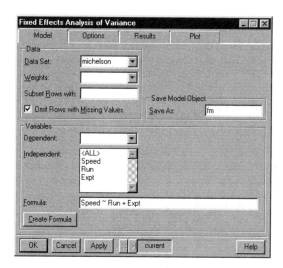

Figure B.6: The dialog box for a fixed-effects analysis of variance in S-PLUS 2000.

Now return to the object browser / explorer, expand the list of data frames / objects in MASS and select `michelson` in the left pane. The right pane will then list the columns: select `Expt` and `Speed` in that order. Then open the **2D Plots** palette and click on the **Box** button. The plot will be generated. Select the x-axis label and edit the text (either 'in place' by double clicking or via the dialog box). Finally, open the annotations palette, select the **Comment** tool and add the title.

Select the `michelson` data frame again, and select `Speed`, `Run` and `Expt` in that order. Then select **Statistics, Analysis of Variance** and **Fixed Effects** from the menus. A dialog box (Figure B.6) will appear, with the appropriate formula already constructed. Give `fm` as the name by which to save the model object. Clicking on **Apply** gives output in the results window, including the analysis of variance table. Now delete `Run +` from the formula, change the name to `fm0` and click on **OK**.

4.0 and **4.5** Select the first page and the class `list` in the browser.

2000 Select the main **Data** folder in the object explorer.

Objects `fm` and `fm0` will have appeared. Right clicking on these will allow them to be summarized and plotted. (Plotting an `aov` object will allow the selection of various diagnostic plots.)

To compare the models, select **Compare Models** from the **Statistics** menu. Select models `fm0`, `fm` in that order from the list of models presented in the dialog box and select **OK**. The analysis of variance table appears in the report window.

Appendix C

Datasets, Software and Libraries

The software and datasets used in this book are available over the World Wide Web. Point your browser at

```
http://www.stats.ox.ac.uk/pub/MASS3/sites.html
```

to obtain a current list of sites; please use a site near you. We expect this list to include

```
http://www.stats.ox.ac.uk/pub/MASS3
http://www.cmis.csiro.au/S-PLUS/MASS
http://lib.stat.cmu.edu/S/MASS3
http://franz.stat.wisc.edu/pub/MASS3
```

The on-line instructions tell you how to install the software under both UNIX and Windows. In case of difficulty in accessing the software please email MASS@ stats.ox.ac.uk.

The on-line complements are available at these sites as well as answers to selected exercises and printable versions of the on-line help for our software.

Note that this book assumes that you have access to the S-PLUS environment. This is a commercial product and if you need to purchase it please see

```
http://www.mathsoft.com/splus
```

for details of its distribution channels.

An 'open source' system has emerged called R which is 'not unlike S' and can be used to explore many of the examples of this book on both UNIX and Windows machines. It and versions of our libraries are available at a collection of sites on the World Wide Web accessible from

```
http://www.ci.tuwien.ac.at/R/mirrors.html
```

Our on-line 'R complements' provide details of what can be done and the changes needed to our examples.

C.1 Our libraries

Our software is packaged as five library sections.

MASS This contains all the datasets and a number of S functions, as well as a number of other datasets that we have used in learning or teaching.

nnet Software for feed-forward neural networks with a single hidden layer and for multinomial log-linear models.

spatial Software for spatial smoothing and the analysis of spatial point patterns. This directory contains a number of datasets of point patterns, described in the text file PP.files (PP.fil under Windows).

class Functions for nonparametric classification, by k-nearest neighbours and learning vector quantization.

treefix Enhanced pruning and prediction routines for tree models.

Under Windows you can make the S objects in a library available by

```
library(name)
```

The help file for a library can be browsed (in the usual Windows help browser) by

```
help(library="name")
```

whether or not the library call has been used.

The libraries may have been installed as system libraries or private libraries; if the latter you need first to issue the command

```
assign("lib.loc", "path to private library directory", w=0)
```

to say where the libraries are. Then S objects in a library are made available by

```
library(name)
```

and

```
library(help=name)
```

gives a short description of the library and a listing of its contents. After attaching the library, help on its objects can be obtained in the same ways as for system objects (page 451).

There are two additional libraries supplied in the Windows bundle. One, helpfix, should be used with first=T under S-PLUS 3.x to make the help more accessible. Library MASSdia provides GUI features such as menus and dialog boxes for parts of the MASS and nnet libraries on S-PLUS 4.5.

Caveat

These datasets and software are provided in good faith, but none of the authors, publishers or distributors warrant their accuracy nor can be held responsible for the consequences of their use.

We have tested the software as widely as we are able but it is inevitable that system dependencies will arise. We are unlikely to be in a position to assist with such problems.

The licences for the distribution and use of the software and datasets are given in the on-line distributions.

C.2   Using libraries

A library in S-PLUS is a convenient way to package S objects for a common purpose, and to allow these to extend the system. A library section is a directory containing a .Data subdirectory, which may contain a .Help subdirectory. (Under Windows it has a _Data directory, and perhaps a _Help subdirectory or a .hlp file.) The directory should also contain a README file describing its contents, and may also contain object modules for dynamic loading.

The structure of a library section is the same as that of a working directory, but the library function makes libraries much more convenient to use. Conventionally libraries are stored in a standard place, the subdirectory library of the main S-PLUS directory. Which library sections are available can be found by the library command with no argument; further information (the contents of the README file) on any section is given by

```
library(help=section_name)
```

and the library section itself is made available by

```
library(section_name)
```

This has two actions. It attaches the .Data subdirectory of the section at the end of the search path (having checked that it has not already been attached), and executes the function .First.lib if one exists within that .Data subdirectory.

Sometimes it is necessary to have functions in a library that will replace standard system functions (for example, to correct bugs or to extend their functionality). This can be done by attaching the library as the second dictionary on the search path with

```
library(section_name, first=T)
```

Of course, attaching other dictionaries with attach or other libraries with first=T will push previously attached libraries down the search path.

Private libraries

So far we have only considered system-wide library sections installed under the main S-PLUS directory, which usually requires privileged access to the operating system. It is also possible to use a private library, by giving `library` the argument `lib.loc` or by assigning the object `lib.loc` in the current session dictionary (frame 0). This should be a vector of directory names that are searched in order for library sections before the system-wide library. For example, on another of our systems we get

```
> assign(where=0, "lib.loc", "/users/ripley/S/library")
> library()
Library "/users/ripley/S/library"
The following sections are available in the library:

SECTION         BRIEF DESCRIPTION

MASS            main library
nnet            neural nets
spatial         spatial statistics
class           classification

Library "/packages/splus3.4/library"
The following sections are available in the library:

SECTION         BRIEF DESCRIPTION

chron           Functions to handle dates and times.
    ....
```

Because `lib.loc` is local to the session, it must be assigned for each session. The `.First` function is often a convenient place to do so (see page 454); see its help page for other ways using `.First.local` or the `S_FIRST` environment variable.

Sources of libraries

Many S-PLUS users have generously collected their functions and datasets into libraries and made them publicly available. An archive of sources for library sections is maintained at Carnegie-Mellon University. The World Wide Web address is

```
http://lib.stat.cmu.edu/S/
```

There are several mirrors around the world.

Among the libraries available from `statlib` are the following which have been mentioned elsewhere in this book.

boot   bootstrap functions from Davison & Hinkley (1997)
delaunay   Dirichlet tessellation and Delaunay triangulation

KernSmooth	kernel density estimation and smoothing
logspline	spline estimation of log-densities
nls2	nonlinear model-fitting
postscriptfonts	additional capabilities for postscript under UNIX
pspline	penalized splines, for estimating derivatives of a smooth fit
rpart	recursive partitioning
sm	kernel density estimation and smoothing
xgobi	XGobi dynamic graphics package (UNIX only)

Version 3 of the nlme library is available from http://nlme.stat.wisc.edu, and version 5 of the survival library from http://www.mayo.edu/hsr/biostat.html.

The convention is to distribute libraries as 'shar' archives; these are text files which when used as scripts for the Bourne shell sh unpack to give all the files needed for a library section. Check files such as Install for installation instructions; these usually involve editing the Makefile and typing make.

Several of these libraries are available prepackaged for Windows users in .zip archives; check the WWW addresses

```
http://lib.stat.cmu.edu/DOS/S
http://www.stats.ox.ac.uk/pub/SWin
```

which also point to tools to help Windows users access 'shar' archives intended for UNIX. (A port of XGobi to Windows is available *via* those sites.) Beware that some libraries packaged for S-PLUS 3.x are incompatible with S-PLUS 4.x since the format of compiled C or FORTRAN code has changed, but those we have ported work with both.

Users are encouraged to share their own efforts; statlib welcomes submissions (see the file submissions from S). Submitters should try to be aware of the potential differences among S-PLUS platforms.

References

Numbers in brackets [ ] are page references to citations.

Abbey, S. (1988) Robust measures and the estimator limit. *Geostandards Newsletter* **12**, 241–248. [131]

Aitchison, J. (1986) *The Statistical Analysis of Compositional Data.* London: Chapman & Hall. [92]

Aitchison, J. and Dunsmore, I. R. (1975) *Statistical Prediction Analysis.* Cambridge: Cambridge University Press. [350]

Aitkin, M. (1978) The analysis of unbalanced cross classifications (with discussion). *Journal of the Royal Statistical Society A* **141**, 195–223. [180, 181, 188]

Aitkin, M., Anderson, D., Francis, B. and Hinde, J. (1989) *Statistical Modelling in GLIM.* Oxford: Oxford University Press. [239]

Akaike, H. (1974) A new look at statistical model identification. *IEEE Transactions on Automatic Control* **AU–19**, 716–722. [185]

Akima, H. (1978) A method of bivariate interpolation and smooth surface fitting for irregularly distributed data points. *ACM Transactions on Mathematical Software* **4**, 148–159. [438]

Analytical Methods Committee (1987) Recommendations for the conduct and interpretation of co-operative trials. *The Analyst* **112**, 679–686. [192, 194]

Analytical Methods Committee (1989a) Robust statistics — how not to reject outliers. Part 1. Basic concepts. *The Analyst* **114**, 1693–1697. [121, 131]

Analytical Methods Committee (1989b) Robust statistics — how not to reject outliers. Part 2. Inter-laboratory trials. *The Analyst* **114**, 1699–1702. [131, 194]

Andersen, P. K., Borgan, Ø., Gill, R. D. and Keiding, N. (1993) *Statistical Models Based on Counting Processes.* New York: Springer-Verlag. [367, 370]

Anderson, E. (1935) The irises of the Gaspe peninsula. *Bulletin of the American Iris Society* **59**, 2–5. [329]

Anderson, O. D. (1976) *Time Series Analysis and Forecasting. The Box-Jenkins Approach.* London: Butterworths. [422]

Atkinson, A. C. (1985) *Plots, Transformations and Regression.* Oxford: Oxford University Press. [72, 161]

Atkinson, A. C. (1986) Comment: Aspects of diagnostic regression analysis. *Statistical Science* **1**, 397–402. [162]

Atkinson, A. C. (1988) Transformations unmasked. *Technometrics* **30**, 311–318. [162, 163]

Azzalini, A. and Bowman, A. W. (1990) A look at some data on the Old Faithful geyser. *Applied Statistics* **39**, 357–365. [119]

Banfield, J. D. and Raftery, A. E. (1993) Model-based Gaussian and non-Gaussian clustering. *Biometrics* **49**, 803–821. [339]

Bates, D. M. and Chambers, J. M. (1992) Nonlinear models. Chapter 10 of Chambers & Hastie (1992). [241]

Bates, D. M. and Watts, D. G. (1988) *Nonlinear Regression Analysis and Its Applications.* New York: John Wiley and Sons. [241]

Baxter, L. A., Coutts, S. M. and Ross, G. A. F. (1980) Applications of linear models in motor insurance. In *Proceedings of the 21st International Congress of Actuaries, Zurich*, pp. 11–29. [239]

Becker, R. A. (1994) A brief history of S. In *Computational Statistics: Papers Collected on the Occasion of the 25th Conference on Statistical Computing at Schloss Reisenburg*, eds P. Dirschedl and R. Osterman, pp. 81–110. Heidelberg: Physica-Verlag. [1]

Becker, R. A., Chambers, J. M. and Wilks, A. R. (1988) *The NEW S Language.* New York: Chapman & Hall. (Formerly Monterey: Wadsworth and Brooks/Cole.). [2]

Bie, O., Borgan, Ø. and Liestøl, K. (1987) Confidence intervals and confidence bands for the cumulative hazard rate function and their small sample properties. *Scandinavian Journal of Statistics* **14**, 221–233. [371]

Bishop, C. M. (1995) *Neural Networks for Pattern Recognition.* Oxford: Clarendon Press. [296]

Bishop, Y. M. M., Fienberg, S. E. and Holland, P. W. (1975) *Discrete Multivariate Analysis.* Cambridge, MA: MIT Press. [226]

Bloomfield, P. (1976) *Fourier Analysis of Time Series: An Introduction.* New York: John Wiley and Sons. [401, 407, 409, 410]

Borgan, Ø. and Liestøl, K. (1990) A note on confidence intervals and bands for the survival function based on transformations. *Scandinavian Journal of Statistics* **17**, 35–41. [371]

Bowman, A. and Azzalini, A. (1997) *Applied Smoothing Techniques for Data Analysis: The Kernel Approach with S-Plus Illustrations.* Oxford: Oxford University Press. [132, 281]

Box, G. E. P. and Cox, D. R. (1964) An analysis of transformations (with discussion). *Journal of the Royal Statistical Society B* **26**, 211–252. [182, 183]

Box, G. E. P., Hunter, W. G. and Hunter, J. S. (1978) *Statistics for Experimenters.* New York: John Wiley and Sons. [122, 123, 179, 180]

Box, G. E. P. and Pierce, D. A. (1970) Distribution of residual autocorrelations in autoregressive-integrated moving average time series models. *Journal of the American Statistical Association* **65**, 1509–1526. [416]

Breiman, L. and Friedman, J. H. (1985) Estimating optimal transformations for multiple regression and correlations (with discussion). *Journal of the American Statistical Association* **80**, 580–619. [294]

Breiman, L., Friedman, J. H., Olshen, R. A. and Stone, C. J. (1984) *Classification and Regression Trees.* Monterey: Wadsworth and Brooks/Cole. [303, 309, 327]

Brent, R. (1973) *Algorithms for Minimization Without Derivatives.* Englewood Cliffs, NJ: Prentice-Hall. [261]

Brockwell, P. J. and Davis, R. A. (1991) *Time Series: Theory and Methods.* Second Edition. New York: Springer-Verlag. [401, 415, 431]

Brockwell, P. J. and Davis, R. A. (1996) *Introduction to Time Series and Forecasting.* New York: Springer-Verlag. [401, 431]

Bryan, J. G. (1951) The generalized discriminant function: mathematical foundation and computational routine. *Harvard Educational Review* **21**, 90–95. [346]

Campbell, N. A. and Mahon, R. J. (1974) A multivariate study of variation in two species of rock crab of genus *Leptograpsus. Australian Journal of Zoology* **22**, 417–425. [356]

Cao, R., Cuevas, A. and González-Manteiga, W. (1994) A comparative study of several smoothing methods in density estimation. *Computational Statistics and Data Analysis* **17**, 153–176. [138]

Chambers, J. M. (1998) *Programming with Data. A Guide to the S Language.* New York: Springer-Verlag. [2, 50]

Chambers, J. M. and Hastie, T. J. eds (1992) *Statistical Models in S.* New York: Chapman & Hall. (Formerly Monterey: Wadsworth and Brooks/Cole.). [2, 216, 289, 474, 475]

Ciampi, A., Chang, C.-H., Hogg, S. and McKinney, S. (1987) Recursive partitioning: A versatile method for exploratory data analysis in biostatistics. In *Biostatistics*, eds I. B. McNeil and G. J. Umphrey, pp. 23–50. New York: Reidel. [308]

Clark, L. A. and Pregibon, D. (1992) Tree-based models. Chapter 9 of Chambers & Hastie (1992). [303, 307]

Cleveland, R. B., Cleveland, W. S., McRae, J. E. and Terpenning, I. (1990) STL: A seasonal-trend decomposition procedure based on loess (with discussion). *Journal of Official Statistics* **6**, 3–73. [418]

Cleveland, W. S. (1993) *Visualizing Data.* Summit, NJ: Hobart Press. [53, 74, 92, 189]

Cleveland, W. S., Grosse, E. and Shyu, W. M. (1992) Local regression models. Chapter 8 of Chambers & Hastie (1992). [437]

Cochrane, D. and Orcutt, G. H. (1949) Application of least-squares regression to relationships containing autocorrelated error terms. *Journal of the American Statistical Association* **44**, 32–61. [429]

Collett, D. (1991) *Modelling Binary Data.* London: Chapman & Hall. [217, 218]

Collett, D. (1994) *Modelling Survival Data in Medical Research.* London: Chapman & Hall. [367]

Copas, J. B. (1988) Binary regression models for contaminated data (with discussion). *Journal of the Royal Statistical Society series B* **50**, 225–266. [269]

Cox, D. R. (1972) Regression models and life-tables (with discussion). *Journal of the Royal Statistical Society B* **34**, 187–220. [380]

Cox, D. R. and Oakes, D. (1984) *Analysis of Survival Data.* London: Chapman & Hall. [367, 368]

Cox, D. R. and Snell, E. J. (1984) *Applied Statistics Principles and Examples.* London: Chapman & Hall. [226]

Cox, D. R. and Snell, E. J. (1989) *The Analysis of Binary Data.* Second Edition. London: Chapman & Hall. [222]

Cox, T. F. and Cox, M. A. A. (1994) *Multidimensional Scaling.* London: Chapman & Hall. [334]

Cressie, N. A. C. (1991) *Statistics for Spatial Data.* New York: John Wiley and Sons. [433, 442]

Cybenko, G. (1989) Approximation by superpositions of a sigmoidal function. *Mathematics of Controls, Signals, and Systems* **2**, 303–314. [298]

Daniel, C. and Wood, F. S. (1980) *Fitting Equations to Data.* Second Edition. New York: John Wiley and Sons. [199]

Darroch, J. N. and Ratcliff, D. (1972) Generalized iterative scaling for log-linear models. *Annals of Mathematical Statistics* **43**, 1470–1480. [213, 230]

Davidian, M. and Giltinan, D. M. (1995) *Nonlinear Models for Repeated Measurement Data.* London: Chapman & Hall. [199]

Davies, P. L. (1993) Aspects of robust linear regression. *Annals of Statistics* **21**, 1843–1899. [170]

Davison, A. C. and Hinkley, D. V. (1997) *Bootstrap Methods and Their Application.* Cambridge: Cambridge University Press. [142, 143, 145, 146, 175, 470]

Davison, A. C. and Snell, E. J. (1991) Residuals and diagnostics. Chapter 4 of Hinkley *et al.* (1991). [217]

Dawid, A. P. (1982) The well-calibrated Bayesian (with discussion). *Journal of the American Statistical Association* **77**, 605–613. [364]

Dawid, A. P. (1986) Probability forecasting. In *Encyclopedia of Statistical Sciences*, eds S. Kotz, N. L. Johnson and C. B. Read, volume 7, pp. 210–218. New York: John Wiley and Sons. [364]

Deming, W. E. and Stephan, F. F. (1940) On a least-squares adjustment of a sampled frequency table when the expected marginal totals are known. *Annals of Mathematical Statistics* **11**, 427–444. [213]

Devijver, P. A. and Kittler, J. V. (1982) *Pattern Recognition: A Statistical Approach.* Englewood Cliffs, NJ: Prentice-Hall. [354]

Diaconis, P. and Shahshahani, M. (1984) On non-linear functions of linear combinations. *SIAM Journal of Scientific and Statistical Computing* **5**, 175–191. [290]

Diggle, P. J. (1983) *Statistical Analysis of Spatial Point Patterns.* London: Academic Press. [433]

Diggle, P. J. (1990) *Time Series: A Biostatistical Introduction.* Oxford: Oxford University Press. [401, 402]

Diggle, P. J., Liang, K.-Y. and Zeger, S. L. (1994) *Analysis of Longitudinal Data.* Oxford: Clarendon Press. [206]

Dixon, W. J. (1960) Simplified estimation for censored normal samples. *Annals of Mathematical Statistics* **31**, 385–391. [129]

Dodge, Y. (1985) *Analysis of Experiments With Missing Data.* New York: John Wiley and Sons. [100]

Duda, R. O. and Hart, P. E. (1973) *Pattern Classification and Scene Analysis.* New York: John Wiley and Sons. [226]

Efron, B. (1982) *The Jackknife, the Bootstrap, and Other Resampling Plans.* Philadelphia: Society for Industrial and Applied Mathematics. [142]

Efron, B. and Hinkley, D. V. (1978) Assessing the accuracy of the maximum likelihood estimator: Observed versus expected Fisher information (with discussion). *Biometrika* **65**, 457–487. [260]

Efron, B. and Tibshirani, R. (1993) *An Introduction to the Bootstrap*. New York: Chapman & Hall. [142]

Ehrlich, I. (1973) Participation in illegitimate activities: A theoretical and empirical investigation. *Journal of Political Economy* **81**, 521–565. [208]

Eilers, P. H. and Marx, B. D. (1996) Flexible smoothing with B-splines and penalties. *Statistical Science* **11**, 89–121. [399]

Ein-Dor, P. and Feldmesser, J. (1987) Attributes of the performance of central processing units: A relative performance prediction model. *Communications of the ACM* **30**, 308–317. [188]

Everitt, B. S. and Hand, D. J. (1981) *Finite Mixture Distributions*. London: Chapman & Hall. [263]

Feigl, P. and Zelen, M. (1965) Estimation of exponential survival probabilities with concomitant information. *Biometrics* **21**, 826–838. [368]

Firth, D. (1991) Generalized linear models. Chapter 3 of Hinkley *et al.* (1991). [211, 213, 215]

Fisher, R. A. (1925) Theory of statistical estimation. *Proceedings of the Cambridge Philosophical Society* **22**, 700–725. [214]

Fisher, R. A. (1936) The use of multiple measurements in taxonomic problems. *Annals of Eugenics (London)* **7**, 179–188. [329, 344]

Fisher, R. A. (1940) The precision of discriminant functions. *Annals of Eugenics (London)* **10**, 422–429. [342]

Fleming, T. R. and Harrington, D. P. (1981) A class of hypothesis tests for one and two sample censored survival data. *Communications in Statistics* **A10**(8), 763–794. [372]

Fleming, T. R. and Harrington, D. P. (1991) *Counting Processes and Survival Analysis*. New York: John Wiley and Sons. [367, 370, 380, 384]

Freedman, D. and Diaconis, P. (1981) On the histogram as a density estimator: L_2 theory. *Zeitschrift für Wahrscheinlichkeitstheorie und verwandte Gebiete* **57**, 453–476. [119]

Friedman, J. H. (1984) SMART user's guide. Technical Report 1, Laboratory for Computational Statistics, Department of Statistics, Stanford University. [290]

Friedman, J. H. (1987) Exploratory projection pursuit. *Journal of the American Statistical Association* **82**, 249–266. [330]

Friedman, J. H. and Stuetzle, W. (1981) Projection pursuit regression. *Journal of the American Statistical Association* **76**, 817–823. [289]

Funahashi, K. (1989) On the approximate realization of continuous mappings by neural networks. *Neural Networks* **2**, 183–192. [298]

Gabriel, K. R. (1971) The biplot graphical display of matrices with application to principal component analysis. *Biometrika* **58**, 453–467. [335, 336]

Gallant, A. R. (1987) *Nonlinear Statistical Models*. New York: Wiley. [241]

Gehan, E. A. (1965) A generalized Wilcoxon test for comparing arbitrarily singly-censored samples. *Biometrika* **52**, 203–223. [368]

Geisser, S. (1993) *Predictive Inference: An Introduction.* New York: Chapman & Hall. [350]

Gelman, A., Carlin, J. B., Stern, H. S. and Rubin, D. B. (1995) *Bayesian Data Analysis.* London: Chapman & Hall. [237]

Golub, G. H. and Van Loan, C. F. (1989) *Matrix Computations.* Second Edition. Baltimore: Johns Hopkins University Press. [100, 101, 436]

Goodman, L. A. (1978) *Analyzing Qualitative/Categorical Data: Log-Linear Models and Latent-Structure Analysis.* Cambridge, MA: Abt Books. [226]

de Gooijer, J. G., Abraham, B., Gould, A. and Robinson, L. (1985) Methods for determining the order of an autoregressive-moving average process: A survey. *International Statistical Review* **53**, 301–329. [413]

Gordon, A. D. (1981) *Classification: Methods for the Exploratory Analysis of Multivariate Data.* London: Chapman & Hall. [336]

Gower, J. C. and Hand, D. J. (1996) *Biplots.* London: Chapman & Hall. [336, 343]

Grambsch, P. and Therneau, T. M. (1994) Proportional hazards tests and diagnostics based on weighted residuals. *Biometrika* **81**, 515–526. [386]

Green, P. J. and Silverman, B. W. (1994) *Nonparametric Regression and Generalized Linear Models. A Roughness Penalty Approach.* London: Chapman & Hall. [281, 284]

Greenacre, M. (1992) Correspondence analysis in medical research. *Statistical Methods in Medical Research* **1**, 97–117. [343]

Haberman, S. J. (1978) *Analysis of Qualitative Data. Volume 1: Introductory Topics.* New York: Academic Press. [226]

Haberman, S. J. (1979) *Analysis of Qualitative Data. Volume 2: New Developments.* New York: Academic Press. [226]

Hampel, F. R., Ronchetti, E. M., Rousseeuw, P. J. and Stahel, W. A. (1986) *Robust Statistics. The Approach Based on Influence Functions.* New York: John Wiley and Sons. [127, 152]

Hand, D. J., Daly, F., McConway, K., Lunn, D. and Ostrowski, E. eds (1993) *A Handbook of Small Data Sets.* London: Chapman & Hall. [149, 199]

Härdle, W. (1991) *Smoothing Techniques with Implementation in S.* New York: Springer-Verlag. [136]

Harrington, D. P. and Fleming, T. R. (1982) A class of rank test procedures for censored survival data. *Biometrika* **69**, 553–566. [372]

Hartigan, J. A. (1975) *Clustering Algorithms.* New York: John Wiley and Sons. [336, 338]

Hartigan, J. A. (1982) Classification. In *Encyclopedia of Statistical Sciences*, eds S. Kotz, N. L. Johnson and C. B. Read, volume 2, pp. 1–10. New York: John Wiley and Sons. [329]

Hartigan, J. A. and Wong, M. A. (1979) A K-means clustering algorithm. *Applied Statistics* **28**, 100–108. [338]

Harvey, A. C. (1989) *Forecasting, Structural Time Series Models and the Kalman Filter.* Cambridge: Cambridge University Press. [431]

Harvey, A. C. and Durbin, J. (1986) The effects of seat belt legislation on British road casualties: A case study in structural time series modelling (with discussion). *Journal of the Royal Statistical Society series A* **149**, 187–227. [431]

Hastie, T. J. and Tibshirani, R. J. (1990) *Generalized Additive Models*. London: Chapman & Hall. [281, 284]

Hauck, Jr., W. W. and Donner, A. (1977) Wald's test as applied to hypotheses in logit analysis. *Journal of the American Statistical Association* **72**, 851–853. [225]

Heiberger, R. M. (1989) *Computation for the Analysis of Designed Experiments*. New York: John Wiley and Sons. [195]

Henrichon, Jr., E. G. and Fu, K.-S. (1969) A nonparametric partitioning procedure for pattern classification. *IEEE Transactions on Computers* **18**, 614–624. [303]

Hertz, J., Krogh, A. and Palmer, R. G. (1991) *Introduction to the Theory of Neural Computation*. Redwood City, CA: Addison-Wesley. [296]

Hettmansperger, T. P. and Sheather, S. J. (1992) A cautionary note on the method of least median squares. *American Statistician* **46**, 79–83. [170]

Hinkley, D. V., Reid, N. and Snell, E. J. eds (1991) *Statistical Theory and Modelling. In Honour of Sir David Cox, FRS*. London: Chapman & Hall. [476, 477]

Hoaglin, D. C., Mosteller, F. and Tukey, J. W. eds (1983) *Understanding Robust and Exploratory Data Analysis*. New York: John Wiley and Sons. [121, 479]

Hornik, K., Stinchcombe, M. and White, H. (1989) Multilayer feedforward networks are universal approximators. *Neural Networks* **2**, 359–366. [298]

Hosmer, D. W. and Lemeshow, S. (1989) *Applied Logistic Regression*. New York: John Wiley and Sons. [222]

Hsu, J. C. (1996) *Multiple Comparison Procedures: Theory and Methods*. London: Chapman & Hall. [192]

Huber, P. J. (1967) The behavior of maximum likelihood estimates under nonstandard conditions. In *Proceedings of the Fifth Berkeley Symposium on Mathematical Statistics and Probability*, eds L. M. Le Cam and J. Neyman, volume 1, pp. 221–233. Berkeley, CA: University of California Press. [260]

Huber, P. J. (1981) *Robust Statistics*. New York: John Wiley and Sons. [127]

Huber, P. J. (1985) Projection pursuit (with discussion). *Annals of Statistics* **13**, 435–525. [330]

Huet, S., Bouvier, A., Gruet, M.-A. and Jolivet, E. (1996) *Statistical Tools for Nonlinear Regression. A Practical Guide with S-PLUS Examples*. New York: Springer-Verlag. [241]

Iglewicz, B. (1983) Robust scale estimators and confidence intervals for location. In Hoaglin *et al.* (1983), pp. 405–431. [127]

Ingrassia, S. (1992) A comparison between the simulated annealing and the EM algorithms in normal mixture decompositions. *Statistics and Computing* **2**, 203–211. [263]

Jackson, J. E. (1991) *A User's Guide to Principal Components*. New York: John Wiley and Sons. [332]

Jardine, N. and Sibson, R. (1971) *Mathematical Taxonomy*. London: John Wiley and Sons. [333]

John, P. W. M. (1971) *Statistical Design and Analysis of Experiments*. New York: Macmillan. [195]

Jolliffe, I. T. (1986) *Principal Component Analysis*. New York: Springer-Verlag. [332, 336]

Jones, M. C., Marron, J. S. and Sheather, S. J. (1996) A brief survey of bandwidth selection for density estimation. *Journal of the American Statistical Association* **91**, 401–407. [137, 138]

Jones, M. C. and Sibson, R. (1987) What is projection pursuit? (with discussion). *Journal of the Royal Statistical Society A* **150**, 1–36. [330]

Journel, A. G. and Huijbregts, C. J. (1978) *Mining Geostatistics*. London: Academic Press. [439]

Kalbfleisch, J. D. and Prentice, R. L. (1980) *The Statistical Analysis of Failure Time Data*. New York: John Wiley and Sons. [367, 368, 369, 378, 386, 391]

Kaufman, L. and Rousseeuw, P. J. (1990) *Finding Groups in Data. An Introduction to Cluster Analysis*. New York: John Wiley and Sons. [333, 337, 340, 341]

Kent, J. T., Tyler, D. E. and Vardi, Y. (1994) A curious likelihood identity for the multivariate t-distribution. *Communications in Statistics—Simulation and Computation* **23**, 441–453. [348]

Klein, J. P. (1991) Small sample moments of some estimators of the variance of the Kaplan-Meier and Nelson-Aalen estimators. *Scandinavian Journal of Statistics* **18**, 333–340. [370]

Knight, R. L. and Skagen, S. K. (1988) Agnostic asymmetries and the foraging ecology of Bald Eagles. *Ecology* **69**, 1188–1194. [236]

Knuth, D. E. (1968) *The Art of Computer Programming, Volume 1: Fundamental Algorithms*. Reading, MA: Addison-Wesley. [25]

Kohonen, T. (1990) The self-organizing map. *Proceedings IEEE* **78**, 1464–1480. [354]

Kohonen, T. (1995) *Self-Organizing Maps*. Berlin: Springer-Verlag. [354]

Kooperberg, C. and Stone, C. J. (1992) Logspline density estimation for censored data. *Journal of Computational and Graphical Statistics* **1**, 301–328. [141]

Krzanowski, W. J. (1988) *Principles of Multivariate Analysis. A User's Perspective*. Oxford: Oxford University Press. [329]

Laird, N. M. and Ware, J. H. (1982) Random-effects models for longitudinal data. *Biometrics* **38**, 963–974. [199]

Lauritzen, S. L. (1996) *Graphical Models*. Oxford: Clarendon Press. [226]

Lawless, J. F. (1982) *Statistical Models and Methods for Lifetime Data*. New York: John Wiley and Sons. [367]

Lawless, J. F. (1987) Negative binomial and mixed Poisson regression. *Canadian Journal of Statistics* **15**, 209–225. [233]

Linder, A., Chakravarti, I. M. and Vuagnat, P. (1964) Fitting asymptotic regression curves with different asymptotes. In *Contributions to Statistics. Presented to Professor P. C. Mahalanobis on the Occasion of his 70th Birthday*, ed. C. R. Rao, pp. 221–228. Oxford: Pergamon Press. [249, 250]

Lindstrom, M. J. and Bates, D. M. (1990) Nonlinear mixed effects models for repeated measures data. *Biometrics* **46**, 673–687. [271]

Lock, R. H. (1993) New car data. *Journal of Statistics Education* **1**(1). See the URL http://jse.stat.ncsu.edu/jse/v1n1/datasets.lock. [208]

Ludbrook, J. (1994) Repeated measurements and multiple comparisons in cardiovascular research. *Cardiovascular Research* **28**, 303–311. [273]

Macnaughton-Smith, P., Williams, W. T., Dale, M. B. and Mockett, L. G. (1964) Dissimilarity analysis: A new technique of hierarchical sub-division. *Nature* **202**, 1034–1035. [342]

MacQueen, J. (1967) Some methods for classification and analysis of multivariate observations. In *Proceedings of the Fifth Berkeley Symposium on Mathematical Statistics and Probability*, eds L. M. Le Cam and J. Neyman, volume 1, pp. 281–297. Berkeley, CA: University of California Press. [338]

Madsen, M. (1976) Statistical analysis of multiple contingency tables. Two examples. *Scandinavian Journal of Statistics* **3**, 97–106. [226]

Mallows, C. L. (1973) Some comments on C_p. *Technometrics* **15**, 661–675. [185]

Mangasarian, O. L. and Wolberg, W. H. (1990) Cancer diagnosis via linear programming. *SIAM News* **23**, 1, 18. [365]

Marazzi, A. (1993) *Algorithms, Routines and S Functions for Robust Statistics*. Pacific Grove, CA: Wadsworth and Brooks/Cole. [171]

Mardia, K. V., Kent, J. T. and Bibby, J. M. (1979) *Multivariate Analysis*. London: Academic Press. [330]

Matheron, G. (1973) The intrinsic random functions and their applications. *Advances in Applied Probability* **5**, 439–468. [442]

McCullagh, P. (1980) Regression models for ordinal data (with discussion). *Journal of the Royal Statistical Society series B* **42**, 109–142. [231]

McCullagh, P. and Nelder, J. A. (1989) *Generalized Linear Models*. Second Edition. London: Chapman & Hall. [202, 211, 213, 214, 231]

McCulloch, W. S. and Pitts, W. (1943) A logical calculus of ideas immanent in neural activity. *Bulletin of Mathematical Biophysics* **5**, 115–133. [297]

McLachlan, G. J. (1992) *Discriminant Analysis and Statistical Pattern Recognition*. New York: John Wiley and Sons. [329]

McLachlan, G. J. and Basford, K. E. (1988) *Mixture Models: Inference and Applications to Clustering*. New York: Marcel Dekker. [263]

McLachlan, G. J. and Jones, P. N. (1988) Fitting mixture models to grouped and truncated data via the EM algorithm. *Biometrics* **44**, 571–578. [280]

McLachlan, G. J. and Krishnan, T. (1997) *The EM Algorithm and Extensions*. New York: Wiley. [280]

Michie, D. (1989) Problems of computer-aided concept formation. In *Applications of Expert Systems 2*, ed. J. R. Quinlan, pp. 310–333. Glasgow: Turing Institute Press / Addison-Wesley. [304]

Milicer, H. and Szczotka, F. (1966) Age at menarche in Warsaw girls in 1965. *Human Biology* **B38**, 199–203. [236]

Miller, R. G. (1981a) *Simultaneous Statistical Inference*. New York: Springer-Verlag. [192]

Miller, R. G. (1981b) *Survival Analysis*. New York: John Wiley and Sons. [367]

Moran, M. A. and Murphy, B. J. (1979) A closer look at two alternative methods of statistical discrimination. *Applied Statistics* **28**, 223–232. [350]

Morgan, J. N. and Messenger, R. C. (1973) THAID: *a Sequential Search Program for the Analysis of Nominal Scale Dependent Variables*. Survey Research Center, Institute for Social Research, University of Michigan. [303]

Morgan, J. N. and Sonquist, J. A. (1963) Problems in the analysis of survey data, and a proposal. *Journal of the American Statistical Association* **58**, 415–434. [303]

Mosteller, F. and Tukey, J. W. (1977) *Data Analysis and Regression*. Reading, MA: Addison-Wesley. [121, 338]

Nagelkerke, N. J. D. (1991) A note on a general definition of the coefficient of determination. *Biometrika* **78**, 691–692. [383]

Nash, J. C. (1990) *Compact Numerical Methods for Computers. Linear Algebra and Function Minimization*. Second Edition. Bristol: Adam Hilger. [261, 436]

Natrella, M. (1963) *Experimental Statistics*. Washington, DC: NBS Handbook 91. [148]

Nelson, W. D. and Hahn, G. J. (1972) Linear estimation of a regression relationship from censored data. Part 1 — simple methods and their application (with discussion). *Technometrics* **14**, 247–276. [369, 378]

Park, B.-U. and Turlach, B. A. (1992) Practical performance of several data-driven bandwidth selectors (with discussion). *Computational Statistics* **7**, 251–285. [138]

Peto, R. (1972) Contribution to the discussion of the paper by D. R. Cox. *Journal of the Royal Statistical Society B* **34**, 205–207. [380]

Peto, R. and Peto, J. (1972) Asymptotically efficient rank invariant test procedures (with discussion). *Journal of the Royal Statistical Society A* **135**, 185–206. [373]

Plackett, R. L. (1974) *The Analysis of Categorical Data*. London: Griffin. [226]

Prater, N. H. (1956) Estimate gasoline yields from crudes. *Petroleum Refiner* **35**, 236–238. [199]

Priestley, M. B. (1981) *Spectral Analysis and Time Series*. London: Academic Press. [401, 407]

Pringle, R. M. and Rayner, A. A. (1971) *Generalized Inverse Matrices with Applications to Statistics*. London: Griffin. [100]

Quinlan, J. R. (1979) Discovering rules by induction from large collections of examples. In *Expert Systems in the Microelectronic Age*, ed. D. Michie. Edinburgh: Edinburgh University Press. [303]

Quinlan, J. R. (1983) Learning efficient classification procedures and their application to chess end-games. In *Machine Learning*, eds R. S. Michalski, J. G. Carbonell and T. M. Mitchell, pp. 463–482. Palo Alto: Tioga. [303]

Quinlan, J. R. (1986) Induction of decision trees. *Machine Learning* **1**, 81–106. [303]

Quinlan, J. R. (1993) *C4.5: Programs for Machine Learning*. San Mateo, CA: Morgan Kaufmann. [303]

Raftery, A. E. (1995) Bayesian model selection in social research. In *Sociological Methodology 1995*, ed. P. V. Marsden, pp. 111–196. Oxford: Blackwells. [208]

Rao, C. R. (1948) The utilization of multiple measurements in problems of biological classification. *Journal of the Royal Statistical Society series B* **10**, 159–203. [346]

Rao, C. R. (1971a) Estimation of variance and covariance components—MINQUE theory. *Journal of Multivariate Analysis* **1**, 257–275. [194]

Rao, C. R. (1971b) Minimum variance quadratic unbiased estimation of variance components. *Journal of Multivariate Analysis* **1**, 445–456. [194]

Rao, C. R. (1973) *Linear Statistical Inference and Its Applications*. Second Edition. New York: John Wiley and Sons. [181]

Rao, C. R. and Kleffe, J. (1988) *Estimation of Variance Components and Applications*. Amsterdam: North-Holland. [194]

Rao, C. R. and Mitra, S. K. (1971) *Generalized Inverse of Matrices and Its Applications*. New York: John Wiley and Sons. [100]

Rawlings, J. O. (1988) *Applied Regression Analysis. A Research Tool*. Pacific Grove, California: Wadsworth and Brooks/Cole. [208]

Redner, R. A. and Walker, H. F. (1984) Mixture densities, maximum likelihood and the EM algorithm. *SIAM Review* **26**, 195–239. [263]

Reynolds, P. S. (1994) Time-series analyses of beaver body temperatures. In *Case Studies in Biometry*, eds N. Lange, L. Ryan, L. Billard, D. Brillinger, L. Conquest and J. Greenhouse, Chapter 11. New York: John Wiley and Sons. [427]

Rice, J. A. (1995) *Mathematical Statistics and Data Analysis*. Second Edition. Belmont, CA: Duxbury Press. [148]

Ripley, B. D. (1976) The second-order analysis of stationary point processes. *Journal of Applied Probability* **13**, 255–266. [445]

Ripley, B. D. (1981) *Spatial Statistics*. New York: John Wiley and Sons. [72, 433, 437, 438, 439, 445, 446, 447]

Ripley, B. D. (1987) *Stochastic Simulation*. New York: John Wiley and Sons. [116, 117]

Ripley, B. D. (1988) *Statistical Inference for Spatial Processes*. Cambridge: Cambridge University Press. [433, 447]

Ripley, B. D. (1993) Statistical aspects of neural networks. In *Networks and Chaos — Statistical and Probabilistic Aspects*, eds O. E. Barndorff-Nielsen, J. L. Jensen and W. S. Kendall, pp. 40–123. London: Chapman & Hall. [296, 298]

Ripley, B. D. (1994) Neural networks and flexible regression and discrimination. In *Statistics and Images 2*, ed. K. V. Mardia, volume 2 of *Advances in Applied Statistics*, pp. 39–57. Abingdon: Carfax. [298]

Ripley, B. D. (1996) *Pattern Recognition and Neural Networks*. Cambridge: Cambridge University Press. [226, 260, 269, 296, 303, 309, 310, 324, 330, 334, 341, 348, 349, 350, 354, 361]

Ripley, B. D. (1997) Classification (update). In *Encyclopedia of Statistical Sciences*, eds S. Kotz, C. B. Read and D. L. Banks, volume Update 1. New York: John Wiley and Sons. [329]

Robinson, G. K. (1991) That BLUP is a good thing: The estimation of random effects (with discussion). *Statistical Science* **6**, 15–51. [194]

Roeder, K. (1990) Density estimation with confidence sets exemplified by superclusters and voids in galaxies. *Journal of the American Statistical Association* **85**, 617–624. [138]

Rosenberg, P. S. and Gail, M. H. (1991) Backcalculation of flexible linear models of the Human Immunodeficiency Virus infection curve. *Applied Statistics* **40**, 269–282. [269]

Ross, G. J. S. (1970) The efficient use of function minimization in non-linear maximum-likelihood estimation. *Applied Statistics* **19**, 205–221. [278]

Ross, G. J. S. (1990) *Nonlinear Estimation.* New York: Springer-Verlag. [241]

Rousseeuw, P. J. (1984) Least median of squares regression. *Journal of the American Statistical Association* **79**, 871–881. [348]

Rousseeuw, P. J. and Leroy, A. M. (1987) *Robust Regression and Outlier Detection.* New York: John Wiley and Sons. [127, 167, 168, 348]

Sammon, J. W. (1969) A non-linear mapping for data structure analysis. *IEEE Transactions on Computers* **C-18**, 401–409. [334]

Santer, T. J. and Duffy, D. E. (1989) *The Statistical Analysis of Discrete Data.* New York: Springer-Verlag. [225]

Scheffé, H. (1959) *The Analysis of Variance.* New York: John Wiley and Sons. [195, 208]

Schoenfeld, D. (1982) Partial residuals for the proportional hazards model. *Biometrika* **69**, 239–241. [385]

Schwarz, G. (1978) Estimating the dimension of a model. *Annals of Statistics* **6**, 461–464. [203]

Scott, D. W. (1979) On optimal and data-based histograms. *Biometrika* **66**, 605–610. [119]

Scott, D. W. (1992) *Multivariate Density Estimation. Theory, Practice, and Visualization.* New York: John Wiley and Sons. [118, 132, 133, 136, 137, 139]

Seber, G. A. F. and Wild, C. J. (1989) *Nonlinear Regression.* New York: John Wiley and Sons. [241]

Sethi, I. K. and Sarvarayudu, G. P. R. (1982) Hierarchical classifier design using mutual information. *IEEE Transactions on Pattern Analysis and Machine Intelligence* **4**, 441–445. [303]

Sheather, S. J. and Jones, M. C. (1991) A reliable data-based bandwidth selection method for kernel density estimation. *Journal of the Royal Statistical Society B* **53**, 683–690. [137]

Silverman, B. W. (1985) Some aspects of the spline smoothing approach to non-parametric regression curve fitting (with discussion). *Journal of the Royal Statistical Society B* **47**, 1–52. [282, 283]

Silverman, B. W. (1986) *Density Estimation for Statistics and Data Analysis.* London: Chapman & Hall. [132, 135, 139]

Simonoff, J. S. (1996) *Smoothing Methods in Statistics.* New York: Springer. [132, 141, 281]

Solomon, P. J. (1984) Effect of misspecification of regression models in the analysis of survival data. *Biometrika* **71**, 291–298. [383]

Stace, C. (1991) *New Flora of the British Isles.* Cambridge: Cambridge University Press. [304]

Staudte, R. G. and Sheather, S. J. (1990) *Robust Estimation and Testing.* New York: John Wiley and Sons. [127, 128, 142, 162]

Stevens, W. L. (1948) Statistical analysis of a non-orthogonal tri-factorial experiment. *Biometrika* **35**, 346–367. [213]

Stone, M. (1974) Cross-validatory choice and assessment of statistical predictions (with discussion). *Journal of the Royal Statistical Society B* **36**, 111–147. [137]

Svensson, Å. (1981) On the goodness-of-fit test for the multiplicative Poisson model. *Annals of Statistics* **9**, 697–704. [239]

Therneau, T. M. and Atkinson, E. J. (1997) An introduction to recursive partitioning using the RPART routines. Technical report, Mayo Foundation. [Distributed in PostScript with the `rpart` package.]. [303, 310]

Thisted, R. A. (1988) *Elements of Statistical Computing. Numerical Computation.* New York: Chapman & Hall. [100]

Thomson, D. J. (1990) Time series analysis of Holocene climate data. *Philosophical Transactions of the Royal Society A* **330**, 601–616. [417]

Tibshirani, R. (1988) Estimating transformations for regression via additivity and variance stabilization. *Journal of the American Statistical Association* **83**, 394–405. [294]

Titterington, D. M., Smith, A. F. M. and Makov, U. E. (1985) *Statistical Analysis of Finite Mixture Distributions.* Chichester: John Wiley and Sons. [263]

Tsiatis, A. A. (1981) A large sample study of Cox's regression model. *Annals of Statistics* **9**, 93–108. [370]

Tukey, J. W. (1960) A survey of sampling from contaminated distributions. In *Contributions to Probability and Statistics*, eds I. Olkin, S. Ghurye, W. Hoeffding, W. Madow and H. Mann, pp. 448–485. Stanford: Stanford University Press. [127]

Upton, G. J. G. and Fingleton, B. J. (1985) *Spatial Data Analysis by Example.* Volume 1. Chichester: John Wiley and Sons. [433]

Vandaele, W. (1978) Participation in illegitimate activities: Ehrlich revisited. In *Deterrence and Incapacitation*, eds A. Blumstein, J. Cohen and D. Nagin, pp. 270–335. Washington: US National Academy of Sciences. [208]

Velleman, P. F. and Hoaglin, D. C. (1981) *Applications, Basics, and Computing of Exploratory Data Analysis.* Boston: Duxbury. [121]

Vinod, H. (1969) Integer programming and the theory of grouping. *Journal of the American Statistical Association* **64**, 506–517. [341]

Wahba, G. (1990) *Spline Models for Observational Data.* Philadelphia: SIAM. [285, 442]

Wahba, G., Wang, Y., Gu, C., Klein, R. and Klein, B. (1995) Smoothing spline ANOVA for exponential families, with application to the Wisconsin epidemiological study of diabetic retinopathy. *Annals of Statistics* **23**, 1865–1895. [285, 288]

Wand, M. P. (1997) Data-based choice of histogram bin width. *American Statistician* **52**, 59–64. [148]

Wand, M. P. and Jones, M. C. (1995) *Kernel Smoothing.* Chapman & Hall. [132, 137, 139, 285]

Weiss, S. M. and Kapouleas, I. (1989) An empirical comparison of pattern recognition, neural nets and machine learning classification methods. In *Proceedings of the Eleventh International Joint Conference on Artificial Intelligence, Detroit, 1989*, pp. 781–787. [323]

White, H. (1982) Maximum likelihood estimation of mis-specified models. *Econometrika* **50**, 1–25. [260]

Whittaker, J. (1990) *Graphical Models in Applied Multivariate Statistics.* Chichester: John Wiley and Sons. [226]

Wilkinson, G. N. and Rogers, C. E. (1973) Symbolic description of factorial models for analysis of variance. *Applied Statistics* **22**, 392–399. [4, 149]

Williams, E. J. (1959) *Regression Analysis.* New York: John Wiley and Sons. [253]

Wilson, S. R. (1982) Sound and exploratory data analysis. In *COMPSTAT 1982, Proceedings in Computational Statistics*, eds H. Caussinus, P. Ettinger and R. Tamassone, pp. 447–450. Vienna: Physica-Verlag. [329]

Yandell, B. S. (1997) *Practical Data Analysis for Designed Experiments.* London: Chapman & Hall. [192]

Yates, F. (1935) Complex experiments. *Journal of the Royal Statistical Society (Supplement)* **2**, 181–247. [195]

Yates, F. (1937) *The Design and Analysis of Factorial Experiments.* Technical communication No. 35. Harpenden, England: Imperial Bureau of Soil Science. [195]

Yohai, V., Stahel, W. A. and Zamar, R. H. (1991) A procedure for robust estimation and inference in linear regression. In *Directions in Robust Statistics and Diagnostics, Part II*, eds W. A. Stahel and S. W. Weisberg. New York: Springer-Verlag. [171]

Index

Entries in `this font` are names of S objects. Page numbers in **bold** are to the most comprehensive treatment of the topic.